FROM FAUST TO STRANGE

D1425540

FROM FAUST
TO STRANGELOVE

Representations of the Scientist

in Western Literature

ROSLYNN D. HAYNES

THE JOHNS HOPKINS UNIVERSITY PRESS
BALTIMORE AND LONDON

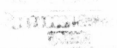

The Johns Hopkins University Press
2715 North Charles Street
Baltimore, Maryland 21218-4319
The Johns Hopkins Press Ltd., London

Library of Congress Cataloging-in-Publication Data will be found
at the end of this book.

A catalog record for this book is available from the British Library.

ISBN 0-8018-4801-6

For Raymond, Nicky, and Rowena

CONTENTS

ACKNOWLEDGMENTS

I am happy to acknowledge the two grants that I received from the University of New South Wales Faculty of Arts to help me carry out the research for this book, but other forms of support have been equally important. The subject has proved a most effective conversation-starter, and I am grateful to the many friends and colleagues who have shared their ideas and favorite examples of fictional scientists with me. The final version has benefited enormously from the advice and comments of Dr. Jon Darius, Director of the National Museum of Science and Industry, London, Dr. Richard O'Grady, of the Johns Hopkins University Press, who read the draft manuscript, and Joanne Allen, who copyedited the manuscript. Most of all, I am grateful to my husband Raymond and my daughters Nicola and Rowena, whose help and enthusiastic encouragement over many years have been essential for the book's completion.

FROM FAUST TO STRANGELOVE

INTRODUCTION

Popular belief and behavior are influenced more by images than by demonstrable facts. Very few actual scientists (Isaac Newton, Marie Curie, and Albert Einstein are the only significant exceptions) have contributed to the popular image of "the scientist." On the other hand, fictional characters such as Dr. Faustus, Dr. Frankenstein, Dr. Moreau, Dr. Jekyll, Dr. Caligari, and Dr. Strangelove have been extremely influential in the evolution of the unattractive stereotypes that continue in uneasy coexistence with the manifest dependence of Western society on its scientists.

Surveys conducted among various social groups to evaluate how scientists are generally perceived have invariably yielded results that indicate an almost wholly negative estimate, in relation to both scientists as a group and the contribution they make to the community. Such an evaluation seems sharply at odds with the self-image cherished by the scientists themselves, the majority of whom regard their activities as both reputable and socially beneficent. On the other hand, the images conveyed by the survey answers bear a remarkable resemblance to literary representations of scientists, many of them centuries old and apparently irrelevant to their modern analogues.

In 1957 Margaret Mead and Rhoda Métraux carried out a pilot study to assess how scientists were regarded among high school students in the United States. They reported that although at an "official level" most students commented positively about the public role of scientists, their answers to questions involving their own personal choices about a career in science or about a scientist as marriage partner conveyed an "overwhelmingly negative" image.[1] More recent surveys of younger children in several different countries have come up with very similar findings. Scientists drawn or described by primary school students are almost invariably male (99.4 percent),[2] middle-aged or older, either bald or having a large mass of hair in the style of Einstein (see fig. 1). What is more disturbing, they are nearly always depicted as working alone and in isolated laboratories; and where the object

FIGURE 1.
The Scientist, a recurrent
image indicated in public
opinion surveys.

of their research is indicated it is usually conspicuously labeled as "secret"
or "dangerous." This deep-rooted suspicion about scientists is not confined
to children. In 1975 surveys measuring the public perception of scientists
were conducted among readers of *New Scientist* and *New Society*. While the
scientist readers who responded considered scientists to be "typically ap-
proachable, sociable, open, unconventional, socially responsible, and popular with broad interests, nonscientist readers saw them as typically the opposite." There was also an overwhelming gender bias in the replies,[3] which
accords with the gender distribution among physics graduates and staff in
physics departments surveyed internationally.[4]

Fictional scientists are expressions of their creators' response to the
role of science and technology in a particular social context and thus are
interesting in their own right; but viewed chronologically they achieve an
additional historical significance both as ideological indicators of the changing perception of science over some seven centuries and as powerful images
that give rise to new stereotypes. These subversive fictional protagonists
have inevitably contributed to Western society's ambiguous love-hate attitude towards science, which has resurfaced in recent decades in the debates
over the use of public money for space research, the benefits and dangers of

nuclear power and state-of-the-art weapons, the responsibility of science and technology for environmental pollution, and the reification of the individual in a postmodern technocracy.

Studying the evolution of representations of scientists in Western literature, and more recently in film, allows us to see how clusters of these fictional images have coalesced to produce archetypes that subsequently have acquired a cumulative, even mythical, importance. The pageant of fictional scientists, from the medieval alchemist to the modern computer programmer, atomic physicist, or cyberneticist, is grouped around six recurrent stereotypes:

—The alchemist, who reappears at critical times as the obsessed or maniacal scientist. Driven to pursue an arcane intellectual goal that carries suggestions of ideological evil, this figure has been reincarnated recently as the sinister biologist producing new (and hence allegedly unlawful) species through the quasi-magical processes of genetic engineering.
—The stupid virtuoso, out of touch with the real world of social intercourse. This figure at first appears more comic than sinister, but he too comes with sinister implications. Preoccupied with the trivialities of his private world of science, he ignores his social responsibilities. His modern counterpart, the absent-minded professor of early twentieth-century films, while less overtly censured than his seventeenth-century prototype, is nevertheless an ineffectual figure, a moral failure by default.
—The Romantic depiction of the unfeeling scientist who has reneged on human relationships and suppressed all human affections in the cause of science. This has been the most enduring stereotype of all and still provides the most common image of the scientist in popular thinking, recurring repeatedly in twentieth-century plays, novels, and films. In portrayals of the 1950s there is an additional ambivalence about this figure: his emotional deficiency is condemned as inhuman, even sinister, but in a less extreme form it is also condoned, even admired, as the inevitable price scientists must pay to achieve their disinterestedness.
—The heroic adventurer in the physical or the intellectual world. Towering like a superman over his contemporaries, exploring new territories, or engaging with new concepts, this character emerges at periods of scientific optimism. His particular appeal to adolescent audiences, deriving from the implicit promise of transcending boundaries, whether material, social, or intellectual, has ensured the popularity of this stereotype in comics and space opera. More subtle analyses of such heroes, however, suggest the danger of their charismatic power as, in the guise of neo-imperialist space travelers, they impose their particular brand of colonization on the universe.
—The helpless scientist. This character has lost control either over his dis-

covery (which, monsterlike, has grown beyond his expectations) or, as frequently happens in wartime, over the direction of its implementation. In recent decades this situation has been explored in relation to a whole panoply of environmental problems, of which scientists are frequently seen as the original perpetrators.

—The scientist as idealist. This figure represents the one unambiguously acceptable scientist, sometimes holding out the possibility of a scientifically sustained utopia with plenty and fulfillment for all but more frequently engaged in conflict with a technology-based system that fails to provide for individual human values.

The majority of these stereotypes (as well as the overwhelming majority of individual characters) represent scientists in negative terms, as producing long-term liabilities for society. Yet these depictions have not only reflected writers' opinions of the science and scientists of their day; they have, in turn, provided a model for the contemporary evaluation of scientists and, by extension, of science itself.

Realization of this ongoing connection permits us to confront the widespread, often unacknowledged, fear of science and scientists in Western society. It also enables us to explore the basis of these fears, to deconstruct what the creators of fictional scientist characters were trying to express, and to ask which of those implicit anxieties are still relevant and should be addressed and which are anachronisms, relics of quite different social circumstances, irrelevant to our own society.

The traditional fear of too much knowledge and the belief that some things should remain hidden have an ancestry far older than the printed word; they can be traced back to biblical texts and Scandinavian mythology. More recently, these attitudes reflect the hostility of medieval Europeans towards the alchemists, whose closely guarded learning derived from the Arab world and was therefore regarded by the Catholic church as distinctly dangerous and all too likely to confirm the Genesis story of the serpent and the apple. Indeed, from the Middle Ages to the twentieth century, scientists as depicted in literature have, with few exceptions, been rated as "low" to "very low" on the moral scale. The early Faustian stereotype of the enchanter, versed in the black arts and most probably in league with the devil, has spawned a series of equally unattractive offspring: megalomaniacs bent on world destruction; absent-minded professors shuffling in slippers and odd socks while disasters befall their beautiful daughters in the next room; inhuman researchers who think only in facts and numbers and are unable to communicate on any other level.

One of the most enduring and influential images of the scientist in both literature and film has been that of Victor Frankenstein, creation of the eighteen-year-old Mary Shelley, who with brilliant perception and foresight analyzed and described the personality traits associated with scientific research and their dangerous implications. Frankenstein's problems begin with his isolation, which leads directly to his suppression of emotional relationships and aesthetic experiences and the delusion that his work is being pursued in the interests of society, when in fact the real goals are power and fame; above all, he fails to foresee and take responsibility for the results of his research.

Of course not all the scientists portrayed in literature have been evil or misguided. Newton himself inspired the most fulsome tributes in verse, and for a short time at the beginning of this century scientists emerged in much popular fiction as adventurer-heroes, saving the Earth from evil space invaders or setting up a utopian society grounded on the principles of science. But this period of hero worship was brief. After Hiroshima the scientists' moral stocks plunged once more; they were depicted as ruthlessly sacrificing individuals, even whole nations, merely to gratify their scientific curiosity.

More recently, this stereotype has been followed by that of the helpless scientist wringing his hands while his computers run amok or poisonous substances from his laboratory destroy the natural environment. In these cases too, writers are reflecting more or less faithfully the attitudes of their society towards actual science and scientists. Those attitudes, in turn, spring from a complex of influences—literary, philosophical, social, religious, and moral—as well as from the personalities, or perceived personalities, of real-life scientists as represented in the media.

The "gulf of mutual incomprehension" between scientists and others received some publicity in the 1960s with the acrimonious "two cultures" debate between the British scientist-turned-novelist C. P. Snow and the influential Cambridge literary critic F. R. Leavis, but the stakes today are higher, and the potential consequences for both scientists and nonscientists more serious, than either of those two antagonists anticipated. By failing to discuss with nonscientists what they are doing, scientists not only endanger society but limit themselves and their research in a number of ways. They may fail to perceive directions that would be profitable to their work; they may fail to convince funding bodies that what they are doing has any economic or social value; they may be left with no control over what is done with their research; and they will almost certainly be diminished as people.

There is also some good news. Although literature has most frequently acted as a mirror, reflecting contemporary attitudes towards science and scientists, it has sometimes pointed the way to new insights. This was certainly true in the case of Francis Bacon's *New Atlantis* (1626), which was directly responsible for the establishment of the Royal Society. It was arguably true in the case of the English Romantic poets of the early nineteenth century, who, in defiance of the whole Newtonian materialist edifice, opted for an understanding of the world in terms of an interactive relationship, a position not dissimilar to that posited by Heisenberg's uncertainty principle and much of subsequent twentieth-century physics.[5] It could also be true today. Some of our most perceptive contemporary writers have described and analyzed the communication breakdown between the disciplines, and a few have even dared to propose solutions. They have pointed to the limitations of the traditional assumption that the scientist must be a detached and wholly rational observer of nature. They have suggested a correlation between this inherently male attitude and the conventional terminology of science, implicitly based on a model of the violation by (male) scientists of a passive, female Nature. These writers are proposing that scientists, who, after all, pride themselves on their mental flexibility, might reexamine their assumptions and their unacknowledged mind-set. They ignore that suggestion at their peril.

This book examines the representation of scientists from the Middle Ages to the present, showing how the recurrent mutual suspicion between scientists and other members of society was developed and reinforced in Western literature and pointing to some of the fictional suggestions for overcoming what is arguably the most pervasive problem of our time, namely, communication failure.

A Note on the Words *Science* and *Scientist*

The word *science* came into English in the Middle Ages, from the French *science,* which meant knowledge in the broadest sense.[6] Subsequently it acquired the connotation of accurate and systematic knowledge, especially that derived from philosophical demonstration, for example, by a syllogism. The Scholastic philosophers used the word *science* to refer to specialized branches of philosophy; thus the seven sciences of medieval learning were grammar, logic, arithmetic, rhetoric, music, geometry, and astronomy. Francis Bacon extended this definition to include knowledge derived from observation and experiment,[7] and hence the study of the natural world came to be called *natural philosophy,* although this term was still regarded as suspect by the Scholastic purists. Even Isaac Newton had taken the precaution of setting out his *Principia* in the format of Euclid's *Geometry* in order to elevate it to the status of a science. Although by 1820 the astronomer William Herschel was rejecting the old Scholastic nexus between science and deductive logic and aligning science exclusively with Bacon's experimental method, the terms *science* and *philosophy* were used synonymously until about 1850. After this time, the term *philosophy* was assigned to theological and metaphysical science, and *science* to experimental and physical science. The relation of social and even biological sciences to the physical sciences remained less clear-cut. Ross considers that the growing prestige of the physical sciences in the nineteenth century explains why they could arrogate to themselves the word formerly used to designate all knowledge, but the hierarchical structure within the sciences is still implicit in much discussion. It has frequently been remarked that, in Ross's words, "a higher status is claimed by, and generally accorded to, the physical and biological sciences, and to physical sciences in particular."[8] There often seems to be an implicit assumption, at least by physical scientists, that mathematical content is an index of scientific status.

The word *scientist* appeared much later, and even when the neologism was coined anonymously by William Whewell,[9] it was greeted with acrimony by most of the doyens of the scientific community. By analogy with *dentist*, it was thought to have connotations of specialization and professionalism distasteful to the heirs of the amateur tradition of British science. The ideal was "that of a man liberally educated, whose avocation was science as an intellectual *cum* philanthropic recreation, to which he might indeed devote most of his time without ever surrendering his claim to be a private gentleman of wide culture. In particular, to be thought to be pursuing science for money was distasteful."[10] (A character such as Lord Hollingshed in Elizabeth Gaskell's novel *Wives and Daughters* epitomizes this ideal.)[11] However, as Ross points out, this superior attitude could not survive the educational reforms of the mid-century, when science became one of the learned professions, along with medicine, theology, and the law. (Roger Hamley in the same Gaskell novel exemplifies this new approach to science.)

In the twentieth century, as science has acquired professional respectability and, perhaps more important, associations of wealth and power, the words *science* and *scientist* have come to bear more positive but scarcely less dangerous connotations of accuracy and even infallibility. Disciplines desirous of demonstrating their intellectual credentials lay claim to the title *science* in much the same way as the terms *philosophy* and *philosopher* were sought-after appellations in the eighteenth century. Terms such as *domestic science, military science, political science,* and *building science* really have meaning only in the original sense of science as knowledge, but their proponents imply in the use of the word a level of rigor that the physical and biological scientists would deny them.

While this would not, perhaps, matter very greatly if it were merely a dispute over semantics, its implications are more disturbing. Basic to the desire to qualify as a science is the public deference to scientific opinion, which assumes an importance in the popular mind not only out of proportion to its likely validity but in violation of the very basis of scientific method. Insofar as scientists exploit this mistaken credulity for their own ends, they are guilty of betraying that tradition of open questioning of all authorities that has allowed science to develop to the position it now holds. Insofar as they themselves believe it, they have returned to the pre-Baconian tradition of the alchemist in search of absolute and unquestionable truth—the philosophers' stone, perpetual motion, and the elixir of life.

CHAPTER ONE

EVIL ALCHEMISTS
AND DOCTOR FAUSTUS

We always met with failure in the end.
Yet though we never reached the wished conclusion
We still went raving on in our illusion,
Sitting together, arguing on and on,
And every one as wise as Solomon.
—Chaucer

Remote as they may seem from twentieth-century atomic physicists or
industrial chemists in white lab coats, surrounded by equipment costing
more than their life earnings, the medieval alchemists were the predecessors
of modern scientists. Not only were they at the cutting edge of experimental
research into the mysteries of nature, but they contributed to the profession
an aura of mystery, secrecy, suspicion, and, at times, irreligion from which it
has never wholly succeeded in detaching itself, either in literature or in the
public perception of scientists. Thus, in order to understand the development of the scientist as a literary character, it is necessary to begin with the
alchemists and the particular historical events surrounding their appearance in Europe.

Abstract concepts consistent with the basic tenets of alchemy were
current in China as long ago as the fifth century B.C., in association with the
Taoist belief that transformation and change are essential aspects of nature.
It was not until the fourth century A.D., however, that the first systematic
treatises, integrating a number of different traditions, were put forward in
Alexandria and Hellenistic Egypt. Aristotle's thesis of the unity of matter
and the interchangeable qualities of the four elements—earth, air, fire, and
water—formed the theoretical basis for the rise of Western alchemy through
its suggestion of the transmutation of matter. Aristotle had taught that everything in nature strives towards perfection and that since gold was considered the perfect and most noble state of matter, it followed that all base
metals must strive to become gold. The alchemist's task therefore was merely to help nature achieve its innate desire.

In Alexandria, the Aristotelian theory of transmutation became linked with the astrological charts of the Babylonians, with Chaldean traditions of magic, and with the practices of the Egyptian metalworkers, who, following the secret recipes of the priests of Isis, were adept at extending a given quantity of gold by producing alloys of gold with silver, copper, tin, and zinc. Thus, from the beginning, practical alchemy was closely associated with the "production" of gold, and it was doubtless this that ensured both its popularity and its prolonged survival.

During the eighth century, alchemy in the sense of metalworking was transmitted to the Arabs, who gave it the name *alkimia* (*al*, the; *Khem*, Egypt) from which the word *alchemy* is derived.[1] To the Alexandrian practice the Arabs, who had begun trading with China in the eighth century, added two more elements derived from Chinese tradition: the search for immortality and the idea of a transforming catalyst, which came to be called the philosophers' stone. Chinese alchemists appear to have been the first to associate the possibility of human immortality with the making of alchemical gold, the immortal substance. One of them, Go-Hung, was believed to have perfected a recipe for a pill of immortality, and the search for this "elixir of life" became another central aspect of alchemy, leading to important contributions to medicine.[2] Applying the Taoist principle of opposites, yin and yang, Chinese alchemists also regarded the primary substance as a necessary factor in the transmutation of metals, an idea that, despite Aristotle's belief in a *prima materia*, had not been postulated in Alexandria. It was the Arabs who integrated these diverse alchemical ideas and associated them with their religion. One of the great Islamic alchemists, Jabir ibn Hayyan, later known in Europe as Geber, affirmed that one could discover the secrets of perfection and the absolute only if one accepted the religious belief of the one God, Allah, the source of all.[3] For the Arabs, then, not only was there no conflict between alchemy and religion, but the Muslim faith was the very basis of theoretical science.

For many centuries alchemy remained under Islamic influence, but following the expulsion of the Moors from Europe and the return of the Islamic schools to Christian direction, the rare manuscripts they held, including those dealing with alchemy, were translated from the Arabic into Latin, thereby providing sourcebooks for medieval alchemists.[4] It was at this point that alchemy inevitably became associated in European thinking with the so-called black arts, with heresy and magic. Alchemists were regarded as being at best sinister and most likely in league with the devil, an impression that was accentuated by the medieval suspicion of knowledge per se. Within the strongly hierarchical hegemony of the Catholic church the story of the Garden of Eden was used to discourage independent thought about the causes and origins of phenomena, lest such knowledge constitute a rival authority. Soon the practice of secrecy, originally evolved to guard the formulas of the initiates, became necessary for sheer survival. Many alchemists lived in isolation, using a cryptic or cabalistic language, to escape persecution similar to that accorded to witches, and for similar reasons. Yet despite this reputation, alchemy exerted a fascination because of its fabulous promises, which came to include not only transmutation of base metals into gold but also other absolutes—eternal youth, perpetual motion, and the creation of life in the form of a homunculus. Apart from the search for gold, the other aims appear at first to be of a philosophical rather than a materialistic nature, but they all involve the prospect of power—over people, over death, over natural laws.

A few alchemists were able to combine their alchemy with a career in the church, the most notable being the Dominicans Albertus Magnus and Thomas Aquinas and the Franciscan Roger Bacon. Not only did these practitioners add a dash of Christian symbolism to the existing brew of traditions in an attempt to make alchemy acceptable to the church; what is more important, they introduced the idea of testing the postulates of alchemy by reason and experiment. It was this questioning and insistence on demonstration, especially on the part of Roger Bacon, that was to form the basis of modern scientific method and to distinguish it from the reactionary alchemy still extant in the seventeenth century.

There was, however, increasing unease among the leaders of the religious orders about friars' involvement in alchemy. A series of acts was passed forbidding it,[5] and in 1380 Charles V of France outlawed alchemy entirely. Thus although alchemy continued to be practiced in defiance of the church, it was forced to become increasingly arcane as a defense against the spies of church and state.

Despite the stereotyped view of alchemists promulgated by the church, there were actually three quite distinct groups: the charlatans, who deliberately deluded others about their ability to make gold but were not themselves deceived; the "puffers," or laboratory assistants involved in the practical problems of making gold but not yet successful in the art; and the adepts, scholarly alchemists who understood the secret language and who really believed that they knew, or were about to learn, the secret of transmutation (see fig. 2). Representatives of all three groups feature in Chaucer's *Canon's Yeoman's Tale.*

By the sixteenth and seventeenth centuries, alchemy had risen in the social scale to acquire royal patronage. The court of Rudolph II, at Prague, was one of the centers where eminent intellectuals gathered from all over Europe to discuss and demonstrate various branches of the occult and their proposed relation to cosmology, medicine, and of course the production of gold by transmutation. Kepler relied on his astrological predictions at Rudolph's court to supplement his stipend as imperial mathematician and finance his astronomical research. Queen Elizabeth I of England was also very receptive to alchemy, employing Cornelius de Lannoy as her personal

FIGURE 2. *An Alchemist and His Assistant,* by Hans Weidetz, ca. 1520

alchemist and encouraging a circle of practitioners about her court. Sir Walter Raleigh and Sir Philip Sidney were both enthusiastic students of alchemy, although there is no record of their experimental results. Dr. John Dee, a mathematician, astrologer, and alchemist, and Edward Kelley, an apothecary-turned-alchemist, were widely believed to conduct transmutations of base metals into gold, and the possibility of a commercial enterprise along these lines was suggested to the queen as a less risky operation than the plundering of Spanish galleons. The description given of one such transmutation allegedly carried out by Kelley bears a striking similarity to the description of deceitful procedures in Chaucer's *Canon's Yeoman's Tale:* "[I] saw him put of the base metal into the Crucible, and after it was set a little upon the Fire and a very small quantity of the Vessel [i.e., the elixir] put in and stirred with a stick of Wood, it came forth in great proportion perfect Gold."[6]

Even by the time of the Renaissance, however, the pursuit of gold had ceased to be the central obsession of alchemy. Theophrastus Bombastus von Hohenheim, better known by his self-bestowed name, Paracelsus, was more concerned with proving his theory of a universal order in the "latent forces of Nature" and with the practice of holistic medicine. He was the first to use the word *chemistry* to express the formulas and techniques of his treatments, which involved chemical drugs rather than the herbal medicines hitherto in favor. This preoccupation of alchemists with medicinal remedies, whether chemical or herbal, was the feature most used to distinguish them in nineteenth-century literature.

The other dream of the alchemists was the production of a homunculus, or minute human being, by artificial means. Of course this was also perceived as a threat to the divine basis of life expounded by the church, and much the same furor attended the attempts to produce a homunculus as obtains today over in vitro fertilization or genetic engineering. The idea of a homunculus was first promulgated in the *Homilies* of Clement of Rome (ca. A.D. 250), which described the alleged conjuring up of a such a being by the sorcerer Simon Magus, but by the early sixteenth century Paracelsus claimed to have a precise recipe for the physical generation of a homunculus from a mixture of semen and blood without resorting to the female uterus.[7] The Faustus legends also claimed that one of the doctor's successful exploits was the production of a homunculus.[8] This interest in solving the mystery of creation was closely related to the legend of the golem, culminating in the story of the Golem of Prague, allegedly created in 1580.[9] The sense of guilt associated with these attempts to emulate the divine prerogative is expressed in the elaborate ritual practices associated with the event. Later developments of mechanical men, or automatons, which would mimic or surpass human reasoning, from Roger Bacon's alleged talking head and Ramon Lull's automatons to the twentieth-century fascination with robots, were the result of variations of the same desire for power over the creation of life. Therefore it is not surprising that the essential features of the golem— the material origin of his body, derived from clay or dust; his role as servant to his human makers; and the implanted divine spark, which endowed the body with life—were retained, at least in the intention of those who sought to make homunculi or automatons, until the twentieth century.

Although it enjoyed a minor revival in the Rosicrucian doctrines of Khunrath, Andreä, Mynsicht, and Robert Fludd, alchemy as conceived in the medieval period had begun to decay in the seventeenth century, as the spirit of empirical enquiry spread throughout Europe. Some alchemists became involved in Hermetic philosophy, but most discarded mysticism and obscurantism, embracing instead the new experimental science of chemistry. There were two main reasons for this decline of alchemy, at least in the form of its traditional preoccupations. One was the accumulation of negative evidence by those who set out in good faith to clarify and codify the diverse claims of alchemists throughout Europe, particularly claims for the transmutation of substances. Ironically, one of the major figures in this process of clarification was the virtuoso and "compleat Gentleman" Sir Kenelm Digby, who himself remained convinced of metallic transformation and the main body of Aristotle's thought. Digby's relatively rigorous procedures were to invalidate so many of the claims of alchemy that the entire body of thought was called into question.[10] The second cause for the decline of

Hermetic doctrines of alchemy was the increasing interest in the experimental procedures and the mechanical philosophy of René Descartes and Thomas Hobbes. The immense attraction of this *philosophia mechanica* was dependent on its apparent explanatory power, its proposal that all events could be readily understood in terms of matter in motion and individual units of matter impacting on each other like billiard balls on a table. After the obscurities and complex procedures of alchemy, this seeming simplicity and universality were intellectually alluring. Some natural philosophers, including Robert Boyle and Isaac Newton, attempted to integrate the older system with the new mechanical philosophy, as exemplified in Boyle's *Origine of Formes and Qualities (According to the Corpuscular Philosophy)* (1666). From a post-Enlightenment perspective, it appears that there was a complete break between alchemy and modern chemistry and that Boyle's earlier publication, *The Sceptical Chymist* (1661), had effectively demolished the Aristotelian system of the four elements, one of the cornerstones of medieval alchemy, and introduced the modern idea of an element, thereby paving the way for chemistry as we know it. However, as we shall see in chapter 4, there was really no such hiatus. Most of the early members of the Royal Society, including the most prestigious, were still eagerly experimenting with the ideas and practices of alchemy. That these activities have been largely suppressed in subsequent accounts of their work is due partly to their failure to establish the truth of these ideas but also, I believe, to the adverse associations of alchemy and in particular the poor image of alchemists. The biographers of Boyle and Newton were concerned to exonerate their subjects from any such implications, and thus later accounts of their work omitted any connections with alchemy. It is only in recent decades that this material has been rediscovered and made widely available.

The diverse traditions and the social status of alchemy determined many characteristics of both the art itself and its practitioners. Because of the intimate relation between alchemy and the alleged production of gold, the presence of charlatans trading on the greed of the populace was almost inevitable. The Hermetic element of secrecy, deriving from the priestly origins of alchemy, was also there from the beginning; indeed, the alchemists often called themselves the sons of Hermes, believing that their art derived from the god Hermes, the Greek counterpart of the Egyptian god Thoth, whom they addressed as Trismegistus, "the Thrice-Great." These religious and magical origins also gave rise to the popular belief that alchemists wielded extraordinary power over nature and transcended the limits of what it was considered proper for man to know. Further, the isolation enforced by both the alchemists' desire to protect their secrets and, later, the threat of persecution enabled them for a long time to retain considerable autonomy; no one asked any questions, because it was too dangerous to know the answers. This attitude, too, has continued to overshadow the depiction of scientists in literature.

Despite the large volume of early alchemical writings, very few alchemists are depicted in these texts.[11] The first fully developed literary portrait of alchemists occurs in Chaucer's *Canon's Yeoman's Tale* (1387), one of the *Canterbury Tales*.[12] This tale is unique in that it presents a range of alchemists of differing ethical status and includes astringent social comment on their clients.

The Canon of the title is a closet alchemist who, while endeavoring to trick others with his promises of turning base metals into gold, has actually deluded himself, since he too believes in the fabulous promises and has lost all his money in the pursuit of his art.[13] When his servant, the Yeoman, asserts

> That all this blessed road we ride upon
> From here as far as Canterbury town,
> Why, he could turn it all clean upside down
> And pave it all with silver and with gold.   ("Prologue ," 475)[14]

his listener, the pragmatic Host, asks pointedly why, in that case, the Canon's clothes indicate his extreme poverty.

> With all those magic powers can't he pay
> For better cloth, if what you say is so?   (476)

The Yeoman admits that the Canon has never delivered the promised gold, either to himself or to others.

> We're always blundering, spilling things in the fire,
> But for all that we fail in our desire
> For our experiments reach no conclusion,   (476–77)

After seven years with the Canon, the Yeoman has lost "Al that I hadde" to "that slippery science" ("Tale," 478). And from his disgruntled account of life in the laboratory, with its accidental breakages and the loss of precious concoctions, it is clear that the Canon is no adept, but only a simple "puffer" who deceives and beggars himself as well as his clients. The Yeoman ruefully admits:

> We always met with failure in the end.
> Yet, though we never reached the wished conclusion
> We still went raving on in our illusion,
> Sitting together, arguing on and on,
> And every one as wise as Solomon.   ("Prologue ," 478)

For these two individuals, caught up in their own illusion, Chaucer inspires more pity mixed with his comedy than condemnation, but the other Canon of the Yeoman's tale within a tale is a very different character. He is depicted unambiguously as a charlatan, making a living out of the delusions and greed of humanity, in this case a gullible priest. After a series of rigged experiments involving the sly introduction of a little silver into the crucible so that it appears to have been produced from mercury, the Canon graciously consents to sell for forty pounds the magic powder allegedly responsible for the transmutation, a preparation that the Yeoman dismisses as

> Possibly chalk, or glass would do as well,
> Or anything else indeed not worth a fly.   ("Tale," 490)

Although the Yeoman's account concentrates on the trickery of the charlatan Canon, Chaucer also makes the point that such fraud could not succeed were it not for the greed of the client, who, already suspiciously wealthy for a priest, desires to multiply his money for nothing. This was the aspect of alchemy on which Ben Jonson was to focus in his play *The Alchemist*.

In addition to expressing trenchant criticism of both self-deluded and fraudulent alchemists, the Yeoman also describes the enthusiasm, amounting to mania, of the sincere devotees who sacrifice everything in the pursuit of truth.

> People will always find it bitter-sweet,
> Or so it seems   .   .   .
> .   .   .   .
> No one can stop until there's nothing left.   ("Tale," 482)

Ironically, the Yeoman himself is clearly of this stamp also. Even while condemning the intellectual pride of those who willfully confuse others with technical jargon, he himself delights in using as many alchemical terms as possible. In fact, the Canon of the title, the Yeoman narrator, and the foolish priest have all been taken in by the false promises of some other practitioner of alchemy. Although he advises his listeners to "Meddle no more with alchemy" and professes to renounce his former master, the Yeoman is too much like him to give alchemy up altogether. After seven years of misery and poverty with the Canon, he is still thoroughly obsessed with the craft. His statement, which is almost a religious testimony, is remarkably similar to the image of the noble devotee of science in the early years of the twentieth century as epitomized in popular treatments of Marie and Pierre Curie.

> And yet for all my misery and grief,
> Long hours and injuries without relief,
> I never could leave the business, any price.   ("Prologue ," 478)

One of the most striking aspects of Chaucer's treatment is that at the end of the tale he proceeds to defend "true" alchemy. Unlike the fraudulent variety, "true" alchemy is concerned, not with making gold or with any materialistic pursuit, but with a "secret of the secrets," which, far from being irreligious, is "very dear" to Christ himself ("Tale," 498). It is interesting that this defense of the "true" alchemy is made in conjunction with a warning to the uninitiated against attempting to discover secrets that are hidden in symbolic language, secrets that it is Christ's will should remain hidden except from the chosen few.[15]

Thus Chaucer's tale condemns false alchemists in terms that almost paraphrase the contemporary ecclesiastical view and presents them as being at best self-deluded; but it affirms a belief in the "true" alchemy, practiced by God's chosen initiates. While later writers were to elaborate enthusiastically upon the former aspect in both comic and tragic terms, Chaucer's defense of the noble alchemist inspired by God was probably the only such claim to be made outside the texts of the initiates and suggests that Chaucer himself may have been something of a closet alchemist.

The best-known alchemist figure in literature, even a stereotype in his own right, is Doctor Faustus. Many of his characteristics, notably his intellectual arrogance, are clearly derived from the older alchemist tradition, but the Faust legend evolved somewhat differently, and the significance attached to it has varied markedly at different periods and under different influences.

The original Doctor Georg Faust, whose title was almost certainly spurious, was born around 1480, perhaps in the German town of Knittlingen. He seems to have been something between the extremes of traveling conjurer-cum-hypnotist and quack doctor, on the one hand, and alchemist, perhaps even serious student of natural science, on the other.[16] With the passage of time, the legends about this Faust became increasingly exaggerated, with anecdotes of magic tricks and familiars predominating. The first written account, the anonymous Spieß edition of *Historia von D. Johann Fausten* published in 1587, presents the story in a highly moralistic light. Faust is a presumptuous man, desirous of overstepping the God-given limits of human knowledge. To achieve this, he makes a pact with the Devil, acquiring, as his part of the bargain, a variety of magical powers, most of which he employs in playing trivial tricks. Predictably, he comes to a suitably drawn-out and gruesome end, while the reader is continually referred to the Scriptures and warned against such intellectual arrogance. Yet, despite the allegedly didactic intention of the author, this Faust has a certain implicit nobility in his determination to acquire knowledge.

It was on the English translation of this chapbook that Christopher Marlowe's play *The Tragical History of Doctor Faustus* (1604) was based. Compared with the German original, the translation stressed the theme of intellectual curiosity, and Marlowe emphasized it even more. From the beginning he presents what has come to be seen as the archetypal dilemma of the Faust figure: his Renaissance-humanist longing to transcend the limitations of the human intellect is accompanied by a medieval awareness that such longing is doomed to failure. Reviewing the branches of current knowledge, Faustus rejects each in turn because of its restricted scope. Like Frankenstein, he complains that medicine cannot bestow eternal life, and he longs instead for "the metaphysics of magicians," since a "sound magician is a demi-god."[17]

Marlowe thus preserves the medieval association between intellectual arrogance and Lucifer's revolt against God, for ultimately Faust seeks knowledge as a means of attaining godlike powers; he desires to investigate not merely neutral knowledge but

> Unlawful things
> Whose deepness doth entice such forward wits
> To practise more than heavenly power permits.   (1.1.78–79; Ridley, 123)

Here we see the explicit association of the alchemist with the overreacher, defying God-given limits, and it is no accident that, as a prelude to his pact with Mephistopheles, Faustus derides both theology and religion and flagrantly dedicates his ceremony to Satan (1.3.5–7; Ridley, 126).[18] Having bargained his soul away, Faustus discovers that he has made as poor a bargain as the priest in Chaucer's tale, for, ironically, he learns nothing of any importance. Mephistopheles' answers to his questions are as evasive and ambiguous as those of a Greek oracle, and the much-vaunted magical powers amount to little more than cheap conjuring tricks, a device that recurs in later satirical treatments of scientists.[19]

These aspects of Faustus's character are taken directly from the German pietist tradition, but Marlowe also introduces a new theme, which is essentially a Renaissance rather than a medieval one: the tragic wasted potential of a gifted man.[20] He therefore gives us a very different conclusion from the grimly exultant tone at the end of Spieß Faust. Paradoxically, as Faustus sinks further into despair, the one unforgivable sin of Christian orthodoxy, his moral honesty and human decency increase. As he insists that his fellow scholars leave him to his fate and faces the full horror of the consequences of his past deeds, he arouses compassion and respect rather than condemnation. Indeed, in the subtext of Doctor Faustus, if not explicitly, the alchemist has effectively attained the moral status of a tragic hero.

The intrinsic ambiguity of the Faust character has inspired a variety of treatments and evoked an even greater variety of responses, all of which have, in turn, attached to scientists. At one extreme Faust may be regarded as an arrogant fool making a bad bargain with the wily Mephistopheles, who outwits him until he finally gets what he deserves. At the other extreme

Faust may be seen as embodying the noblest desire of man to transcend the limitations of the human condition and to extend his powers, for good as much as for evil, a Promethean figure who asserts the rights of man over a tyrannical order that seeks to enslave him. This latter characterization of Faustus is the one that predominates in German Romantic literature. Somewhere in between is the Faust of Lutheran piety, a kind of minor Satan, who rebels against God's restrictions and refuses to repent and whose inevitable punishment is presented with all the didactic stops pulled out.

The aspects of the Faust stereotype that predominate at any particular time or place vary with the relative status accorded to man and his intellect, compared with the value placed on obedience to the prevailing hegemony, whether of church or state. During the medieval period, the pact with Mephistopheles was central, and Faust was predictably dragged off to hell with the full approval of the audience. However, during the Renaissance and preeminently in the period of German Romanticism, the nobility of Faust's quest (whether for knowledge or experience or for some more mystical truth) was stressed, and the significance of the satanic connection was left ambiguous. Whichever light he was seen in, however, Faust, like the alchemists, was a figure of both fascination and dread, providing an awful example of the moral dangers of intellectual aspirations and pride.

The intimate connection in the popular mind between alchemy and magic was to issue in a steady stream of works featuring some kind of magus who possessed special powers, often implicitly satanic, over nature. For the most part, such magicians are presented as evil or, at best, awesome figures. A rare exception is Shakespeare's Prospero, the seemingly benign magus of *The Tempest* (1611), who not only has natural forces under his control but possesses magic of a similar (though superior) kind to that of the powerful witch Sycorax. Yet even Prospero is depicted with some ambiguity. The morality of causing a shipwreck that, although it produces no actual casualties, nevertheless elicits grief and suffering in the minds of the survivors, who assume their relatives drowned, remains problematic. Moreover, Prospero's final admission that he himself, through preoccupation with his books and the study of magic, was partly responsible for the mismanagement of his former kingdom sets a retrospective qualification on both his magical powers and the morality of his revenge. Magic emerges therefore as a doubtful good, possibly benign in the right hands (Prospero is virtually unique in renouncing his powers when justice appears to have been done) but suggesting also the potential for manipulation (such as Prospero exercises over the innocent spirit Ariel) and, in other hands, evil— the shipwreck *could* have been a real one.

Very different in its assumptions is Ben Jonson's play *The Alchemist* (1610). Whereas Chaucer had warned against the serious threat posed by charlatan alchemists, Marlowe had depicted the tragic grandeur of man's attempt to transcend his limitations, and Shakespeare had, through the character of a benign magus, suggested the ethical dangers implicit in magical control over nature and spirits, Jonson, the Renaissance man of reason, presents the quest for transcendent knowledge as a ridiculous illusion that flourishes only because of man's arrogance and greed. It is significant that in *The Alchemist* there *is* no alchemist, only low-life characters Subtle and Face pretending, through a facility with jargon, to have alchemical knowledge.[21] Jonson makes it plain that their temporary "success" is due not to their cleverness but to the stupidity, avarice, and vanity of their willingly deluded victims.

The Alchemist was part of Jonson's more general intention to ridicule those whose pretensions led them to seek or pretend to transcendence in any sphere, and his concern, arising from an actual situation, was more than merely academic. The prototypes for Subtle and Face were most probably the well-known contemporary practitioners of magic and alchemy, Dr. John Dee, Edward Kelley, and Simon Forman,[22] all of whom had gained considerable influence over the superstition-prone Elizabethan court even though alchemy was officially outlawed.[23] In Jonson's play, these educated and elevated personages are stripped of any claim to greatness and reduced to the level of a cunning servant and an itinerant rogue. While his master Lovewit flees from London to escape from the plague, Jeremy, the butler, enters into partnership with the fraudulent "alchemist" Subtle and his prostitute friend Doll Common, assuming the role of his assistant, Face (the names are, of course, indicative of their moral status). The house is then turned into a reception center for the greedy and pretentious, who flock there only to be duped by the syndicate's necromancy, horoscopes, alchemical mumbo-jumbo, and the sundry gold-making operations allegedly taking place in the "laboratory" offstage, but seen by none of the clients.

Their leading dupe, and indeed the play's chief exponent of alchemy, is Sir Epicure Mammon, who, predictably, is concerned to transmute everything into easy gold by means of "the [philosophers'] stone." So gullible is Mammon that he virtually swindles himself, believing any claptrap he is told and promising his skeptical friend Surly "irrefutable evidence":

> I'll show you a book where Moses and his sister
> And Solomon have written of the art;
> Ay, and a treatise penn'd by Adam
>
> . . . .
>
> I have a piece of Jason's fleece too,
> Which was no other than a book of alchemy.[24]

The absence of any actual alchemist from the play is indicative of Jonson's estimate of the claims of alchemy. There *are* no alchemists, he suggests, except insofar as deceiving rogues pretend to be such and play upon

FIGURE 3.
*The Alchemist in Search
of the Philosopher's Stone
Discovers Phosphorus,*
by Joseph Wright,
1771–95. Derby
Museum and Art
Gallery.

the avarice, folly, and pretensions of society. What is more important, unlike Chaucer and Marlowe, Jonson has a more powerful "hero," Lovewit, the urbane, unpretentious man of true reason, who sees the charlatans for what they are and is therefore immune to their "power." He expels the syndicate of pseudo-alchemists from his house and demotes Face to his proper position as butler. For Jonson, the proper exercise of the intellect, what we might call common sense, and a morality immune to the lures of avarice are sufficient protection against the tricks of alchemy or any form of magic, which relies for its power on the pretension and greed of stupid people.

Jonson's major innovation in the depiction of alchemists, then, was to strip them of their exotic aura and show them as no more than ordinary cheats, play-acting to deceive their willing victims. Through them he satirizes the very basis of magic, namely, the idea that man is potentially a demigod who can manipulate the universe and achieve magical control over time, place, or matter.

It is not accidental that in *The Alchemist* Jonson links the "magicians" with a superficially very different group, the extreme Puritans. He considered both to be equally arrogant in regarding their personal revelations as superior to those of other people. Their claims to divine inspiration are the root cause of the unintelligible jargon designed to keep others in subjection, whether it be the occult vocabulary of alchemy or the apocalyptic prophecies of Broughton that Doll pours forth, which delude the gulls into believing they can usher in a new utopian order, either material or religious.[25]

As we shall see in chapter 3, Jonson's weapon of satiric deflation was so widely adopted that it became almost the norm in the late-seventeenth- and eighteenth-century representation of scientists, even in depictions of the serious founding members of the Royal Society. Indeed, the extraordinary and enduring fame of Sir Isaac Newton was at least partly contingent on the omission of this major aspect of his work from the Newtonian iconography transmitted to successive generations through literary and popular culture.

Perhaps because of the diverse responses it has elicited at different times, the alchemist stereotype has proved the most resilient of any. It has recurred in representations of the scientist with both good and evil implications, from medieval times to the present, sometimes with specific allusion to magic, as in Maugham's *Magician* (1908), but more frequently in oblique suggestions about looks and attitudes, as in the case of Nebogipfel, the original version of H. G. Wells's Time Traveller.[26] Because this stereotype is associated with a mystique that sets the alchemist figure apart from ordinary people, it is often taken to suggest communion with supernatural powers (see fig. 3). Such powers may be ennobling, as in the case of the Curies in real life and countless fictional examples of the noble scientist sacrificing his life to science, but they are more often darkly satanic, as in the Faustian tradition, causing their devotees to seek knowledge harmful to humanity.

It is not difficult to see how much of the popular image of the scientist derives from the alchemist tradition. The aspiration to know more than others about the causes of natural processes is inevitably associated with the desire for power over nature and hence with the attempts to manipulate or modify some aspect of nature for one's own convenience. Like their alchemist predecessors who held out the lure of gold, modern scientists often feel compelled to justify their research in terms that promise their funding bodies monetary gain or social benefits.

It is from the alchemists, too, that science has derived its tradition of involvement with processes that ordinary people cannot understand. The symbols and formulas of modern physics and chemistry, a convenient shorthand for their practitioners, represent an almost insuperable barrier to nonscientists wishing to understand it. The alchemist's cave may have become the chemist's laboratory, but the principle is the same.

CHAPTER TWO

BACON'S NEW
SCIENTISTS

We have consultations, which of the inventions and experiences,
which we have discovered, shall be published, and which not: . . .
we do publish such new profitable inventions, as we think good.
—Bacon

Prior to the seventeenth century, the conceptual traditions of Europe were still fundamentally inimical to the progress of science and, equally, to the revaluation of the scientist as a respectable member of society. First, the Old Testament taught that the desire for knowledge was both dangerous and evil, for the Eden myth associated the search for knowledge with disobedience to God, with pride and presumption, and, ultimately, with the Fall of Man. The pursuit of any branch of knowledge other than theology was therefore regarded with suspicion and fear as being likely to engender atheism, even diabolism, as portrayed in the various versions of the Faust legend. Further, the Renaissance admiration for the classical writers reinforced the idea that mankind, so far from improving, was embarked on a decline, intellectual as well as ethical. The golden age was firmly located in the distant past and regarded as being beyond recall either in the present or in the foreseeable future. There was thus a marked degree of pessimism about the future of mankind, which militated against that belief in progress that underlies the pursuit of science in the modern sense. In addition, there was no prevailing confidence that the individual had the ability to determine his own or society's development and overcome his problems; that role was still firmly assigned to Providence. Such resignation of initiative was incompatible both with a belief in causality and with the development of scientific method.

Again, while the first flush of Renaissance humanism had temporarily promoted the physical world (as opposed to a spiritual dimension) to a position where it was considered worthy of study, by the last decades of the sixteenth century there was a growing sense of disenchantment with the high hopes and promises of the Renaissance and the Reformation: neither had ushered in the millennium, and man remained as much a victim of his earthbound state as ever. In England the economic and religious troubles, as well as the political intrigues that had dogged the last decade of Elizabeth's reign, fostered a widespread disillusionment with the power of reason to resolve the difficulties of the age. Despite the impetus given to scientific investigation by Copernicus's *On the Revolutions of the Heavenly Bodies* (1543) and Kepler's *New Astronomy* (1609), Sir Francis Bacon, Lord Verulam and sometime Lord Chancellor of England, actually feared a total collapse of learning and the onset of a second Dark Age. His precarious personal situation after falling into disfavor with Queen Elizabeth may have contributed to his pessimism, but, for whatever reason, he wrote in 1603:

> I see in the present time some kind of impending decline and fall of the knowledge and erudition now in use. Not that I apprehend any more barbarian invasions . . . but from the civil wars which may be expected I think . . . to spread through many countries, from the malignity of religious sects, and from those compendious artifices and devices which have crept into the place of solid erudition I have augured a storm not less fatal for literature and science.[1]

In his determination to stem the tide of ignorance and darkness and institute a complete reform of learning that he called the "Great Instauration," Bacon's first step was to change the unfavorable image of scientific study and break the link with the Faust legend. This he did by what might be seen as a clever theological ruse, although there is no evidence to suggest disingenuousness on Bacon's part. By neatly locating the basis of science in God's laws as embodied in nature, he at one stroke made the study of such laws not only ethically respectable but virtually mandatory for the believing Christian.[2] Indeed, Bacon inverted the traditional implication of the story of the Fall to suggest the possibility of a glorious restoration "of man to the sovereignty and power (for whensoever he shall be able to call the creatures by their true names he shall again command them) which he had in the first state of creation."[3] Thus natural science, so far from being an instrument of the Devil, is presented as a means of overcoming man's postlapsarian limitations.

Bacon's program involved not only a new conception of the proper goals of study and an apparently new methodology—empirical observation and experiment—whereby they could be implemented, but also a new image of the scientist, who could be entrusted to carry out the great plan. This last premise was fundamental to the success of the whole undertaking, for Bacon was all too aware of that "sort of discredit . . . that groweth unto learning from learned men themselves" and "those errors and vanities which have intervened among the studies themselves of the learned."[4]

For men have entered into a desire of learning and knowledge, sometimes upon a natural curiosity and inquisitive appetite; sometimes to entertain their minds with variety and delight; sometimes for ornament and reputation; and sometimes to enable them to victory of wit and contradiction; and most times for lucre and profession; and seldom sincerely to give a true account of their gift of reason to the benefit and use of men.[5]

Bacon therefore spent considerable effort in recasting the image of his man of science; apart from his many philosophical works, his fictional *New Atlantis,* published in 1626, the year of his death, was devised specifically to popularize his ideas about the place of science and scientists in society. In this work a party of European travelers, representing Bacon's contemporaries, comes by accident to the utopian island of Bensalem (the good Jerusalem?), where they have the opportunity to investigate the college of natural philosophy, called Salomon's House, "dedicated to the study of the works and creatures of God . . . the noblest foundation (as we think) that ever was upon the earth, the lantern of this kingdom."[6] In Bensalem, scientists are respected and revered in the community as a new moral elite committed to socially directed research.[7]

In characterizing his scientists, Bacon specifically rejected the traditional ideal of contemplation, adopted by both classical scholars and the medieval church, on the grounds that "it is reserved only for God and the Angels to be lookers on."[8] Instead he proposed an ideal of active service, *philanthropia:* the needs of the community were to determine what research should be pursued. Thus a primary characteristic of Bacon's new scientist is compassion for "the sorrows of mankind and the pilgrimage of this our life," and his chief desire is to alleviate these through the fruits of his learning. In his preface to the *Magna Instauratio* Bacon declares:

> Lastly I would address one general admonition to all; that they consider what are the true ends of knowledge, and that they seek it not either for pleasure of the mind, or for contention, or for superiority to others, or for profit or fame, or power or any of these inferior things; *but for the benefit and use of life; and that they perfect and govern it in charity.*[9]

It is clear from what follows that Bacon's premise is essentially utilitarian, interpreting human welfare in almost wholly material terms of comfort and technological competence. This was in sharp contrast to the contemporary view of compassion as secondary to evangelism and looked forward to the modern welfare state based on humanist values.

Given this materialist premise, it was inevitable that Bacon's new scientists would display attitudes very different from those of their predecessors, whether classical or medieval. With hindsight we can see that these attitudes were necessary to the development of modern science, and it is arguable that without Bacon's analysis its systematic development might have been much slower. To start with, in seeking to ameliorate the conditions of mankind in this world, his scientists are basically secular and future-oriented in their thinking. No longer is the golden age located in the classical past; it is waiting to be ushered in through the efforts of noble and altruistic scientists. This leads to two innovatory effects on both the methodology of the new science and the ethical behavior of the scientists. As distinct from the precedent-oriented system of the past, where some earlier authority, usually Aristotle, was invoked to determine the validity of all claims, Francis Bacon, like his earlier namesake, Roger Bacon, stresses the importance of experimentation and observation of the particular as the necessary basis of the general. Indeed, he explicitly accuses Aristotle of bending the facts to suit his hypotheses:

> Nor let any weight be given to the fact that in his [Aristotle's] books on animals, and his problems, and other of his treatises, there is frequent dealing with experiments. For he had come to his conclusion before; he did not consult experience, as he should have done, in order [to proceed] to the framing of his decisions and axioms; but having first determined the question to his will, he then resorts to experience, and bending her unto conformity with his placets leads her about like a captive in a procession.[10]

Similarly, Bacon rejects the popular medieval sport of logic-chopping because "it commands assent therefore to the proposition, but does not take hold of the thing."[11] Instead, he claims that his own methodology is grounded in observation, experimentation, and induction.[12] Not content with examining natural phenomena under normal conditions, these scientists have extended the scope of their research and are investigating the behavior of substances at immense altitude, depth, and temperature in order to be able to generalize. This necessarily involves new equipment; indeed, almost one-quarter of *New Atlantis* is devoted to a description of the "Preparations and Instruments" of Salomon's House.[13] The preoccupation with equipment was to characterize the development of British science from the early days of the Royal Society and played a significant role in forwarding the industrial revolution in England in advance of its European neighbors, who were still conducting science largely along theoretical lines. Bacon also introduced the notion of the international community of science as his "Merchants of Light," who, unlike that other contemporary hero figure, the patriot, transcend national boundaries and allegiances, traveling the world to collect and share their knowledge with all races.[14]

Undeterred by the fact that the accomplishment of these elaborate experiments and international travels was impossible for individuals or even for one generation of scientists, Bacon proposed the idea of a team effort, whereby scientists would labor not for their own individual goals but as part of a long-term communal venture. They are therefore free to specialize in various areas and procedures, happy that their results will be absorbed into the growing corpus of knowledge.[15] In thus replacing the jealously guarded isolation of the alchemists with a group effort, Bacon was, of course, also demystifying scientific knowledge and making it more acceptable.

This team concept presupposes, in turn, that the scientists concerned have no need to become involved in the competitive scramble for individual success and honors. In answer to the objection that this is to ask for the impossible, for a cessation of human nature, Bacon argues that the motivation of *philanthropia* can overcome the elements of pride and self-interest that had been apparent in even the most highly esteemed scholars of the past. So effective was Bacon's publicizing zeal that since his time this belief in the essential altruism of science has been reaffirmed whenever scientists have been perceived as particularly heroic or necessary to the community.

Being part of a team also frees Bacon's scientists from the ultimate sin of despair. Whereas Faust and the alchemists endeavored to discover all the secrets of the universe in isolation and inevitably lost hope of completing such a task in one lifetime, Bacon's scientist is released from such a burden. He knows that his own particular discoveries will become part of an ongoing stream of knowledge and rests content in having contributed to the future store.[16]

Although he insisted on the need for an empirical basis for knowledge, Bacon did not underestimate the importance of theory; rather, he evolved a rationale of research and development that had to wait some three centuries for implementation. He realized that pure research, which he called *experimenta lucifera* (experiments bearing light), if based on correct observation, could in the long term be more productive of human good than could a narrow, technologically based science, *experimenta fructifera* (experiments bearing fruit), which sought only immediate and obvious gains.

Again, despite their long-term perspective, Bacon's scientists are far from lethargic in their efforts. It is not accidental that Bacon should be best remembered for the aphorism Knowledge is Power, for his scientists are passionately committed to an active program of subduing nature to the will of man. The Father of the House of Salomon asserts: "The End of our Foundation is the knowledge of Causes and secret motions of things, and the enlarging of the bounds of Human Empire, to the effecting of all things possible."[17] It is for this reason that in recent years Bacon has been held responsible for the mechanistic and reductionist rationale of science and in particular for the patriarchal and imperialist attitude toward nature implicit in the notion of ordering, taming, and dominating the natural world by technological means.[18]

Unlike the leaders of Sir Thomas More's Utopia, the scientists of New Atlantis are treated with pomp, even reverence,[19] and dressed with princely richness and finery. This splendor has been seen as an indication of Bacon's conservatism in the social and political realm, as distinct from his innovation and reforming zeal in the intellectual sphere,[20] but an alternative and more likely explanation is that Bacon was endeavoring to make the role of the scientist acceptable to an aristocratic class that did not readily identify with austerity. In effect, the scientists of Salomon's House have become a new elite class, a new priesthood, revered by the people and performing special religious ceremonies.[21] In *New Atlantis* the scientists exert their own censorship concerning which of their discoveries shall be made public on the basis of the common good. "We have consultations, which of the inventions and experiences, which we have discovered, shall be published, and which not: And take all an oath of secrecy, for the concealing of those which we think fit to keep secret."[22]

If Bacon was optimistic about the readiness with which society would embrace science, he seems to twentieth-century readers (who have every reason to be hypersensitive to such matters) even more so in regard to the moral qualities of his scientists. "Only let man regain his right over Nature, which belongs to him by the gift of God; let there be given to him the power: right reason and sound religion will teach him how to apply it."[23]

That Bacon made no attempt to accommodate his enterprise to the church was both unusual and highly significant. Although the ethical values of Bacon's scientists are parallel to those of Christianity, their basis is secular rather than religious, pragmatic rather than devotional. The scientist is to cultivate charity and eschew pride, not because he has a Christian obligation to do so or because he expects a heavenly reward for his altruism, but because such behavior is ultimately more effective in the pursuit of science. Indeed, Bacon has been suspected of opportunism, if not of hypocrisy, in his claims that the study of nature leads to a knowledge of the existence, character, and omnipotence of God,[24] especially since he carefully avoids any reference to the argument from design so enthusiastically embraced by the next generation of natural philosophers, who sought an alliance with the church. On the contrary, he actively advised that "we do not presume by the contemplation of nature to attain to the mysteries of God" and "do not unwisely mingle or confound these learnings [i.e., natural knowledge and religion] together."[25] In carefully distinguishing between scientific and religious knowledge, Bacon was not intending to denigrate the pursuit of science; perhaps, rather, the contrary. Whatever his reasons, it was good strategy, insofar as it forestalled any charge that scientists were bent on undermining the power of the church, a charge to which the alchemists, with their cult of mystery and offer of magical powers, had laid themselves open. In this assertion of two independent truths, sacred and secular, Bacon paved the way for the future divergence of science from religion and the autonomy of the former.[26]

Despite his innovative program and methodology, however, Bacon himself was no scientist. Although he stressed the importance of experimental method, he ignored the importance of precise measurement and of mathematics; indeed, his view of science was almost wholly qualitative. He also seems to have had a naive faith in the use of inductive method, believing that if sufficient observational data were collected, a general scientific law would inevitably emerge. As we shall see in the next chapter, this was to have a profound effect on the early years of the Royal Society, encouraging indiscriminate collection of anything and everything as potentially useful. However, Bacon was extremely important as an entrepreneur of the new science, expounding and explaining and, above all, making its practitioners respectable.[27] He was the first to free scientists from the cloud cast upon them by the medieval church and to hold them up as the natural and acceptable leaders of a utopian state in which material and ethical progress seemed to develop in harmony.

The next fictional work to incorporate Bacon's ideas was *The Man in the Moone: or a Discourse of the Voyage thither by Domingo Gonsales, the Speedy Messenger*, published anonymously in 1638 by Francis Godwin, bishop of Hereford. Gonsales, arguably the first individual experimentalist in fiction, ignores Aristotelian dogma, building instead on the results of his controlled observations, as Bacon had advocated. Indeed, it seems likely that the basic idea for Gonsales's flying machine was suggested by Bacon's description of an experiment for flying in *Sylva Sylvarum*. Both have in common the use of birds connected to the flying contraption by strings and pulleys and the initial strategy of taking off from a high cliff over the water to make use of the updraft. In defiance of Aristotle, who had taught that a body is always attracted to its ordained place in the universe, Godwin adopted Bacon's idea that the force of the earth's gravity decreases with distance.[28] Thus Gonsales, drawn upwards by his twenty-five birds, discovers that after about an hour the birds need to strain less to bear his weight.

Being a conscientious Baconian experimentalist, Gonsales also dismisses Aristotle's view that the upper atmosphere must be fiery hot and moist (since it was the home of the element of fire and since the attributes of air were heat and moisture) on the grounds that "my experience found [it] most untrue."[29] This manner of repudiating Aristotle by appeal to experience—"mine eyes having sufficiently informed me"—is peculiarly Baconian and constitutes the basis of his experimental method. Gonsales also adduces "experience" to "prove" that the earth rotates below him like a "huge Mathematicall Globe."[30]

Gonsales's discovery of his flying mechanism is itself the result of Bacon's methodology, the chain effect of observation, hypothesis, experimentation, and induction. A chance observation of the characteristics of the native birds or gansas suggests a means of sending messages across the island; this is then linked with his knowledge of physics (pulleys) to suggest a mechanism for lifting materials. Gonsales then tests the device with a sheep before trying it himself. His subsequent voyage to the moon, itself the result of this earlier scientific process, leads to observations that in turn overthrow old hypotheses and suggest new explanations of phenomena. This was exactly the procedure Bacon had outlined in the House of Salomon.

Like his predecessor, Godwin never doubts that the pursuit of science will be an unmixed blessing, for the scientist is an instrument of God's purpose, and certainly God will prevent the discovery of whatever might prove harmful. When Gonsales asks the Lunars whether they have discovered how to make themselves invisible, they answer that "God would not suffer it to be revealed to us creatures subject to so many imperfections, being a thing so apt to be abused to ill purposes."[31] Thus, by the simple expedient of invoking God's power to keep hidden whatever may prove harmful, Godwin, like Bacon, claims divine blessing on all scientific enterprise and vindicates whatever discoveries may come to light. It is a neat piece of theological argument, for to doubt the conclusion is to be caught doubting the implicit major premise of God's omnipotence.

The most influential of Bacon's contemporaries to adopt his agenda was Samuel Hartlib (1600–1662), who was interested in reforming education and promoting the open dissemination of knowledge. Hartlib not only collected about him a circle of prominent intellectuals from various disciplines who were interested in integrating their ideas but set himself up as a collection center for all kinds of information and attempted to obtain government funds for this undertaking. In this latter intention he was unsuccessful, but at his death the role was taken over by Henry Oldenburg, who, as secretary of the Royal Society, at first maintained it as a personal initiative and then, in 1665, formalized such dissemination in the *Philosophical Transactions of the Royal Society*.[32] This venture exemplified Bacon's ideal of the widespread collation and open communication of scientific knowledge and contributed significantly to canceling the associations of secrecy and mysticism that had characterized the image of the alchemist.

Bacon's House of Salomon inspired not only the "Philosophical Colledge" of Abraham Cowley's imagination[33] but, in more concrete terms, the institution in London of the College of Philosophy (1645) which after the restoration of the monarchy became the Royal Society (1662), acknowledged as being "due to the prophetic scheme of 'Salomon's House' in the *New Atlantis*." The mathematician John Wallis and the chemist Robert Boyle, two founding members of the Royal Society, explicitly acknowledged Bacon's model, while Thomas Sprat wrote in his *History of the Royal Society of London*

(1667): "I shall onely mention one great Man who had the true Imagination of the whole extent of this Enterprize, as it is now set on foot, and that is Lord Bacon. In whose Books there are every where scattered the best arguments that can be produc'd for the defence of Experimental Philosophy: and the best directions that are needful to promote it. . . . if my desires could have prevail'd . . . there should have been no other Preface to the History of the Royal Society but some of his Writings."[34] The frontispiece of Sprat's *History* features an engraving of Charles II and Bacon in which Charles II is described as "Author et Patronus," and Bacon as "Artium Instaurator" (see fig. 4), while Abraham Cowley's ode "To the Royal Society," printed in the same volume, compares Bacon to Moses in rescuing mankind from the wilderness of ignorance.[35]

The Royal Society's indebtedness to Bacon rested chiefly on its adoption of his interlocking propositions of experimentalism, open communication, and usefulness. Sprat, contrasting the new science with the former scholastic philosophy, stressed this aspect of usefulness,[36] and from its inception, the Royal Society enjoined its members to share information. In his introduction to the first issue of the *Philosophical Transactions* (1666), Henry Oldenburg, the then secretary, pointed out that the new periodical was being published "to the end that such Productions being clearly and truly communicated, desires after solid and useful knowledge may be further entertained . . . and those addicted to . . . such matters, may be invited and encouraged to try and find out new things, impart their knowledge to one another, and contribute what they can to the Grand design of improving Natural knowledge."

Bacon's proposal for the responsible censorship of knowledge by the members of the House of Salomon was also adopted by some members of the Royal Society. The eminent chemist Robert Boyle published in the *Philosophical Transactions* an account of an experiment whereby he had succeeded in combining a particularly "noble" form of mercury with gold to produce an exothermic reaction that he believed would be therapeutic. (Such an incalescent reaction had long been the dream of alchemists.) However, Boyle was uncertain about the possible "political inconveniences" that might attend the recognition of such mercury if it should "fall into ill hands"; he was therefore using his article to consult with the learned world about the relative merits of keeping silent about such a substance. Isaac Newton had no doubt that Boyle should maintain "high silence" and informed Secretary Oldenburg of this.[37]

New Atlantis also inspired both the founders of other academies in Europe and the French Encyclopedists in their resistance to attempts by the church to limit and control science. The Encyclopedists adopted with enthusiasm Bacon's program of cataloguing all knowledge about the natural

FIGURE 4. Frontispiece from Thomas Sprat's *History of the Royal Society of London* [1667], showing, *from left to right:* William, viscount Bruckner, president of the Royal Society; Charles II as "Author et Patronus"; and Sir Francis Bacon as "Artium Instaurator."

world, as Diderot was to acknowledge in the "Prospectus" to his *Encyclopé-die:* "If we have come to it successfully, we shall owe most to the Chancellor Bacon."[38]

    When we recall the secrecy associated with the work of the alchemists and the way they protected the mystery of their activities in order to preserve and heighten their magical power over the uninitiated, it is immediately apparent how completely Bacon recast the scientist as an idealized figure that has been rivaled in only a handful of utopian proposals and that has continued to exert influence over modern science. It is therefore somewhat ironic that Bacon derived many of the tenets of his natural philosophy, including suggestions for transmuting other metals into gold, from alchemy, with which he was well acquainted.[39] This, however, is not apparent in *New Atlantis,* and it is significant that this text has been selected, more or less in isolation, as a model for the scientific utopia that not only appealed to Bacon's century but was to reemerge in almost the same terms in the early twentieth century and to endure until the Second World War.

Bacon's call for empirical procedures, the preeminence of experimental results over dogma and authority, his ideal of an international cooperative effort by scientists devoted to the ongoing pursuit of truth, the requirement that research be open and its results communicated to other scientists, and, perhaps more contentiously, his stipulation that science should serve the needs of the community rather than its own purposes have all become accepted principles of modern science, even when, as in wartime, for instance, it pays only lip service to them. Moreover, although Bacon's account of the methodology of science as empirical observation, hypothesis, experiment, and inductive generalization has been challenged by modern philosophers of science, such as Karl Popper, Thomas Kuhn, and Paul Feyerabend, who deny that this is how science in fact operates, it still corresponds most closely to the popular conception of how scientists work and has thus been extremely influential in determining how scientists are presented in fiction, which in most cases is written by nonscientists.

It is apparent that all these ideals hinge on the moral caliber of the scientists involved. Hence it is not surprising that the proposals of Bacon, like those of H. G. Wells, his twentieth-century successor in proposing a scientifically based utopia, depend almost entirely on the assumption that scientists have a preeminent sense of dedication and morality. This was perceived by their critics as the weakest point in the utopian argument; but in Bacon's case this reaction was longer in coming, for with the institution of the Royal Society, it did indeed seem to many that the "New Atlantis" would soon emerge in England, if not throughout Europe. Bacon's infectious belief in the glorious future of learning, if men would but sail beyond the Pillars of Hercules, first cast the scientist in the role of adventurer and world savior, an idea that was to reemerge with more problematic implications in the twentieth century.

In elevating and idealizing scientists, however, Bacon simultaneously degraded nature to the status of a relatively impotent opponent to be subdued: "If a man endeavour to establish and extend the power and dominion of the human itself race over the universe, his ambition (if ambition it can be called) is without doubt a more wholesome thing and a more noble than the other two [those who seek individual or national prestige]."[40] Prior to Bacon's rallying call "Knowledge is Power," natural philosophers, whether alchemists, physicians, or herbalists, perceived themselves as an integral part of a (fallen) nature; but Bacon's ideas hinge on a revival of the prelapsarian concept of man as set over nature, to dominate and control it. This assumption was to survive longer even than that of the noble scientist, and except for a brief period coinciding with the Romantic movement (discussed in chapter 6), it has continued to dominate both the methodology of science and the attitudes of Western society until the present day. However, the checkered estimate that characterized Bacon's unfortunate political life has been reflected in the evaluation of his philosophical ideas. After enjoying in the early twentieth century the adulation of H. G. Wells, who effectively adopted the scheme of New Atlantis for his scientific utopias, Bacon is again under review. In the view of many environmentalist philosophers of science today, and especially of ecofeminists, Bacon bears a major part of the responsibility for the deep division between science and nature and for the perception of nature as a passive object available for exploitation, manipulation, and domination.[41] He is thus, by extension, regarded as responsible for the environmental crisis of the late twentieth century. His immediate successors, however, were troubled by no such problematic considerations.

CHAPTER THREE

FOOLISH
VIRTUOSI

The smallest worm insults the sage's hand;
All Gresham's vanquish'd by a grain of sand.
—Henry Jones

It now seems a strange irony that the first systematic satires of scientists should have been elicited by the founding members of the Royal Society, that is, by the first people we may consider scientists in the modern sense. Yet they, along with their contemporaries the amateur virtuosi, were ridiculed on the Restoration stage as pursuing useless, irrelevant, and usually disgusting research into topics with which no gentleman should concern himself. Even Charles II, their first and most prestigious patron, was not above mocking their endeavors.[1] Later they were rapped on the knuckles by the leading moralists of the day for intellectual arrogance and, perhaps worse, for the "Dulness" of the factual and unadorned style of writing they espoused. The complex reasons underlying this irreverent reception of some of the great names in the history of science are still debated,[2] but the following historical considerations were certainly major factors in producing that climate of opinion.

In the latter half of the seventeenth century the ideas put forward in Bacon's *New Atlantis* suddenly gained momentum. Of particular importance was his proposal that the new learning should reject authority per se and substitute for it the inductive method based on observation and experimentation. This procedure was overwhelmingly vindicated by the English physician William Harvey's discovery of the circulation of the blood and his concept of the heart as a mechanical pump. Harvey had trained at the famous medical school in Padua under the anatomist Fabricius, who also founded the science of embryology. Fabricius had discovered valves in the large veins but had not realized their significance because current ideas about the circulation of the blood, based on the second-century theories of Galen, could not accommodate them.[3] Harvey returned to England to study the circulatory system without any preconceptions, experimenting with tying the blood vessels of animals to examine the flow of blood. He soon established that blood flows away from the heart through the arteries and returns through the veins, directed by the valves. He professed "to learn and teach anatomy not from books but from dissections, not from the tenets of Philosophers but from the fabric of Nature."[4] Harvey was also the first to apply mathematical calculations to the volume and flow of blood, deducing that the blood had to circulate continually. From this he concluded that the heart acted as a pump to keep the blood in constant circulation. Harvey was thus one of the first scientists to evolve a theory based on experimental observation, and his book *De Motu Cordis* (1628) revolutionized physiology by overthrowing centuries of authority.

The founding of the Royal Society of London fulfilled another of Bacon's recommendations, namely, that scientists should collaborate and pool their results in the common cause of producing a cumulative natural history, from which the fundamental laws governing the universe would eventually be deduced. The "Invisible Colledge," which met in the rooms of Sir Christopher Wren, professor of astronomy at Gresham College, was one of several scientific groups to spring up in London and Oxford during the Interregnum, and in 1660 their members amalgamated. Two years later this larger group was granted a royal charter, thereby becoming the Royal Society of London for the Improving of Natural Knowledge.[5] True to the ideals of their spiritual founder, Sir Francis Bacon, the members of the Royal Society took as their motto *Nullius in verba*, accepting nothing simply from report. The physicist Robert Hooke, one of the founding members, described the aims of the Royal Society in terms that read like a summary of the practices in New Atlantis: "This Society will not own any Hypothesis, System or doctrine of the Principles of Naturall Philosophy, . . . nor dogmatically define, nor fix axioms of scientific Things, but will question and canvass all till by mature debate and clear argumentism, chiefly such as are deduced from legitimate Experiments, the Truth of such Experiments be demonstrated."[6]

Such a high value was placed upon the actual performance of experiments before the gathered members that nothing was considered demonstrated "till the whole Company has been fully satisfi'd of the certainty and constance; or, on the other side, of the absolute impossibility of the effect."[7] Theoretical science was viewed skeptically by the early members of the Royal Society, for it savored too much of the clean-hands approach of Aristotle and the ancients and was thus open to the abuses of the logicians. They voiced their reservations even of Descartes, because he "did not perfectly tread in his [Bacon's] Steps, since he was for doing too great a part of his work in his Closet, concluding too soon, before he had made Experiments enough."[8]

This emphasis on observation and experimentation and the resulting skepticism about hypotheses not grounded on sense data seemed to imply a distrust of pure reason at a time when, through a combination of factors, the temper of the age was more strongly rationalistic than at any other time in British history.[9] Further, the Royal Society's emphasis on mechanical rather than theoretical science, in accordance with Bacon's insistence that research should be socially useful, appeared to the urbane Restoration wits a continuation of the unlovely utilitarianism of the Puritans.

The association of science with Puritanism was, indeed, more than coincidental, although the exact nature of the connection is hotly disputed.[10] Whatever the details, it remains clear that during the Interregnum, Puritan enthusiasts had tried to instigate revolutionary changes in the university curriculum, severely curtailing humanities subjects in order to make way for experimental science on the Baconian model, that is, vocational and immediately practical knowledge of such matters as planting orchards and grinding lenses. Christ College, Oxford, was proposed as the first institution to be radically "reformed" along these lines. It is small wonder that when the danger passed, along with the Commonwealth, the threatened faculties retaliated with attacks on the Royal Society, seemingly the inheritor of these abhorrent views. The bishop of Chichester repeatedly directed sermons against the Royal Society and even objected to the publication of a book of verses because it contained one poem in praise of this body.[11]

Even more implacable enemies of Puritanism were the wits of the Restoration stage. The Puritans had closed the theaters on the grounds that they were the seedbeds of immorality, so it is not surprising that playwrights were in the vanguard of the scientist-bashing movement. In his *History of the Royal Society*, Thomas Sprat records a real fear of the attacks of the "Wits and Railleurs": "For they perhaps by making it [the New Philosophy] ridiculous, becaus it is *new*, and becaus they themselves are unwilling to take pains about it, may do it more injury than all the Arguments of our severe and frowning and dogmatical *Adversaries*."[12]

Another serious liability of the scientific cause was the charge of atheism arising from its perceived association with Descartes's mechanical philosophy. Descartes had developed his *philosophia mechanica*, purporting to explain all natural phenomena mathematically on the basis of matter and motion, rather than teleologically, as Aristotle had done. As a result of Harvey's work, even animate nature came to be understood mechanically in terms of pumps and levers, differing from inanimate nature only in containing a spirit, an immaterial substance as well as matter. Despite their reservations about Descartes's nonexperimental method, Bacon's followers initially welcomed the Cartesian dualism as being conducive to their aim of ultimately explaining all phenomena in terms of a few basic physical laws, and at first this alliance with a respectable rationalist philosopher worked to their advantage. However, when the atheistic English philosopher Thomas Hobbes ridiculed the immaterial aspect of the dualistic system, it began to appear that the Cartesian model was fundamentally irreligious and that the scientists were trying to dispose of God, leaving only a mechanical creation. In a climate of atheist-hunting, this made the scientists a ready target. Edmund Halley gained a reputation for mechanistic views, from which he had to extricate himself,[13] and had the Royal Society not numbered so many bishops among its founding members, it might have fared much worse. Hobbes's enthusiastic support became a liability the Royal Society could ill afford, and in their zeal to deny any such alliance and show that they gave due weight to spiritual, "immaterial substance,"[14] some of its members even attempted to gather data in support of ghosts and witches.[15] Joseph Glanvill, a devoted Baconian, encouraged by Robert Boyle the acclaimed physicist, was almost obsessive in collecting allegations of witchcraft, which he published in *Saducismus Triumphatus: A full and plain Evidence Concerning Witches and Apparitions,* a work that ran into four editions from 1681 to 1726. Boyle himself published *The Excellencey of Theology, Compared with Natural Philosophy* (1674) to counter the fact that "under-valuation . . . of the study of things sacred . . . is grown so rife among many (otherwise ingenious) persons, especially studiers of physicks."[16] The most enduring vindication of the pursuit of science in relation to religious faith was John Ray's *Wisdom of God Manifested in the Works of the Creation* (1691), in which he, like Boyle, argued for the concept of the scientist as the priest of natural theology. Westfall has suggested that the amount of time spent by these eminent naturalists in vindicating religion was related to their own unacknowledged fears: "More than answering hypothetical atheists, they were trying to satisfy their own doubts."[17]

As suggested in earlier chapters, it would be a mistake to assume that these Royal Society members had entirely rejected the profession of alchemy. Many, including some of the most eminent names, such as Boyle and Newton, were still deeply involved in the riddles of alchemy, although they attempted to resolve them in a different way, inclining towards the mechanical approach of Descartes.[18] Thus much of the opprobrium associated with alchemists still attached to them, at least in the early years of the society. These scientifically doubtful excursions into the occult furnished another stick with which the rationalists could beat members of the Royal Society; indeed, the ghost-hunting activities of scientists feature in several contemporary satires along with the escapades of the virtuosi.

In terms of their representation in literature, however, the worst liability of the Royal Society members was their association with the virtuosi. An interesting feature of Restoration science was the strong bond of collaboration between the recognized scientists (equivalent to today's trained, professional scientists) and the much larger body of "enthusiasts" (corresponding to today's amateurs), who had less educational background in science but unbounded enthusiasm for miscellaneous scientific projects, believing that they were thereby contributing important data to a cumulative natural history. These wealthy amateurs, frequently called virtuosi to distinguish them from the scientific elite or natural philosophers,[19] were in many ways the successors of the earlier, less sophisticated alchemists and, like them, included among their ranks the whole gamut of charlatans, gullible fools, and serious seekers after knowledge. They were certainly a doubtful asset to the Royal Society, for although some of them amassed collections of scientific value, most were indiscriminate hoarders of anything and everything. Their private museums, or "cabinets," were carefully cataloged and absurdly overvalued, yet many of the major scientists of the time were reluctant to discard them, in case they should prove important in Bacon's long-term scheme (see fig. 5).[20] With their huge but indiscriminate collections, the virtuosi effectively parodied one of Bacon's fundamental ideas, the amassing of as many separate facts as possible in order eventually to understand the whole. This emphasis on analytical procedures, which was to characterize science for the next three centuries, involved a consequent rejection of the holistic approach favored by the Greek and medieval philosophers, who had attempted to understand the part teleologically in terms of the whole.

The virtuosi, however, fulfilled other, more useful functions. Many contributed considerable funds to the Royal Society, served as officers, and eagerly undertook the expensive and time-consuming collection of vast amounts of data for the "scientists" to examine and interpret. Moreover, without the enlarged market provided by the virtuosi, it is unlikely that scientific apparatus and instruments would have developed as rapidly, or become as affordable, as they did.[21]

Many of the virtuosi were merely dilettantes, though, jumping on any fashionable bandwagon and dabbling in sensational curiosities, without any serious scientific purpose. These characteristics were a gift to the satirist, and it is not surprising that a large number of Restoration plays feature foolish, would-be scientists grubbing among useless or disgusting trivia, oblivious to what is going on around them. These characters were, indeed, the predecessors of the absent-minded professor, a type that was to bulk large in less biting comedies about scientists in the early twentieth century.

Much of the satire was directed as indiscriminately as the name *virtuosi*, so that in 1697 William Wotton expressed the opinion that the Royal Society had suffered more from "the sly Insinuations of the *Men of Wit* . . . That every Man whom they call a *Virtuoso* must needs be a *Sir Nicholas Gimcrack*" than from allegations of atheism. Indeed, the sources of funding were drying up as wealthy patrons came to be persuaded that the Royal Society was a haven for charlatans and incompetents.[22]

Another feature of contemporary science to evoke ridicule was the development of the telescope. Attempts to improve on Kepler's refracting telescope by using weaker convex objective lenses led to the production of

FIGURE 5. *The Museum of a Virtuoso*, by Ferrante Imperato, 1672. Historia Naturale, Venice.

longer and longer instruments. Johannes Hevelius produced his *Seleno-graphia* (1647), the first atlas of the moon, using telescopes from 8 to 10 feet long; and his 150-foot instrument erected in Danzig proved so difficult to point, and so unstable in even the slightest breeze, that its usefulness was severely limited.[23] It was little wonder that news of it provided ammunition for visual caricature. Stage telescopes were invariably so long and cumbersome as to require several bearers, and astronomers were represented as either gullible, prey to wild speculations, or devious, insofar as they attempted to deceive others. Since astronomy had played an important role in the researches of the Royal Society from its inception in Wren's Invisible Colledge, such satirical representations necessarily rebounded upon its members.

It is against this background that we must look at the representations of scientists in the literature from the Restoration to the mid-eighteenth century.

Early satiric representations of scientists in this period show them as wholly comic, almost invariably stupid, and out of touch with the real world. They imagine that they have great power over nature or are about to acquire it through some new discovery, but they pose no threat to society because, not having the sense to realize their own stupidity, they are easily outwitted or manipulated. One of the first such satires was *The Description of a New World called the Blazing World* (1666), by Margaret Cavendish, duchess of Newcastle, an early feminist who prided herself on her scientific interests and knowledge. Somewhat to the consternation of the Royal Society, she insisted on honoring this hitherto exclusively male province with a visit, an event that elicited several comic ballads as well as a caustic description by Samuel Pepys of the lady's eccentric style of dress.

Cavendish was intent on exposing the pettiness and self-seeking nature of contemporary scientific debate, and *The Blazing World* is essentially a moral beast fable set in a new world where the scientists combine human qualities with the characteristics of various animal species.[24] The Bear-men are experimental philosophers, the Bird-men are astronomers, the Fly-worm- and Fish-men are natural philosophers, the Ape-men are chemists, and the Spider- and Lice-men are mathematicians. The Empress of this fabled world soon finds that, instead of proceeding with the accumulation of knowledge according to her Baconian plan, her subjects have resorted to disputes and quarrels. The Bird-astronomers argue fiercely over contemporary points of scientific contention—the cause of the sun's light, why the sun and moon change their shape and size, and why there are spots on the sun. When the Empress requests her experimental philosophers to resolve the dispute by observing the appropriate bodies through their telescopes, she finds that they cannot agree either about what they see or about what it means.

> After they had thus argued, the Empress began to grow angry at their Telescopes, that they could give no better Intelligence; for, said she, now I do plainly perceive, that your Glasses are false Informers, and instead of discovering the Truth, delude your senses; Wherefore I Command you to break them, and let the Bird-men return to their natural eyes, and examine Celestial objects by the motions of their own sense and reason. . . . Nature has made your sense and reason more regular than Art has made your Glasses, for they are meer deluders, and will never lead you to the knowledge of Truth; Wherefore I command you to break them; for you may observe the progressive motions of Celestial bodies with your natural eyes better than through Artificial Glasses.[25]

This passage involves pointed criticism of Francis Bacon, who had affirmed the need for continual invention of new tools, since unaided sense and reason were incapable of discovering the truth. However, the response Cavendish attributes to the astronomers is even more satirical: they plead to be allowed to keep their telescopes *in order* to prolong their disputes.

For, said they, we take more delight in Artificial delusions, then in natural truths. . . . for were there nothing but truth, and no falsehood, there would be no occasion for to dispute, and by this means we should want the aim and pleasure of our endeavours in confuting and contradicting each other; neither would one man be thought wiser than another, but all would either be alike knowing and wise, or all would be fools; wherefore we most humbly beseech your Imperial Majesty to spare our Glasses, which are our onely delight, and as dear to us as our lives.   (28)

This early satire introduces many of the criticisms that were to be leveled against the virtuosi, criticisms of their delight in disputation as an end in itself, their concern with "delusions" rather than with the truth, and their preoccupation with instruments for their own sake rather than with any attempt to develop their usefulness to society.

It seems unlikely that Margaret Cavendish intended these comments to be applied to the members of the Royal Society, but certainly the so-called "Ballad of Gresham Colledge" (its actual title was more explanatory: "In Praise of that choice Company of Witts and Philosophers who meet on Wednesdays weekly att Gresham Colledge") was specifically directed against this august body. The poem satirizes the scientists' preoccupation with quantitative measurement to the exclusion of qualitative considerations and their determination to reduce natural complexity to simple demonstration. Today, when most of these scientific procedures are taken for granted, we may well miss the intended irony concerning these men who "take nothing uppon trust," a clear reference to the motto of the Royal Society.

> Att Gresham Colledge a learned Knott
> Unparallel'd designes have laid
> To make themselves a Corporation
> And knowe all things by Demonstration.
> . . . .
> By demonstrative Philosophy
> They playnly prove all things are bodyes,
> And those that talke of Qualitie
> They count them all to be meer Noddyes.
> Nature in all her works they trace
> And make her as playne as nose in face.
> . . . .
> To the Danish Agent[26] late was showne
> That where noe Ayre is, there's noe breath.
> A glasse this secret did make knowne
> Where[in] a Catt was put to death.

Out of the glasse the Ayre being screwed'
Pusse dyed and ne're so much as mewed.

The selfe same glasse did likewise cleare
Another secret more profound:
That nought but Ayre unto the Eare
Can be the Medium of Sound,
For in the glasse emptied of Ayre
A striking watch you cannot heare.

. . . .

These men take nothing uppon trust
Therefor in Counsell sate many howres
About fileing Iron into Dust
T'experiment the Loadstone's powers.[27]

The ballad proceeds to describe other investigations being carried out by Royal Society members, all of whom are occupied in practical research designed to improve the material quality of life.[28] Because we now know that the projects mentioned were ultimately successful, it is at first difficult to realize that the poem was intended to be highly satirical, but the author clearly implies that the projects described are either so obvious as to be apparent to common sense (a cat dying when deprived of air) or else ridiculously far-fetched. To appreciate the caustic tone, it is necessary to recall that a century was to pass before longitude could be measured accurately, since it depended on the perfecting of the chronometer, and two hundred years before the diving bell was a practicality.[29]

Samuel Butler also ridiculed the virtuosi as stupid and self-deceived, but in addition he stressed their underlying moral deficiencies. Incapable of perceiving the beauty and wonder of ordinary nature, they were intent only on novelty and, in their insatiable desire for fame, were prepared to sacrifice mere truth. For Butler, astronomy was only a more fashionable form of astrology, and in his astrologer character Sidrophel (star lover) of *Hudibras* (1664) he therefore felt free to ridicule indiscriminately all the scientific preoccupations of his day—Descartes's vortexes, the atomic theory in physiology, blood transfusion, the stentrophonic tube, botany, pendulum watches, chemistry, and bottled air. In *The Elephant in the Moon* (1676) Butler's satirical net was cast more widely still, and he appeared bent on including the whole membership of the Royal Society, since characters corresponding to Sir Paul Neal, Sir John Evelyn, Robert Hooke, Antonie van Leeuwenhoek, and even Robert Boyle have been identified.[30] Butler's charge against the new philosophers was the same one he had leveled at the virtuosi: that they sought marvels rather than the truth. In the poem, which has clear echoes of Margaret Cavendish's satirical fantasy, a group of astronomers gathers to observe the moon, constructing elaborate theories of lunar politics and sociology on the basis of no observable evidence. Indeed, they are determined in advance to observe a "wonder." Soon they are overjoyed to find, as they think, a battle in process on the moon and, in the midst of the fray, an elephant. Delighted at this visual "evidence," they enthusiastically plan an article for the next issue of *Philosophical Transactions*, the journal of the Royal Society, until it is suggested by a servant, the representative of common sense, that the "elephant" is a mouse that has been trapped in the telescope (see fig. 6). The telescope thus symbolizes the distorted vision of the scientists, who are interested only in the hypothetical elephant, not in an actual mouse. Even when apprised of their mistake, Butler's astronomers are anxious to suppress the truth in order to further their scientific reputations, but eventually they open the telescope and find, along with the "elephant" mouse, a swarm of flies and gnats, the "combatants" in the lunar battle. Butler's moral is heavily underlined.

> Those who greedily pursue
> Things wonderful instead of true;
> That in their speculations choose
> To make discoveries strange news;
>
> . . . .
>
> In vain strive Nature to suborn,
> And for their pains are paid with scorn.[31]

Thus Butler sees the scientists as not merely foolish but hypocritical; they profess to seek truth but actively suppress it if it interferes with their pursuit of fame. Instruments designed to improve man's vision—the telescope and the microscope—are of no use against moral blindness and the refusal to see.

Not satisfied with this tour de force, Butler apparently contemplated a more extended satire on contemporary scientists. The incomplete fragment "A Satire on the Royal Society" reduces the scientific debates of the day to ridiculous irrelevancies:

> And all their constant occupations:
> To measure wind, and weigh the air,
> And turn the circle to a square;
> To make a powder of the sun,
> By which all doctors should b'undone;
> To find the Northwest passage out,
> Although the farthest way about; . . .
> To stew the Elixir in a bath
> Of hope, credulity, and faith;[32]

FIGURE 6. *The Astronomer,* colored aquatint by Thomas Rowlandson (1756–1827), published by Rudolph Ackermann, 1 May 1815. The caption reads: "Why I was looking at the Bear: / But what strange Planet see I there!" Reproduced by kind permission of a private collector, to whom the author extends her thanks.

The belief that the virtuosi were interested only in the unnatural, the monstrous, was still current some seventy years later. The novelist Henry Fielding, describing the would-be serious young men of his day, includes among their amusements "natural philosophy, or rather unnatural, which deals in the wonderful and knows nothing of Nature, except her monsters and imperfections."[33]

The first fully developed satirical portrait of the new scientist, and certainly the most influential in the period, was that of Sir Nicholas Gimcrack in Thomas Shadwell's box office success *The Virtuoso* (1676). Although it has been widely assumed that Shadwell was poking fun at the Royal Society members, insofar as many of Gimcrack's experiments and theories were only slightly changed from reports in the *Transactions of the Royal Society* or in Robert Hooke's *Micrographia,* this was a simplified estimate both of the play's main character and of its overall place in the tradition of Restoration satire.[34] Not only is it explicitly stated at the beginning of the play that Sir Nicholas has been refused membership in the College[35] but, so far from mocking the practices of the Royal Society, Shadwell actually uses them as the yardstick with which to beat Gimcrack, the dilettante virtuoso who flits from wonder to wonder without method or mental discipline. Nevertheless, in the character of Gimcrack, Shadwell effectively deconstructs contemporary perceptions of scientists' pretensions, their cultivated marks of difference, and their arrogance towards their fellows. There is thus an implicit moral beneath the overt satire.

Like all virtuosi, Gimcrack is an indiscriminate collector of oddities, an easy prey for charlatans. He has been duped into paying ten shillings each for eggs alleged to have been laid with hairs in them—from a tradesman who had inserted the hairs through a fine hole—and has spent two thousand pounds on microscopes to find out the nature of eels in vinegar, but the conclusions he draws from his observations are ludicrous. Perhaps the most damning indictment of the virtuoso, and one that was to be echoed by nearly all the eighteenth-century satirists and moralists, is that voiced by Gimcrack's niece Miranda. She describes him as "one who has broken his brains about the nature of maggots, who has studied these twenty years to find out the several sorts of spiders, and never cares for understanding mankind" (1.2.11–13). Indeed, he has traveled in Italy without taking any interest in its culture: "Tis below a virtuoso to trouble himself with men and manners. I study insects" (3.3.86–89). This aspect of the virtuoso was a persistent source of humor well into the eighteenth century, and Gimcrack was as well known as any contemporary celebrity. Indeed, in number 216 of the *Tatler* (1710) Addison published what was alleged to be the will of Sir Nicholas, whereby he bequeathed to his family such valuable items as "a dried cockatrice . . . my receipt for preserving dead caterpillars, . . . my preparations of winter May-dew and embryo pickle . . . my rat's testicles and whale's pizzle."

Gimcrack is satirized on a number of other counts as well, most frequently for the uselessness of his experiments, a factor that Shadwell exploits brilliantly for comic purposes. Gimcrack is first seen lying awkwardly on a table holding between his teeth a string attached to a frog in a bowl of water. By imitating the movements of the frog, he is allegedly learning to swim, not, it is emphasized, because he ever intends to do so, for he hates water. "I content myself with the speculative part of swimming; I care not for the practic. I seldom bring anything to use; 'tis not my way. Knowledge is my ultimate end" (2.2.83–86). This manifesto of the cult of uselessness is elaborated throughout the play, as Gimcrack frolics from one experiment to the next, without any system or consideration of their implications. He repeatedly shuns any thought of practical applications, insisting that his only motive is the pursuit of knowledge for its own sake—"so it be knowledge, 'tis no matter of what" (3.3.26–27).

Of course, unlike the members of the Royal Society, Gimcrack never follows through on the results of his experiments. Robert Boyle measured the weight of air at different altitudes in order to determine the relation between mass, pressure, and volume; Gimcrack collects bottles of air from different areas merely for the novelty, "to let it fly in my chamber." Similarly, Gimcrack's blood transfusions between men and sheep are ludicrous, because they are done merely out of curiosity and are wholly destructive; however, Shadwell is careful to have two other characters point out that the practice of transfusion per se (a topical issue in the Royal Society)[36] is by no means ridiculous.

> BRUCE: . . . excellent experiments may be made in changing one creature into the nature of another.
>
> LONGVIL: Nay, it may be improved to that height to alter the flesh of creatures that we eat, as much as grafting and inoculating does fruits. (2.2.225– 29)

Sir Nicholas is also typical of the virtuoso rather than of the serious scientist in that he makes excessive claims for his experiments merely in order to make them appear more marvelous and unnatural, for, like Butler's astronomers, he has no interest in the normal. He must claim not only to swim but also to fly: "I am so much advanc'd in the art of flying that I can already out fly that ponderous animal called a bustard" (2.2.30–31). Not content to describe the habits of spiders, he insists that he has a tame spider called Nick that follows him around—"the best natur'd, best condition'd spider that ever was met with . . . he was of the spaniel breed, sir" (3.3.75–78). He does not merely claim to have accomplished a successful blood transfusion between a man and a sheep; he must assert that the man thereupon assumed an ovine character and grew a tail and wool. He has also, he claims, improved the stentrophonic tube, so that he can hear things being said eight miles away (5.2.54–56), yet, typically, he does not hear his wife and nieces plotting against him in the next room. These exaggerations result partly from his pretentiousness (it is not accidental that his best friend is the ridiculous orator Sir Formal Trifle)[37] and class consciousness but also from his inability to understand the difference between cause and effect, between science and superstition. We discover that, like many of the early members of the Royal Society, he is also a Rosicrucian, absorbed in magic and spiritualism.

Finally, although most of the comedy arises from the ridiculousness of his pseudo-scientific antics, Gimcrack stands condemned on moral and social grounds as well, notably for his hypocrisy, which is seen in his closet lechery and his belief that mankind is beneath his notice. These exaggerated vices are responsible for the highly successful comedy of the play. Sir Nicholas *is* the play to such an extent that the usual intrigues of Restoration comedy are subsidiary to his experiments and his ludicrous theories. Indeed, Shadwell's portrait gave rise to a trend in Restoration satire, and Sir Nicholas Gimcrack became the literary ancestor of a whole tribe of virtuosi of both sexes, all having traits modeled on his.[38] Like him, Dr. Boliardo, of Aphra Behn's *The Emperor in the Moon* (1687), who in line with the current fashion for astronomy has studied the moon through his telescope, is ridiculed for his pretensions, which include wearing all manner of mathematical instruments hanging from his belt and having a servant bear after him a thirty-foot-long telescope; like Gimcrack, he is easily manipulated by the more astute gallants courting his pretty wards, thereby demonstrating that intellectual and moral pretensions are only a façade for the universal failings of the human condition.

Given the gender imbalance of scientists then as now, it is interesting that not all the virtuosi characters were male. For the first—and last—time in literary history, there was a rash of female "scientists" as, in the wake of Margaret Cavendish, scientific ladies made their debut on the stage. Usually they were first ridiculed and then, if they were young and attractive, reformed, ending up as sensible wives in the socially acceptable mold. Shadwell's *The Sullen Lovers* (1688) presents Lady Vaine Knowall, a would-be virtuosa who is really only a pedant with little if any accurate knowledge and, like Gimcrack, totally ignorant of the world around her. In Thomas Wright's *The Female Virtuoso's* (1693), an adaptation of Molière's *Les Femmes Savantes*, the female wits prepare to form a learned society, the Academy of Beaux Esprits, but Lady Meanwell, the leading virtuosa, is full of useless projects, all totally impractical, and (clearly a far worse charge) is an unsatisfactory wife.

In *The Basset-Table* (1706) Susannah Centlivre created a somewhat more appealing female counterpart in the young virtuosa Valeria, "a Daughter run mad after Philosophy," whose passion is collecting insects and examining them microscopically. She makes her first entrance pursuing "a huge Flesh Fly," which she has just received from her suitor, Mr. Lovely, for dissection. Indeed, Valeria goes further than Gimcrack in her devotion to science. Unhampered by any unscientific, feminine tenderness, she is an ardent vivisectionist, eagerly dissecting her pet dove to test "whether it is true that doves lack gall," opening up a dog to find a tapeworm, and examining a fish under her microscope. "Can Animals, Insects or Reptiles be put to a nobler Use than to improve our Knowledge?" she asks rhetorically, in an age when animal experimentation in the cause of science was not publicly acceptable.[39] However, she is eventually made to see the error of her ways and marry Mr. Lovely. It is clearly implied that her aspiration to scientific learning is unnatural and she must be reeducated to fit the accepted female role.

A more vigorous attack on the female virtuosa is the character of Lady Science in James Miller's anonymously published comedy *The Humours of Oxford* (1726). Lady Science has none of Valeria's pert charm; she is "a Female Bookworm," "a Pretender to Learning and Philosophy" who suffers from Sir Nicholas Gimcrack's fault of being out of touch with life. A precursor of Oscar Wilde's Lady Bracknell, she interrogates her prospective son-in-law not about his morals or religion but about his cosmology—whether he holds to the Ptolemaic or the Copernican hypothesis and whether he believes it "ever possible to find out the Longitude." The main condemnation of Lady Science, however, hinges on her daring to aspire to a province ordained for men. "The Dressing-Room, not the Study, is the Lady's Province—and a Woman makes as ridiculous a Figure poring over Globes, or thro' a Telescope, as a Man would with a Pair of *Preservers* mending Lace."[40] Clearly it is not scientists per se who are being satirized here, only those who have dared to surrender their femininity to science.

Scientist-bashing was good theater, especially when the ridicule could be supplemented with preposterous visual props and farcical situations, and most of the ridiculous scientists in Restoration drama derived from no more serious purpose than novelty and a backlash against the Puritans who had closed the theaters. This early wave of satire directed against scientists reflects the accumulated resentments of the previous century, rather than any immediate threat. It was only when the research of the Royal Society came to be perceived as socially and economically important that the satire became more serious and vicious, more sternly moral than comic in intention. Ironically, the person who, more than any other, was responsible for elevating the status of science in England and who became a figure of national eulogy was also to provoke the most extreme reaction against the premises of scientific materialism and the reductionist procedures associated with it. That person was, of course, Isaac Newton, one of the few actual scientists to feature as a literary character *in propria persona*.[41]

CHAPTER FOUR

NEWTON:

A SCIENTIST FOR GOD

Nature and Nature's laws lay hid in night;
God said, "Let Newton be!" and all was light.
—Pope

It was primarily through the influence of one person, Isaac Newton (1642–1727), that the popular image of the scientist changed from that of either a stupid or a sinister character to that of a highly respected man of genius, representing the highest attainments of reason (see fig 7). At the time of his death and in the years immediately following, Newton was honored with the highest literary praise ever accorded a scientist. This adulation was the more unusual in being applied explicitly to an actual scientist rather than to a fictional character, although, as we shall see, the representations of Newton became increasingly mythologized. It was not until the first decade of the twentieth century that another such wave of literary adulation was offered to scientists, and then, unlike the praise of Newton, it manifested itself in a spate of heroic fictional scientist characters rather than in acclamation of actual scientists. Compared with our own time, the eighteenth century was less cynical and less self-conscious about praising public figures, and the tributes accorded to Newton were as open as they were fulsome. Newton was perceived as Britain's national treasure, and to praise him became an act of patriotism, even of piety.

Without in any way detracting from Newton's real and enduring achievements, it is important to realize the extent to which Newtonian iconography, bordering on hagiography, was constructed through a rigorous selection of the physical and mathematical elements of his work and a simultaneous exclusion of other aspects of his thought, notably his extensive study of alchemy, to produce the image that was intellectually, politically, and even psychologically congenial to the climate of opinion in England around the time of his death. Whether unconscious or not, this exercise in censorship and propaganda, involving the suppression of Newton's strongly alchemical interests, was so effective that his persistent attempts to link the physics of the *Principia* to his alchemical research and thereby evolve a

FIGURE 7.
Sir Isaac Newton, president of
the Royal Society, 1703–27.
Portrait by Sir Godfrey
Kneller, 1688. National
Portrait Gallery, London.

vast unifying theory linking all matter and reactions throughout the universe remained virtually unrecognized until the mid-twentieth century. The first indication of the disparity between the person and the legend was John Maynard Keynes's paper "Newton, the Man," produced for the Royal Society's tercentenary celebrations of 1946. Keynes, who had tracked down and acquired some fifty-seven of Newton's alchemical manuscripts from 121 auctioned lots, concluded from his study of the manuscripts that "Newton was not the first of the age of reason. He was the last of the magicians, the last of the Babylonians and Sumerians, the last great mind which looked out on the visible and intellectual world with the same eyes as those who began to build our intellectual inheritance rather less than 10,000 years ago."[1] More recently, Betty Jo Teeter Dobbs has traced Newton's overriding concern with demonstrating the unity of Truth grounded in the unity of God.[2] It is this prevailing belief that links his alchemy, his Arian theology, and his writings on the causal relations of matter and forces.

In comparison with the height of his subsequent fame, the recognition of Newton's work was slow in coming. Intended by his widowed mother to be a farmer, Isaac Newton showed no inclination for this career and in 1661 had the good fortune to be sent to Trinity College, Cambridge, by a maternal uncle who had studied there. At Cambridge Newton encountered the intellectual ferment elicited by Descartes's *philosophia mechanica*, by Kepler's optics and his laws of planetary motion, by Galileo's mechanics, and by the new mathematics of his own teacher, Isaac Barrow, who became the first Lucasian Professor of Mathematics while Newton was at Cambridge. In 1664 the university was closed because of plague, and the newly graduated Newton was forced to return to his mother's house in Lincolnshire. During this year of "exile" from Cambridge, he worked out his three laws of motion and his theory of gravitation, although he did not publish these results until 1687. His earliest scientific paper on light and color, read on his behalf at the Royal Society in 1672, was accorded a far from favorable reception,[3] and the ensuing controversy caused him to eschew publication of his work. It was only at the urging of Edmund Halley, later Astronomer Royal, that he consented to write down the elaborate proofs (which he had mislaid!) for his theory of planetary motion, including the infinitesimal calculus he had had to derive for this purpose. This work on planetary motion, together with the treatise *On the System of the World*, was at Halley's instigation published as *Philosophiae Naturalis Principia Mathematica* (universally known as the *Principia*) in 1687. The work was almost immediately recognized as vindicating a new method of analysis, one that quickly came to be seen as the necessary and sufficient instrument for scientific progress. The first book of the *Principia* deals with definitions of space, mass, and time and motion in a nonresistant medium. The second book considers motion in a resisting medium and forms the basis of hydrostatics and hydrodynamics. But it was the third book, entitled *De systemate mundi*, or *On the System of the World*, that was to seize the popular imagination. Its introduction confidently affirms, "Superstat ut ex iisdem principiis doceamus constitutionem systematis mundani" (It remains that from the same principles we demonstrate the form of the system of the world), and it did indeed provide the complete framework for the celestial mechanics of the next two centuries. "The true system of the world has been perceived, developed, and perfected," wrote the French historian D'Alembert, and even Laplace was moved to exclaim that "it [the *Principia*] has all the certainty which can result from the immense number and variety of phenomena, which it rigorously explains, and from the simplicity of the principle which serves to explain them."[4] Certainly the *Principia*, ending with the "General Scholium," in which Newton asserted, "Hypotheses non fingo" (I do not create hypotheses), was taken as the ultimate vindication of Bacon's experimental science against the theoretical constructions of Descartes. The first words of Newton's *Opticks* (1704) were later to reaffirm this in English: "My Design in this Book is not to explain the Properties of Light by Hypotheses, but to propose and prove them by Reason and Experiments."[5]

Although the *Principia* was eagerly bought, it was less widely understood. Newton, in order "to avoid being baited by little smatterers in mathematics," had designed its obscure presentation "to be understood only by able mathematicians," and it has been claimed that only a handful of New-ton's contemporaries, including one woman, the marquise de Châtelet, were able to understand its complex geometrical arguments.[6] It was not long, however, before popular, predigested versions of the main points of the work were current in England. On the Continent, acceptance was slower, partly because of French allegiance to the Cartesian system, but Voltaire's *Eléments de la philosophie de Newton* eventually ensured the popularization of the work, even in France. Indeed, Voltaire remarked in 1735: "Verses are hardly fashionable any longer in Paris. Everyone begins to play the mathe-matician and the physicist. Everyone wants to reason. Sentiment, imagina-tion and charm are banished."[7] He was referring to the increasing influence of scientific discourse not only on the topics but even on the language of literature.

In the light of the specificity, not to mention the difficulty, of the *Princi-pia*, it is necessary to ask why a mathematical text should have acquired such widespread public veneration and mystique. There were at least two major reasons and a number of minor ones. Most important was the unified world-view that it offered, an image of nature that typified order, simplicity, and harmony and that was represented as eminently reasonable and predict-able. Taken together, these qualities engendered in Newton's contempor-aries a sense of power and a belief that man, far from being of no account in the expanding universe being revealed by the telescope, was elevated to a position of superiority, since he alone was capable of understanding the working of the whole celestial system.

Almost as important as the implied order and harmony of the solar system was the universality inherent in Newton's treatment. The term *natu-ral philosophers*, used almost exclusively for scientists during the seventeenth and eighteenth centuries, is indicative of the contemporary belief in the unity of knowledge. It was considered necessary to integrate each new discovery into the existing philosophical framework, and Newton's schema provided a unifying model of the cosmos for hitherto diverse areas of re-search, a model based not on controversial teleological principles, as Aris-totle's was, but on irrefutable mathematical laws. The very word *laws* carried with it a suggestion of authority and the implication that the maker of the laws, in this case Newton, had imposed them on an unruly universe. The fact that the so-called laws of science are actually no more than the descrip-tions that best fit the observed phenomena was submerged in the trans-ferred sense of power conveyed by the idea of the law enforcer.[8]

What is less well realized is the extent to which both these attributes, harmony and universality, were central goals of alchemy and certainly of Newton's own alchemical research. In her comprehensive study *The Founda-tions of Newton's Alchemy* Betty Jo Teeter Dobbs has shown conclusively that Newton was essentially a chemical alchemist, drawing on the researches of the Hartlib circle and heavily influenced by Boyle's belief in "one Catholic Matter," from which transmutations could be performed in the manner asserted by alchemists. Throughout his life Newton had a strong desire to prove nature "consonant" and "conformable to her self" and looked for the structure of the world in alchemy—a system of the small world to match with his system of the greater. This longing is expressed in the preface to the *Principia*—"I wish that we could derive the rest of the phenomena of Nature by the same kind of reasoning from mechanical principles"—and it is expressed even more strongly in the draft preface.[9] Dobbs concludes that "Newton's alchemical thoughts were so securely established on their basic foundations that he never came to deny their general validity, and in a sense the whole of his career after 1675 may be seen as one long attempt to integrate alchemy and the mechanical philosophy."[10] Newton never revoked his early idea that all matter was generated by fermentation and condensation from some common material. In the *Principia* he suggested that vapors from the sun, stars, and comets might be condensed into "water and humid spirits" and then, "by continued fermentation," into all the more dense substances. He did not remove these passages from later editions of the *Principia*, but in the *Opticks* he substituted *Light* for *vapours*—not a major change, in fact, since Newton believed that light was corpuscular. "Are not gross Bodies and Light convertible into one another . . . ?"[11]

The second major attraction of the *Principia* for Newton's contemporaries was its vindication of quantitative science, in particular the experimental methodology expounded by Bacon, as a powerful tool applicable to all spheres of operation. The laws of motion exactly defined the concepts of mass, inertia, and force and their relation to velocity and acceleration. The *Principia* demonstrated the predictive abilities of science and released scientists from the need to seek mystical causes for phenomena they had previously been unable to understand, such as gravity. It tamed the hitherto random and hostile universe into a well-oiled and entirely reliable clockwork model. Newton showed that three relatively simple laws of motion applied to all observable movements of inanimate nature, from the motion of waves to invisible air. Together with the inverse square law, they explained not only terrestrial motion but also the gross movements of the planets and of comets, hitherto regarded as so capricious as to be attributable only to divine retribution, and even the precession of the equinoxes, which formerly had not been considered mathematically explicable. In brief, it can be said that before Newton's work most scientists felt obliged to invoke ad hoc principles and occult causes based on human and divine analogies for all but the simplest terrestrial phenomena, while after the assimilation of the *Principia* it was assumed that all terrestrial and celestial movements were explicable in precise and numerical terms, by calculations based on a few general laws.

Again, there is a link with contemporary preoccupations in alchemy, insofar as it was then assumed by the Hartlibians (see chapter 2 above) that the riddle of alchemy could be solved by experimental method.[12] Even something as fundamental as Newton's ideas about gravity was related to alchemical concepts. His original explanation of gravity as the result of the pressure of a descending shower of aether had to be modified when, as the result of an experiment with pendulums, he was forced to conclude that the aether did not exist. He therefore had recourse to the concepts of attraction and repulsion, which he called "Powers" and "Virtues," whereby particles of matter act upon each other at a distance. This was no longer mechanism in Descartes's sense; indeed, it was strikingly similar to the theories of sympathies and antipathies to be found in the occult literature of the Renaissance,[13] and when the *Principia* was first published, Newton's critics were not slow to brand it as occult. This aspect of Newton's intellectual ancestry was conveniently ignored by the image-makers of the Enlightenment, it being no longer respectable or expedient to recall it.

Descartes had firmly believed that mathematization of the natural world and even of human society would reveal the order underlying apparently unpredictable events. Thus the success with which Newton applied the same basic methods of Galileo, Descartes, and Bacon to explain universal gravitation and the movements of the planets in the solar system was perceived not only as establishing a universal law but as exalting the methodology that could encompass so vast a cosmic order. Further, it inevitably suggested the possibility that all branches of knowledge, not just physical science, might be susceptible to the same universal laws. Thus it was not considered anomalous that in 1728, the year after Newton's death, J. T. Desaguliers produced a ponderous application of Newtonian physics to politics in *The Newtonian System of the World: The Best Model of Government, an Allegorical Plan*. In a period when the religious and political turmoils of the previous century were still fresh in living memory, it was alluring to believe that scientific principles could resolve these deep divisions once and for all. R. G. Olson writes:

> The scientific-mathematical-empirical method seemed to receive almost pontifical authority by its association with Newton's accomplishments in natural philosophy; and while it may be an oversimplification, I think it is not really a major distortion to say that the enlightenment of the eighteenth century was in large measure an attempt to extend the domain of application of scientific method to include not only inanimate nature, but also the laws of human thought, morality, society, and religion.[14]

Obviously, in such a climate, the idea that Newton had attempted to perform the same unifying function, but from the starting point of alchemy, was less than welcome. Indeed, it is doubtful that it could have been comprehended in the later decades of the eighteenth century, for the extreme rationalist position of the Enlightenment, ironically referable in considerable part to Newton's own contribution, would have been unable to reconcile such an idea with the image of the great mathematical physicist.

The success of Newton's *Principia* at both a practical and a theoretical level was also instrumental in propelling Western culture towards the scientism that has largely characterized it ever since. On the one hand, Newton's mechanistic and rationalistic procedures destroyed, at least temporarily, the climate of opinion that had allowed the Faust figure to flourish. The quest for knowledge was no longer associated with sin, but regarded as the highest pursuit of mankind, and the optimistic Baconian belief that scientific investigation would be infallibly beneficial became the accepted philosophy of the age. Samuel Johnson, who in many ways epitomized the educated Englishman of the Enlightenment, asserted categorically, "Sir, a desire for knowledge is the natural feeling of mankind; and every human being, whose mind is not debauched, will be willing to give all that he has to get knowledge."[15]

Again, the manifest practical and economic success of Newtonian mechanics quickly put an end to the notion, explored in the preceding chapter, that science was useless. Apart from the broadly applicable laws of motion, Newton's lunar tables, from which, given an accurate chronometer, longitude could be calculated, and his lunar theory of tides were important improvements in naval science and cartography. Among his other activities, Newton also devised the reflecting telescope, which overcame both the problem of chromatic aberration and the inconvenient length required for adequate magnification in the older refracting telescopes. The new, shorter telescopes were much easier to use and of course cheaper to construct and mount. As they increased in popularity, more people were able to observe for themselves the greatly expanded universe they revealed. In such a universe, the individual was no longer important enough to draw the fire of devils and angels, even if they were assumed to exist. Magic, sin, and spiritual aspirations seemed increasingly irrelevant in this orderly, mechanical cosmos.

These perceptions of the significance of Newton's work are crucial to an understanding of the extraordinary tributes showered upon Newton by contemporary English writers, especially poets. The eulogies elicited, especially at his death, far exceed those accorded any other scientist before or since. E. N. da C. Andrade has summed up the impact of Newton on his contemporaries in the following terms:

> If we are to try to represent Newton's achievements by some modern analogy, to construct some imaginary figure who should be to our times what Newton was to his, we must credit this synthetic representative with, I think, the whole of relativity up to, and somewhat further than, the stage

at present reached—we must suppose our modern Newton to have satis-
factorily completed a unitary field theory. In light we must credit him both
with having established the existence of spectral regularities and with
their explanation in terms of the quantum theory. Possibly, too, we must
give him the Rutherford atom model and its theoretical development, a
simple astronomy in little to correspond to the solar system. Let us, then,
think of one man who, starting in 1900, say, had done the fundamental
work of Einstein, Planck, Bohr and Schrödinger, and much of that of
Rutherford, Alfred Fowler and Paschen, say, by 1930, and had then become, say, Governor of the Bank of England, besides writing two books of
Hibbert lectures and spending much of his time on psychical research, to
correspond with Newton's theological and mystical interests. Let such a
man represent our modern Newton and think how we should regard him.
Only so, I think, can we see Newton as he appeared to his contemporaries
at the end of his life.[16]

Impressive as it is, this list does not convey perhaps the most important attribute that Newton embodied for his contemporaries, namely, a confidence in the harmony and order of the universe, which was expanding
before their eyes with every advance in the construction of telescopes.
Whereas in 1611 Donne had written, "The new Philosophy calls all in
doubt," Newton had seemingly restored order and predictability through
the application of a very few basic principles that appeared to explain all
natural phenomena. Thus the response was not merely, or even primarily, to
Newton the person so much as to the picture of nature that he spread before
them, a nature magnificent and expansive, yet governed by simplicity and
uniformity, a system that, by epitomizing the classical values of order and
harmony, domesticated for them the expanding universe that geographers
and astronomers were uncovering. One indication of this contemporary
fascination with order in nature was the popularity of the orrery, an intricate
clockwork model of the solar system named after Charles Boyle, fourth earl
of Orrery. Although its original purpose was to explain the principles of
astronomy, it became a fashionable acquisition, suggesting an attraction
beyond the pedagogic. Joseph Wright's painting *A Philosopher giving a Lecture on the Orrery* (ca. 1764–66) conveys vividly both a sense of devout
contemplation on the part of the onlookers, ranging from young children to
middle-aged men and women, and the power of the scientist-figure who
explains this complex system (see fig. 8).[17]

In his formulation of the basic rules for scientific procedure, Newton
had stated, "We are to admit no more causes of natural things than such as
are both true and sufficient to explain their appearance," since "Nature is
pleased with simplicity." It was the apparent simplicity of Newton's system
that took hold of the popular mind, cutting like Occam's razor through

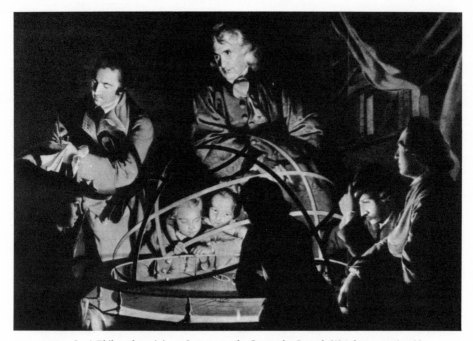

FIGURE 8. *A Philosopher giving a Lecture on the Orrery,* by Joseph Wright, ca. 1764–66. Wright's painting conveys vividly a sense of devout contemplation on the part of the onlookers, ranging from young children to middle-aged men and women. Derby Museum and Art Gallery.

Cartesian vortexes and other complex theories adduced to account for observed phenomena. There is a definite sense of relief in James Thomson's lines on Newton:

> The heavens are all his own; from the wild Rule
> Of whirling VORTICES, and circling SPHERES,
> To their first great simplicity restored.[18]

This simplicity of the Newtonian scheme was extolled by many other contemporary poets, as in this stanza by William Tasker:

> Th'eccentric comet's course he knew,
> From principles sublimely few,
> Explain'd all Nature's laws.
> Th'Attractive and repulsive force,
> He taught to solve the planets' course
> Encircling thee, O Sun![19]

The anonymous *A Philosophic Ode on the Sun and the Universe* (1750) also praises Newton specifically for revealing the order of the universe:

Newton, immortal Newton rose;
This mighty frame, its orders, laws,
His piercing eyes beheld:
That Sun of Science pour'd his streams,
All darkness fled before his beams,
And Nature stood reveal'd.[20]

We have seen that, partly as a result of his desire to vindicate the universality enshrined at the heart of alchemy, Newton had insisted that we should "to the same natural effects . . . as far as possible assign the same causes so that the pull of gravity or the reflection of light on earth and on the planets would be attributed to the same causes."[21] Such a requirement for regularity and consistency was seen by his contemporaries as both comforting and aesthetically pleasing. The importance of the aesthetic considerations may be seen in James Thomson's "Poem Sacred to the Memory of Sir Isaac Newton," where he praises Newton's ability to see "Motion's simple Laws" determining the course of the universe. Newton saw "every Star" as

the living Centre each
Of an harmonious System: all combin'd
And rul'd unerring by that single Power,
Which draws the Stone projected to the Ground.

Even Light itself is reduced to obeying the unifying laws of optics:

Nor could the darting Beam, of speed immense,
Escape his swift Pursuit, and measuring Eye.
Even LIGHT ITSELF, which every thing displays,
Shone undiscover'd, till his brighter Mind
Untwisted all the shining Robe of Day.

And, in a fine example of sexist language infiltrating the discussion of science,

Nature herself
Stood all subdu'd by him, and open laid
Her every latent Glory to his View.[22]

Almost as important to Newton's contemporaries as the revelation of order and harmony was their confidence in the certainty of his mathematically and geometrically couched conclusions, which seemed to admit of no dispute.[23] Voltaire's *Micromégas* (1752) is not unusual in extolling the exactitude of mathematical and physical sciences as opposed to the vagueness and confusion of philosophical systems.[24] Even Newton's work on light, although less obviously related to the mechanical scheme, was characterized by measurement and calculation rather than by generalities and thus seemed to carry an inherent guarantee of truth. The distinguished astronomer Edmund Halley, who was responsible for the publication of the *Principia*, was moved to preface the volume with his own Latin "Ode to the Illustrious Man, Isaac Newton," which, in translation, concludes:

> Come celebrate with me in song the name
> Of Newton to the Muses dear; for he
> Unlocked the hidden treasuries of Truth:
> So richly through his mind had Phoebus cast
> The radiance of his own divinity.
> Nearer the gods no mortal may approach.[25]

Moreover, despite the mechanistic basis of Newtonian physics, a large part of Newton's appeal to his contemporaries resulted from the religious implications they saw in his work, for although the new science in effect excluded purpose (divine or otherwise) as a legitimate consideration, it was not at first seen as anti-religious.[26] On the contrary, Newton affirmed the need for the continuing involvement of God in the universe "to provide a cause for gravitational attraction and to adjust certain imperfections in the system."[27] As occult and the supernatural explanations of the world were dispensed with and replaced by indisputable mechanics, so the former image of the godless scientist was replaced by that of a wise religious teacher who interpreted the ways of God to man. The respected physician Sir Thomas Browne, who, like Harvey, had studied medicine in the rationalistic ambience of Padua, reaffirmed Bacon's assertion that the study of God's works was almost as pious as the study of God's word: "There are two Books from whence I collect my Divinity; besides that written one of GOD, another of His servant Nature, that universal and publick Manuscript, that lies expans'd unto the Eyes of all."[28] And the poet John Dryden had included in his *Annus Mirabilis* (1666) an apostrophe to the Royal Society couched in similar terms:

> This I fore-tel, from your auspicious care,
> Who great in search of God and Nature grow:
> Who best your wise Creator's praise declare,
> Since best to praise his works is best to know.[29]

In the first edition of the *Principia* (1687), Newton had not explicitly mentioned God; nor did the first edition of the *Opticks* (1704) say anything about God as the creator and perceiver of the world. However, Roger Cotes, Plumian Professor of Astronomy at Cambridge and editor of the second edition of the *Principia* (1713), urged Newton to "add something by which your Book may be cleared from some prejudices which have been industriously laid against it." He referred to a letter from Leibniz to Harsöker remarking adversely on the omission of God from a work that presumed to present the system of the world. In compliance, Newton added the "General Scholium" at the end of book 3 as a tacit acceptance of the limitations of knowledge, beyond which faith is the only appropriate attitude.[30] Similarly, he later added to the *Opticks* two more Queries, which treat of God as creator and perceiver of the physical world. Query 28 concludes: "And these things being rightly despatch'd, does it not appear from Phaenomena that there is a Being incorporeal, living, intelligent, omnipresent, who in infinite Space, as it were in his Sensory, sees the things themselves intimately, and thoroughly perceives them, and comprehends them wholly by their immediate presence to himself: . . . And though every true step in the Philosophy brings us not immediately to the Knowledge of the first Cause, yet it brings us nearer to it, and on that account is to be highly valued."[31]

At the conclusion of the *Opticks,* in Query 31, Newton makes explicit the connection between natural and moral philosophy, so important to his contemporaries: "And if natural Philosophy in all its Parts, by pursuing this Method, shall at length be perfected, the Bounds of Moral Philosophy will be also enlarged. For so far as we can know by natural Philosophy what is the first Cause, what Power he has over us, and what benefits we receive from him, so far our Duty towards him, as well as that towards one another, will appear to us by the Light of Nature." This is very different from the assumption of self-sufficient wisdom attributed to earlier scientists. Wisdom is now associated with humility and reverence for the works of the Creator, and what was then known of Newton's religious and mystical tendencies only served to encourage the image of the high priestly scientist, coworker with God in the revelation of his wisdom. If God had created the universe physically, then Newton had recreated it intellectually for his contemporaries, and his interest in mysticism would not have seemed inappropriate.

In this regard, Newton's work can be seen as the culmination of many efforts by natural theologians to establish the presence of God in the world. Where John Ray discovered the wisdom of God in the adaptation of his creatures,[32] Richard Lower in the design of the heart, and Nehemiah Grew in botanical structure, Newton found divinity in the system of the universe and thereby seemed to his contemporaries to have given new meaning to the words of the Psalmist, "The Heavens declare the glory of God." Ray had argued that the study of God's handiwork was not merely a luxury but a Christian duty, and both he and Robert Boyle had affirmed a belief in "the Scientist as Priest." Richard Blackmore's lengthy and popular poem *Creation, a philosophical Poem in Seven Books* (1712) set out to demonstrate the presence of God in every aspect of the world, from the anatomical wonders revealed by Harvey to the wonders of astronomy revealed by Newton, and pays specific tribute to Newton's inspirational effect on his successors.[33]

Neither Newton nor his contemporaries realized that a God derived from nature, especially a nature viewed from the perspective of mechanics, could be no more than a reflection of the implicit values of physics, a Great Mechanic. With hindsight we can see that Newton was projecting his own criteria of the superiority of order and design onto the solar system and then using what he saw to vindicate his belief in the presence of a like-minded God. Thus the simplicity and order of the solar system, with the planets circling the sun, are evidence of divine intelligence only if we assume that these values, rather than complexity and maximum degrees of freedom, represent the divine purpose. The reason for the immediate appeal of the former set of values is that it generates a sense of predictive power through understanding.

The poet-mystic William Blake was the first to recognize this fundamental flaw in Newton's natural theology, but even if Newton's contemporaries had perceived the circularity of his arguments, they would still have applauded them. In the fight against atheism a slight bending of scientific principles troubled no one's conscience, certainly not Newton's. He enthusiastically helped Richard Bentley prepare the first Boyle lectures, entitled *A Confutation of Atheism,* which made use of Newton's cosmology as evidence for the theological argument from design.[34] Newton's letter to John Harrington of 30 May 1698 outlines one of his cherished theories, namely, that the ratios between the spaces occupied by the colors in the spectrum are the same as the harmonic ratios of the octave. The fact that he endeavored for years, unsuccessfully, to prove this indicates how strongly he desired to find evidence for his religious presuppositions.[35]

So widespread was the mythology associated with Newton that after his death his soul was popularly supposed to have winged its way straight to the throne of God, and in many poems of the period he was depicted as a space traveler embarking on a cosmic voyage to learn from God those few truths that had eluded him on earth. Alan Ramsay's fulsome "Ode to the Memory of Sir Isaac Newton: Inscribed to the Royal Society" includes such stanzas as the following:

> The god-like man now mounts the sky,
> Exploring all yon radiant spheres;
> And in one view can more descry
> Than here below in eighty years.
>
> . . . .
>
> Now with full Joy he can survey
> These Worlds, and ev'ry shining Blaze,
> That countless in the *Milky Way*
> Only through Glasses show their Rays.[36]

Dr. Samuel Bowden also published "A Poem Sacred to the Memory of Sir Isaac Newton," which ends with Newton voyaging through space to learn more than he could on earth.[37] And Mrs. Brereton's poem "Merlin," which contrasts the magician unfavorably with the representative of true science, portrayed Newton as heaven's favorite, rewarded by a bird's-eye vision of the universe.[38] James Thomson, Samuel Bowden, and Edward Young all made use of the idea of Newton pursuing a postgraduate course in celestial physics as an appropriate heavenly pastime.

Thus, after his death Newton became an important symbol for scientists in the climate of ambivalence that developed during the eighteenth century. On the one hand, there was acclaim for the discoveries of science, particularly for the demonstration of harmony in the universe, which could be interpreted in religious terms as evidence of the greatness of the Creator; on the other hand, there was a wariness about ascribing too much praise to any mortal and thereby encouraging a godless pride in the human intellect. Newton's death conveniently resolved this dilemma, allowing him to be universally endorsed as a national treasure. Reverence for his achievements could now be combined with humility before the omniscience of God.

Indeed, in Thomson's poem, through a complex system of analogy, Newton is implicitly cast, apparently without any sense of blasphemy, as the new Adam, realigning man to the Creation. Just as the first Adam destroyed the harmony of the Garden of Eden and bequeathed chaos to his descendants, so Newton has reversed that process by restoring man's perception of the order and beauty of the universe.[39] Thomson emphasizes Newton's work on light, astronomy, and the tides, which he implies to be the counterparts of the first elements of the Creation described in Genesis, and contrasts fallen, "erring" man with the "all-piercing Sage."[40] Henry Grove expressed much the same popular myth of the great soul illuminating the darkness of ignorance and the religious revelation implicit in scientific discovery:

> How doth such a Genius as Sir *Isaac Newton*, from amidst the Darkness that involves human Understanding, break forth, and appear like one of another Species! The vast Machine, we inhabit, lies open to him, he seems not unacquainted with the general Laws that govern it, and while with the Transport of a Philosopher he beholds and admires the glorious Work, he is capable of paying at once a more devout and more rational Homage to his Maker.[41]

From Newton's fame it was a short step to Britain's fame, and a great part of the reverence accorded Newton by his compatriots, especially in the spate of eulogies occasioned by his death, was nationalistic in inspiration. The British Newton had overthrown the French Descartes, and no European could stand against him.[42] Even Leibniz and Huygens, though critical, were finally unable to refute Newton's conclusions. David Mallet's poem *The Excursion* (1728), written a year after Newton's death, is an example of such nationalistic, self-congratulatory verse: "To thee, great Newton! Britain's justest pride, / The boast of human race."[43] The same nationalistic senti-

ment may be found at the close of Thomson's "Poem Sacred to the Memory
of Sir Isaac Newton":

> O Britain's Boast! whether with Angels thou
> Sittest in dread Discourse,
>
> .   .   .   .
>
> Or whether, mounted on cherubic Wing,
> Thy swift Career is with the whirling Orbs,
>
> .   .   .   .
>
> O'er thy dejected Country chief preside,
> And be her GENIUS call'd; her Studies raise,
> Correct her Manners, and inspire her Youth.
> For, tho' deprav'd and sunk, she brought thee forth,
> And glories in thy Name; she points thee out
> To all her Sons, and bids them eye thy Star.[44]

Poets were not the only writers to apotheosize Newton. The tribute to Newton in *Mist's Weekly Journal* of 8 April 1727 expatiated upon Newton's charismatic effect upon the nation, leading to a surge of patriotic fervor:

> We are inspired into a generous, tho' distant, Emulation of those transcen-
> dent Virtues, which even on this Side the Grave, exalted the Possessors of
> them to a Degree beyond Mortality. When such great Examples as these
> are set before us, the Passions are necessarily quickened and excited, and
> the Soul with a becoming Pride dilates and extends it self, pleased as it
> were to behold the Dignity of human Nature.
>
> I have been led into these common Reflections by the Death of the late
> Sir Isaac Newton, the greatest of Philosophers, and the Glory of the *British
> Nation*.

Although Newton was the prototype of the great scientist, other lesser
lights were also included in the general enthusiasm for British science. Dryden's poem "To my Honour'd Friend, Dr. Charleton" (1663) paid tribute to
the recipient as well as to Bacon, Gilbert, Boyle, and Harvey in tones of
restrained applause. Sprat's *History of the Royal Society* (1667) was prefaced
by Abraham Cowley's ode "To the Royal Society," in which the members of
that body are compared in rapid succession to Moses's Israelites led by
Bacon to the promised land, to Gideon's picked band, and to the infant
Hercules.[45] Thomson linked Bacon, Boyle, and Locke with Newton as prototypes of man's deliverance from ignorance by the revelation of God in
nature,[46] and in 1705 Edmund Halley, the second Astronomer Royal, captured the public imagination with his synopsis of known comets.[47] After his
death in 1729 Halley was rewarded with a cosmic tour of discovery similar
to the one Newton was popularly supposed to have undertaken, and Rich-

ard Savage confidently asserted that the heavenly tourist would thereby have his faith confirmed:

> Hence Halley's soul etherial flight essays;
> Instructive there from orb to orb she strays;
> Sees, round new countless suns, new systems roll!
> Sees God in all! and magnifies the whole.[48]

This liaison between astronomy and religious faith was, however, always fragile if subjected to scrutiny, and it is not surprising that it failed to survive long. Inevitably, after such excesses of adulation a reaction set in, and scientists fell under attack from literary moralists and satirists, this time on the three main charges of arrogance, delusion, and irreligion.

CHAPTER FIVE

ARROGANT AND GODLESS: SCIENTISTS IN EIGHTEENTH-CENTURY SATIRE

We nobly take the high Priori Road,
And reason downwards till we doubt of God

. . . .

Thrust some Mechanic Cause into his place:
Or bind in Matter, or diffuse in space.
—Pope

Although it was slow by the standards of the twentieth century, when fame automatically elicits a desire to expose the deficiencies of public figures, there was inevitably a reaction against the excessive adulation of Newton. Not all his contemporaries were as easily persuaded of the god-fearing humility of scientists as the writers quoted in the preceding chapter. After the witty Restoration satires, which emphasized the comic foolishness of the virtuosi, the mood of the age became more serious, and the charges leveled against scientists changed also. In particular, those educated exclusively in the humanities and steeped in the literature of the past had little sympathy with a world-view that denigrated the past and looked instead towards a glorious future to be ushered in through the power of science. The scientists were accused of ignoring the moral dimension, of failing to see that, in Pope's words, "the proper study of mankind is man."

The eminent classical scholar Méric Casaubon was one of the first to deny explicitly that a knowledge of science could civilize a man or teach him ethical responsibility, but major literary figures, such as Alexander Pope, Jonathan Swift, Samuel Butler, and Samuel Johnson, became increasingly critical of the moral failings they saw as being engendered by the very success of the new science. First, resurrecting the Faustus image, they believed that scientists were attempting to discover more than it was proper for humanity to know. Second, they shared a barely suppressed anger at what they saw as the arrogance of scientists, especially the proponents of

Baconian method, with its assumption that eventually man will fully understand and exploit the mysteries of the universe. Here, again, there are echoes
of Faustus the overreacher, particularly in the implications of Bacon's dictum that knowledge is power.

The third and essentially new component in the eighteenth-century
criticism of scientists was the fear that science might indeed succeed in
deriving a self-sufficient, purely mechanistic system, with no moral dimension and no need of God. Newtonian physics, by explaining the movement
of all bodies, great and small, terrestrial and celestial, in one universal principle of gravitation led insidiously to the demand that all scientific explanations be couched in terms of particles and atoms moving in space—the so-
called billiard-ball universe of Laplace, who claimed that, given the original
position and momentum of every particle in the universe, he could predict
every subsequent event without recourse to the "hypothesis" of God.[1] God
could be accommodated in the cast of the cosmic drama only in the walk-on,
walk-off part of First Cause. It will be recalled from the preceding chapter
that Newton himself was far from subscribing to this view, but after his
death the Newtonian system was increasingly regarded among scientists as
almost exclusively mechanistic. Moreover, if primary properties constituted
the only reality, then the moral qualities of beauty, truth, and goodness were,
by implication, unreal, nonexistent, and irrelevant. It was no wonder that
the eighteenth-century moralists feared the advances of science on moral as
well as theological grounds and attacked the enemy with the strongest
ammunition that an age steeped in satire could contrive.

Alexander Pope's social satire *The Dunciad* (1728–42) includes an attack on the intellectual arrogance of contemporary scientists, who Pope
believed were deposing God from the universe and replacing him with
scientific laws. One of the scientists in his poem declaims:

> All-seeing in thy mists, we want no guide,
> Mother of Arrogance, and Source of Pride!
> We nobly take the high Priori Road,
> And reason downwards till we doubt of God:
>
> .   .   .   .
> Thrust some Mechanic Cause into his place:
> Or bind in Matter, or diffuse in space.[2]

Pope's later *Essay on Man* (1733), while freely admitting the religious
value of science in illustrating the wisdom of God in the Creation, points out
the limitations of natural philosophers, who cannot explain the causes of the
wonders they observe, and warns against the pride of those who profess to
know the secrets of God. Because of its immense scale and because of the
double connotation of the phrase "the heavens," astronomy came to epito-

mize for many moralists the arrogance and pretensions of science. "Is the great chain, that draws all to agree, / And drawn, supports, upheld by God, or thee?"[3]

After Newton's death, the grandiloquent flights of Newtonian eulogies precipitated more vituperative censure on the part of contemporary moralists against living scientists for their hubris in replacing ethics and religion with science and reason and for effectively installing Newton in the place of the deity. Whereas Pope in rebuking contemporary worshipers of Newton had exonerated Newton himself,[4] Jonathan Swift made no such concessions. Swift wrote, "It is hard to assign one Art or Science which has not annexed to itself some Fanatic Branch: such are, The Philosophers' Stone; The Grand Elixir; The Planetary Worlds; The Squaring of the Circle."[5] He ridiculed the pretentious and absurd claims of those who, like Bacon, set out to take all knowledge for their province: "The whole school of the Greshamites [the Royal Society] are too wild in their claims; the whole realm of human knowledge is too broad for human nature to conquer."[6]

It was already clear in his early work *The Battle of the Books* (1697) that Swift had more sympathy for the morals and values of the ancients than for those of the moderns, represented chiefly by those sympathetic to the new science; and by the time he published *Gulliver's Travels* (1726) his emphasis had changed from humor to bitter condemnation. In part this reflects the changed social status of scientists, who by the eighteenth century were no longer merely isolated eccentrics, figures of fun to be irreverently mocked, but economically and politically influential in the state.[7] Like Pope, Swift saw that, whatever their intellectual attainments, they were prey to the same moral failings as other men, and thus the very success and power they enjoyed made them potentially more dangerous and sinister. He therefore created a new satirical image of the scientist: one who had the power to wreak havoc and destruction on society and who was morally responsible for the havoc caused by his experiments. These powerful scientists Swift called "Projectors," a term loaded with contemporary significance. Projectors were speculators whose fantastic schemes threatened innocent people with financial ruin, and the more frequent use of the word pertained to precarious commercial speculations such as the notorious "South Sea Bubble." By applying it to scientific projects, Swift emphasized the widespread economic destruction that could ensue from the schemes of irresponsible scientists.

Like Pope's *Essay on Man, Gulliver's Travels* is concerned with man's pride and reaffirms, in effect, that the proper study of mankind is not nature but man, regarded in moral terms.[8] In book 3, which describes the ivory-tower abstractions of Laputa and the ludicrous excesses of Balnibarbi, Swift deals with man's pretensions to reason and more specifically with his intellectual pride. To exemplify this state, he concentrates mainly on astronomers, who typify for him man's perverted preoccupation with the physical world to the exclusion of his own moral and spiritual condition.[9]

What does Swift have against astronomy? In the first place, for Swift, a churchman, it exemplified the sin of pride, insofar as it represented a desire to transcend human limitations and aspire to the wisdom of God by trying to understand the universe; it was in the spirit of the Faustian overreacher or, in biblical terms, of the Tower of Babel. The close parallels between the experiments of Swift's Academy of the Projectors in Lagado and those reported in the *Philosophical Transactions* of the Royal Society have been well documented.[10] This theme is developed with increasing venom throughout book 3 as Gulliver encounters The Flying Island of Laputa, whose dimensions, significantly, are proportional both to those of the earth as calculated by Newton and to the terella of William Gilbert's experiments in magnetism.[11] Laputa thus symbolizes the astronomers' determination to force nature into a rigorous mathematical framework of their own devising. Newton's entire treatment of space and time would have been abhorrent to Swift, who accepted no absolutes in human affairs and regarded the formulation of universal laws as only one step from the assumption that man could usurp the role of God. It is significant, therefore, both that the Laputans are convinced that they can hear the music of the spheres (traditionally heard only by angels) and that the music they themselves produce in response to this is a cacophony, indicating their delusion.[12]

In Swift's view, related to this arrogant aspiration to acquire the knowledge of absolutes were the scientists' emphasis on abstract thinking, their obsession with other planets rather than their own, and their consequent contempt for everyday experience. The Laputans no longer even touch the ground, but live on a flying island, and their minds "are so taken up with intense speculations that they can neither speak nor attend to the discourses of others, without being rouzed by some external taction upon the organs of speech and hearing" (bk. 3, chap. 2, 169), a parody of John Locke's idea that knowledge is acquired only through stimuli impacting on the senses. Indeed, in their preoccupation, the Laputans, having no instinct of preservation, have to be forcibly restrained from falling over a cliff. It is not accidental that their chief interests are mathematics and music, the main occupations of the members of the Royal Society and also widely regarded as the purest expressions of absolute truth. As the chapter title "A Phenomenon solved by modern Philosophy and Astronomy" forewarns, the Laputans' abstract speculation about flight and its allegedly practical application, the complicated movement of their flying island "in an oblique Direction," are an intended parody of Newton's "resolution of any one direct force . . . into two oblique forces" in the plotting of planetary orbits and the paths of comets.

In Swift's satire, this preoccupation with the abstract is exacerbated by the Laputans' obsession with elaborate scientific instruments, another gibe at the Royal Society. Gulliver's Laputan tailor constructs an ill-fitting suit because he insists on taking his client's measurements with "scientific" instruments—a quadrant and compass—and on surveying everything, terrestrial and lunar, in the process. Again, because of their contempt for practical geometry, which they despise as "vulgar and mechanic," the Laputans' houses are as ramshackle as their clothes, for, like many of Newton's calculations, their complex mathematics cannot be followed by the builders. Some of the effects of their willful abstraction are far from harmless. The Flying Island proceeds at the whim of the King, who is either oblivious of the damage done to his subjects down below in Balnibarbi when the island comes between them and the sun or else uses this as a threat to suppress them. In Balnibarbi itself the Projectors, fresh from training in Laputa, look with one eye upward to the stars and with the other into their own minds (their major studies are astronomy, mathematics, and music), and in their mania to usher in a utopian future of their own design they take no account of nature. They have devastated the once fertile land with projects designed to change the naturally useful into its useless opposite—producing naked sheep, soft marble, and tangible air, sowing the fields with chaff, reducing excrement to its original food, and petrifying the hooves of a living horse. This too is an attack on the experiments conducted by the Royal Society, a more vitriolic version of Shadwell's gibes at the virtuosi.

It is clear, then, that Swift, so far from succumbing to the optimism of eighteenth-century science, believed pessimistically that scientific discoveries would be at best a temporary benefit: even if they led to increased wealth, this would only hasten luxury, corruption, and social decay. He also attacked the notion that knowledge, especially of astronomy, brings peace and happiness. The Laputans, who have cataloged ten thousand stars, discovered two moons of Mars, and charted the path of ninety-three comets, live in a state of perpetual anxiety.

> These people are under continual disquietudes, never enjoying a minute's peace of mind; and their disturbances proceed from causes which very little affect the rest of mortals. . . . [They fear] that the earth by the continual approaches of the sun towards it, must in course of time be absorbed or swallowed up. That, the face of the sun will by degrees be encrusted with its own effluvia, and give no more light to the world. That, the earth very narrowly escaped a brush from the tail of the last comet, . . . and that the next, which they have calculated for one and thirty years hence, will probably destroy us. . . . They are so perpetually alarmed with the apprehensions of these, and the like impending dangers, that they can neither sleep quietly in their beds, nor have any relish for the common pleasures or amusements of life. (Ibid., 175–76)

FIGURE 9. *Bedlam: A Rake's Progress Plate VIII*, by William Hogarth, 1735. Hogarth's engraving of the Bethlehem Hospital (Bedlam) for mental patients shows two victims of science. The man peering at the ceiling through a roll of paper imagines he is an astronomer. Behind him, the man who has drawn a ship, the earth, the moon, and various geometric patterns is attempting to work out a means of calculating longitude. Sir John Soane's Museum, London.

This is an unmistakable reference to the predicted return of Halley's comet. There are also allusions to the contemporary fascination with sunspots, seen here as a pox on the face of the sun, and the obsession with cataloging the increasing number of stars observable with telescopes of ever-bigger aperture.

Although only book 3 of *Gulliver's Travels* explicitly focuses on scientists as the representatives of intellectual pride, it is important also to see it in the context of book 4, where the rationalistic, horselike Houyhnhnms represent a much more general image of the disembodied intellect. The Houyhnhnms have many admirable qualities, particularly when they are contrasted with the barbaric Yahoos, but Swift ridicules the notion that we can escape our human limitations and become Houyhnhnm-like. The more

Gulliver tries to do so, the more ridiculous he becomes. There is, moreover, a sinister suggestion of the cruelty and inhumanity contingent upon the passionless state of the Houyhnhnms, motivated by pure reason and impervious to emotional considerations, a view that was to be developed extensively by the Romantic writers of the next century. Thus, ultimately, Swift's purpose, like Pope's, is to redirect mankind from an excessive and optimistic reliance on rationalism, from abstract schemes that emanate from pride of the intellect, and to reaffirm the limitations of the human condition, to which the only proper response is humility. That he should have chosen the figure of the scientist to typify the arrogant intellect is an indication of both the increasing power and security of scientists compared with their counterparts a century earlier and the extent to which natural philosophy had come to represent the boundaries of contemporary knowledge.

The next stage after the acute anxiety experienced by the Laputans over the projected arrival of comets is madness, and eighteenth-century scientists did not escape this charge. In the final plate of his series of engravings *The Rake's Progress* (1735) William Hogarth depicted a scene from Bedlam (Bethlehem Hospital for the reception of lunatics), where two of the inmates have been rendered insane by their pursuit of astronomy or its application (see fig. 9). One astronomer is shown peering at the ceiling through a roll of paper that he imagines to be a telescope, while another inmate, who has drawn on the wall a ship, the earth, the moon, and various geometric figures, is evidently attempting to square the circle and to use astronomy to calculate longitude—another pressing preoccupation of the period.

A similar delusion about his power over the heavens besets the nameless astronomer in Samuel Johnson's moral essay *Rasselas* (1759). This learned man "has spent forty years in unwearied attention to the motions and appearances of the celestial bodies, and has drawn out his soul in endless calculations."[13] His long years of isolation and contemplation have deluded him into believing that the weather and the seasons are dependent on him, and the strain of such a responsibility has proved too much for his mind. Johnson diagnoses his case as resulting from too much solitude, with consequent overindulgence in the imagination. (This charge is particularly interesting in the light of the Romantics' accusations, less than half a century later, that science destroyed the imaginative faculty.) Since astronomy demands many hours of uninterrupted solitude, in which such fantasies can take root, it would seem almost impossible, according to Johnson's calculations, to combine astronomy and sanity.

Severe as this eighteenth-century moral censure was, it paled beside the fundamentally new and much more radical attack on the Newtonian system mounted by William Blake, who accepted none of the assumptions of the scientific method. For him, Newton, revered not fifty years before as the champion of religion, shared with Bacon and Locke the doubtful distinction of constituting an infernal trinity, and his dismissal of scientists generally as deficient human beings was to be reiterated and elaborated upon by later Romantic writers of the nineteenth century.

CHAPTER SIX

INHUMAN SCIENTISTS:
THE ROMANTIC PERCEPTION

> But the Spectre, like a hoar-frost and a mildew, rose over Albion,
> Saying: "I am God, O Sons of Men! I am your Rational Power!
> Am I not Bacon and Newton and Locke, who teach Humility to
>   Man,
> Who teach Doubt and Experiment? . . ."
>
> . . . May God us keep
> From single vision and Newton's sleep!
> —Blake

The eighteenth-century satirists discussed in the preceding chapter attacked the scientists of their time on the grounds of arrogance, for believing that they could transcend the human condition. Yet their criticism had little effect in view of the manifest practical spin-offs from science, especially in England, where the industrial revolution, based on technological innovation, was seen as the lucrative and inevitable outcome of scientific method. In theoretical terms as well, the unprecedented success of Newtonian physics in explaining the behavior of physical entities, from billiard balls to planets, seemed to guarantee the imminent realization of the Cartesian dream that reason, exemplified in mathematics, would ultimately resolve all the problems facing mankind.[1] Mathematics became the theoretical keystone of the Enlightenment, and its methods were widely emulated in other branches of knowledge. The codification of events with a view to establishing accurate causal predictions of the behavior of material entities and the reductionist procedures of analysis were regarded as the preeminent, if not the only, paths to knowledge.

In such a climate, it was inevitable that a mechanistic model of the animate world would be evolved to parallel that of the inanimate. Descartes had proposed that the bodies of animals could be described and understood as complex machines, and although he was careful not to say so explicitly, it is clear that he believed human beings were no exception, for his description of the human body closely parallels accounts of the mechanical statues of his time.[2] By the eighteenth century another French philosopher, Julien Offray de La Mettrie, had no such reservations. Basing his theory on analogy with the complex clocks of his time, La Mettrie declared in his provocative *L'Homme machine* (1747) that all animals, including human beings, are only complex animal-machines.[3] Already in England the philosopher John Locke, a Fellow of the Royal Society and an ardent admirer of Boyle and Newton, had embarked on an empirical enquiry into the powers and functioning of the human mind and concluded that all thought and knowledge were derived from external sensations and subsequently built up into a system of knowledge by an essentially mechanistic process, analogous to the processes described by physics. Joseph Ben-David has argued that Europe was swept by a scientistic movement, which became the new orthodoxy, whereby scientifically based procedures infiltrated every branch of knowledge, from economics to political thought, providing a new vocabulary for literature as well as science and influencing even the conventions of eighteenth-century European art.[4]

The Romantic writers' attack on this entrenched position of science was more radical than that of their predecessors. The earlier charge that scientists were attempting to transcend human limitations was far from inimical to the Romantics; indeed, the Prometheus figure, representing the desire to understand the universe, recurs throughout Romantic literature. Instead, these writers focused on the limitations rather than the triumphs of science. Their case against the classical mechanics, which formed the basis of Enlightenment science, was that by limiting the universe to the sum of separate, measurable entities it limited man as well, denying the validity of emotions, nonrational experiences, spiritual longings, and individuality. Consequently, the image they presented of the scientist was not that of an overreacher but rather that of someone less than human, not so much arrogant as deficient in human qualities.

Unlike the eighteenth-century satirists, whose criticism of science was almost exclusively negative, many of the Romantics offered alternatives to the Newtonian view of the universe, proposing a vitalistic science to replace the mechanistic one, or a life force with which Man could communicate. Whatever form it took, the Romantic writers had no doubt that their vision represented the Truth about the universe and would eventually encompass all knowledge and experience.

There was much in the actual science of their time to encourage these views. The cutting edge of science had moved away from Newtonian systems and botanical classification to chemistry and the little-understood phenomena of electricity and magnetism, areas where the mysterious concept of force at a distance was observed but not yet accounted for.[5] The work of Henry Cavendish and Joseph Priestley on electricity and the composition of gases in England, of Antoine Lavoisier in chemistry in France, and the experiments with electricity of Benjamin Franklin in America and Galvani in Italy seemed, in the popular mind, to reveal a world not far removed from alchemy and magic. Electrolysis appeared to fulfill the alchemists' dream of depositing pure gold; Galvani's experiments with static electricity applied to a dead frog's leg, causing it to twitch, were suggestive of restoring, if not creating, life and held promises of breaking down the barriers between living and nonliving. In chemistry the theory of elective "affinities," evolved to account for the relative tendencies of elements to form stable compounds, had clear analogies with human behavior and suggested a unitary principle applicable to both the animate and the inanimate world. Even the isolation of the gas oxygen and demonstrations of its necessary role in sustaining life were seen as related to vitalism.

Despite the similarity of scientific knowledge throughout Western Europe, the attitudes towards science and scientists adopted by Romantic writers show national variations. Basic to the relationship of Romanticism with the sciences in Germany was the *Naturphilosophie* evolved by the philosopher Schelling, who, in his desire for a unified world-view, proposed a new physics based on the premise that Nature was one huge living organism. The goal of science, in Friedrich Schelling's view, was therefore to discover the *Weltseele*, or world-soul, of this organism. He even proposed a hypothesis about the nature of this soul, namely, that it was characterized by polarity between opposing forces to produce a kind of cosmic heartbeat. Magnetic and electric polarity and chemical "affinities" were all seen as manifestations of this universal *Urpolarität*, which could be traced throughout Nature from crystal structures to the spiritual states of man.[6] The essential difference between *Naturphilosophie* and Enlightenment science was the insistence of the former on *Geist*, or spirit, and on a continuity between the spirit of man and a spiritual dimension in Nature. *Naturphilosophie* was therefore opposed to the Newtonian model of the universe as astronomical clockwork synchronized according to the laws of mechanics. The very term *mechanisch* became, for Schelling and the circle of Romantic writers who collected around him in Jena, *the* polemic adjective of abuse, the antithesis of their organic imagery.[7] On the other hand, many of the German Romantic writers had a considerable scientific background for their ideas.[8] Novalis had studied mineralogy, physics, chemistry, mathematics, and physiology; Friedrich Schlegel had studied physics; Goethe had specialized in botany but had read widely in all the contemporary sciences; and J. W. Ritter was a pharmacist, physiologist, chemist, and physicist.[9] In their diverse but comparable ways, these writers were intent on "romanticizing" the world in all its aspects, including science. They not only believed that science yielded material for poetry but were equally confident that their Romantic perspective would lead to a breakthrough in science. Goethe's theory of light and color perception, based on his own observation that shadows had color, the phenomenon of the "after image," was consistent with his belief that the observer had some active part in the perception of colors and was not merely a passive recipient of light particles falling upon the eye, as Newton and Locke had postulated. So passionately did Goethe feel about this that he conducted a lifelong battle, involving considerable personal expense, against the Newtonian theory of light.[10] Ritter's interest in electrophysiology led him to study the response of mimosa leaves to electric stimuli, a careful and highly regarded piece of experimental research but one that arose from his Romantic belief that the whole of Nature partook of sensitivity to some degree. Similarly, Ritter's best-known scientific work, his discovery of ultraviolet rays in the spectrum of the sun, was a direct result of his theory of polarity, which for aesthetic reasons required something to balance the infrared rays at the other end of the spectrum.

Given this close involvement with contemporary science and the interest of the German Romantics in portraying psychological and philosophical truths about scientists, the figure of the isolated alchemist pursuing truth and power became a useful archetype. It is therefore not surprising that they revived and reinterpreted the Faust figure, endowing him with nobility and heroic stature.[11] In these Romantic Faust portrayals there is no suggestion that knowledge is evil, only that it is insufficient to satisfy the aspirations of the heroic genius, who then embarks on other schemes. In F. M. Klinger's novel *Fausts Leben, Taten und Höllenfahrt* (1791) the protagonist begins with an intellectual enquiry into cosmic principles and problems of purpose, but by the end of book 1 he has discarded this program in favor of an investigation of evil and injustice. Like the traditional Faust, he is a rebel, but less against the restrictions of the human condition than against the moral limitations of the universe. His hubris lies in his assumption that he himself can right these injustices. But far from being condemned for this, he is portrayed as a moral idealist, albeit misguided and doomed to failure, a Promethean figure denouncing not only the inhumanity of man but, by implication, the necessary immorality of the Creator of such a universe.[12] For Maler Müller's Faust also, the craving for knowledge is only one of many goals, and again the protagonist is depicted as a heroic character, whose desire to transcend his humanity is endorsed. In Müller's treatment, moral strictures applicable to ordinary men are irrelevant.

The most famous Romantic Faust is Goethe's (part 1, 1805; part 2, 1832). He is depicted as an eminent scholar who, having mastered all the separate branches of learning, shares the alchemists' desire to know the unifying principle of Nature: "That I may know what the world contains in its innermost being."[13] In his opening monologue, on the vanity of learning and the charm of magic, Faust explains that he has turned to magic only because he believes it will disclose more of the mystical secrets of nature than orthodox science does. Knowledge qua rational explanation of the theoretical laws of the universe is of little interest to him, because from the beginning of *Faust* he already knows everything science has to teach, and in his second conversation with Mephisto he describes himself as "healed" of his striving for knowledge (line 1768).[14] He desires, rather, an intuitive understanding of Nature and ultimate experience of reality through the senses. Thus Goethe transforms Faust's traditional desire for knowledge into a quest for a spiritual unity with Nature, a kind of pansophism, and aligns this quest with sensuous experience. He thereby manages to accommodate two seemingly contradictory traits of the medieval Fausts—their desire for limitless knowledge and their apparently debased actions. Faust himself articulates this dualism in his lament, "Two souls, alas! dwell in my breast,"[15] and the uniting of these two contraries in the drama itself becomes part of the Romantic quest for unity out of diversity.

Goethe thus dispensed with the earlier trappings of the Faust story— alchemy and the desire for abstract knowledge—and presented instead a scientist who seeks confirmation of his theoretical ideals not in the laboratory but in the experiences of life itself.[16] As J. W. Smeed points out, "What distinguishes the *Sturm und Drang* Fausts from Lessing's Faust and, to some extent, from the Faust of the original chapbook is that they do not so much want to *know* more than other men as to *be* more than other men."[17]

German Romanticism produced more than a dozen works dealing with the Faust character,[18] but Goethe was the last to enunciate the ideal of a genuinely universal world-view, one embracing the intellect, the emotions, and transcendent experience. After Goethe, the growth of information in all scientific disciplines led to an inevitable specialization in science that is reflected in the literature. Experimental scientists and natural philosophers are increasingly depicted as divergent in their interests, so that there was no longer any expectation that a unifying *Weltanschauung* might be possible, or even desirable.[19]

Unlike their German counterparts, the English Romantic poets, few of whom had any training in science,[20] showed little faith in scientists' ability to evolve an organic world-view. Only Shelley and Byron responded positively to the scientific implications of the Promethean myth, and although Coleridge, with his interest in German philosophy, was not unsympathetic to the concept of science as the search for some universal principle, none of these writers created a character who could reasonably be regarded as a scientist.[21] It is true that Wordsworth professed to look forward to the time when "the Man of Science" and the poet would cooperate on a joint project, but for the most part, the English Romantics emphasized the essential conflict between intuition and the emotions on the one hand and scientific rationalism and mechanism on the other.

This is again a reflection of the diverse developments of science in different countries. In England, technology and, by extension, science were closely associated with the industrial revolution, which had brought not only wealth to the nation but also destruction of vast tracts of the countryside and appalling working conditions for those who had been forced into the towns to find work in factories. Compared with their German philosopher-scientist counterparts, the heroes of contemporary English science were essentially engineers and technicians, men such as James Hargreaves, Richard Arkwright, James Watt, and Thomas Newcomen, who were concerned with solving the practical problems of the manufacturing industry rather than with enunciating a universal hypothesis or positing a life force. It is therefore not surprising that the English Romantic writers identified science with technology, with the ugliness of the factories invading the countryside, and with the reification of the individual as an extension of the machine at which he worked. In Germany and France, on the other hand, the effects of the industrial revolution came much later; thus for their Romantic writers, science and its practitioners are more closely associated with theoretical issues.

The attack of the English Romantic writers on the scientific edifice was two-pronged, being directed against both its materialist and its rationalist aspects, that is, both the content and the method of science. In place of materialism, which they regarded as a limiting and fundamentally untrue account of experience, they asserted the primacy of a transcendent, metaphysical reality, affirming an organic view of nature in which the whole could not be reduced to the sum of its parts, nor the animate explained in terms of the inanimate.[22] Their protest was directed not primarily against the mechanistic procedures of science per se but rather against the quasi-religious article of scientific faith that the theoretical model constituted the whole reality. Related to this protest against mechanism was the Romantic rejection of rationalism as the only valid means of knowing. Instead, the English Romantics stressed the centrality of nonrational modes of experience—visionary perception, intuition, the subconscious as accessed in dreams and even madness, and, above all, the imagination. All these modes of knowing were antithetical to the procedures of science. In addition, they asserted the value and uniqueness of the individual over the generalizations and abstractions of scientific laws.[23]

Virtually all these aspects of English Romanticism can be seen as reactions against the figure of Newton, who inevitably represented for them the prototypical scientist. With his compact, internally coherent model of the universe as a machine, he epitomized the materialistic view of the world, denying reality to whatever could not be weighed and measured in terms of so-called primary characteristics. Although, as we saw in chapter 4, Newton was by no means a thoroughgoing mechanist and was even, in his own way, a visionary,[24] his public image came to represent, for later generations, analytical and reductionist procedures, which the Romantic writers regarded as actively destructive of the essential oneness and integrity of the universe.

William Blake, the earliest and the most extreme of the English Romantics, was also Newton's most uncompromising critic. Whereas Newton had been welcomed by the eighteenth-century deists as an ally in the demonstration of the existence of God—who else, they argued, could have invented a system so mathematically perfect and artistically satisfying as that encompassed in Newton's laws?[25]—Blake saw him, with his three-dimensional, mechanistic model of the universe, as a dangerous heretic, blinded by the error of materialism from seeing the complexity of truth.

> Now I a fourfold vision see,
> And a fourfold vision is given to me.
> . . . May God us keep
> From single vision and Newton's sleep![26]

In Blake's view, Newton the archmechanist, Bacon the exponent of experimentalism, and Locke, representing the philosophy of the five senses, constituted an infernal trinity of intellectual arrogance:

> But the Spectre, like a hoar-frost and a mildew, rose over Albion,
> Saying: "I am God, O Sons of Men! I am your Rational Power!
> Am I not Bacon and Newton and Locke, who teach Humility to Man,
> Who teach Doubt and Experiment? and my two wings, Voltaire,
>   Rousseau?"[27]

Blake further identifies them with the industrialism that sprang from the success of scientific principles.

> For Bacon and Newton, sheathed in dismal steel, their terrors hang
> Like iron scourges over Albion. . . .
> And there behold the Loom of Locke, whose Woof rages dire,
> Wash'd by the Water-wheels of Newton;[28]

In the *Opticks*, Newton had explained light as a stream of particles, and Blake equated these with the atomic theory of Democritus, insofar as both theories assumed a mechanical universe explicable in terms of smaller physical entities. Against this reductionist view, Blake again asserts the primacy of a transcendent spiritual reality.

> You throw the sand against the wind,
> And the wind blows it back again.

> And every sand becomes a Gem
> Reflected in the beams divine;
> Blown back they blind the mocking Eye,
> But still in Israel's paths they shine.

The Atoms of Democritus
And Newton's Particles of Light
Are sands upon the Red Sea shore,
Where Israel's tents do shine so bright.[29]

Paradoxically, however, Blake credited Newton with having achieved his unitary view of the universe by an extraordinary feat of the imagination. His engraving *Newton* (1795) reflects the complexity of his own response (see fig. 10). It shows a muscular youth measuring with dividers the base of an equilateral triangle drawn on a scroll. The curve of the youth's bent back parallels the arc inscribed in the triangle, and the triangular forms of his right leg, left wrist, and stretched fingers of both hands mimic the angle of the dividers and of the drawn triangle, implying that his humanity has been reduced to a mathematical parody of itself.[30] However, the youth, characterized by great intensity and energy (invariably positive attributes in Blake's view), draws his design, not on a stone tablet or in a book, but on a scroll, which in Blake's iconography invariably signifies imaginative creation.[31] Indeed, it was precisely because he also recognized Newton's genius that Blake was so critical of him. In the final Apocalypse, represented in his *Jerusalem* (1804), Bacon, Newton, and Locke, who, whatever their errors, were sincere seekers after truth, appear in the heavens as the foremost representatives of Science, counterbalancing Milton, Shakespeare, and Chaucer, the preeminent champions of Art.[32] The apparent anomaly is resolved when we realize that Blake regarded true science as eternal and essential; what he repudiated was a science divorced from humanity, denying the validity of the emotions, generalizing, abstracting, and denigrating individual experience.[33]

William Wordsworth showed a similar ambivalence towards scientists. On the one hand he elevated Newton, or at least the Newton represented in the statue in Trinity College, Cambridge, as an imaginative discoverer of far-off worlds, making him, in effect, an honorary Romantic hero, if not poet:

... Newton with his prism and silent face,
The marble index of a mind for ever,
Voyaging through strange seas of Thought alone.[34]

But he regarded science as an essentially lonely and anti-social pursuit in contrast to the universal experience of poetry. "The Man of Science seeks truth as a remote and unknown benefactor; he cherishes and loves it in his solitude: the Poet, singing a song in which all human beings join with him, rejoices in the presence of truth as our visible friend and hourly companion." Scientists could contribute to the imaginative subject matter of poetry only when their discoveries were shown to be relevant to the human condition.

FIGURE 10. *Newton,* by William Blake, 1795. For Blake, Newton symbolized the dangers of rationalism. Obsessively absorbed in his diagram, which his body has come to emulate, Newton thinks the whole of life can be measured mathematically. Tate Gallery, London.

"If the labours of men of Science should ever create any material revolution, direct or indirect, in our condition and in the impressions which we habitually receive; the Poet . . . will be ready to follow the steps of the Man of Science . . . carrying sensation into the midst of the objects of Science itself."[35] But for Wordsworth the partnership is strictly conditional, and it is clear that he regards the role of the poet as the more important of the two.

In addition, Wordsworth, like Blake, was opposed to narrow, analytical procedures; he deplored the man who would "peep and botanise on his mother's grave." Whereas for the scientist analysis was not only justified but obligatory, for the poet it was limiting and indeed falsifying. In lines that may now be read as a prescient formulation of the uncertainty principle, Wordsworth rejects the then-current scientific dissociation of subject and object:

> Our meddling intellect
> Misshapes the beauteous forms of things—
> We murder to dissect.[36]

This is a condemnation, not, as it is often made to seem when quoted in isolation, of the intellect per se, but of the *meddling* intellect, which regards nature as an alien object to be investigated rather than as an organic entity with which it can enter into a relationship.[37] It is a direct rejection of the so-called objective methodology and of the Baconian view that man should dominate Nature and force her to yield up her secrets. The English painter Joseph Wright, of Derby, vividly captured this spectrum of responses to science in his painting *An Experiment on a Bird in the Air Pump* (see fig. 11). The demonstrator who presides over the air-pump experiment, designed to show that air is really necessary to support life, stands poised ready to reintroduce air into the jar so that the bird may revive, but in order to dramatize the effect, he delays this until the critical moment when the bird is on the point of death. His obsessive expression and anachronistic garb suggest the alchemist totally preoccupied with his experiment. The man in the foreground, who carefully records the time taken for the bird to lose consciousness, is also concerned only with the mathematical results. By contrast, the children, whose values are always, for the Romantics, a better guide to Truth, are concerned chiefly for the suffering of the bird.[38]

Percy Bysshe Shelley, although himself sympathetic to many aspects of science and particularly of chemistry, also rejected the imperative to dominate Nature, on the grounds that, in so doing, man would diminish himself. "The cultivation of those sciences which have enlarged the limits of the empire of man over the external world has, for want of the poetic faculty, proportionally circumscribed those of the internal world; and man, having enslaved the elements, remains himself a slave."[39] It was for similar reasons that a late Romantic American poet, Walt Whitman, expressed his contempt for the astronomer, who, imprisoned physically by his lecture room and mentally by charts, figures, and columns, presumes to explain the unconfined immensity of the stars. He is contrasted with the Romantic poet, who, dissatisfied with the limitations of reason, escapes into the solitude and freedom of the night and, through a mystical rather than an intellectual experience, discovers the meaning of the stars.

> When I heard the learn'd astronomer;
> When the proofs, the figures, were ranged in columns before me;
> When I was shown the charts and the diagrams, to add, divide, and
> measure them;
> When I, sitting, heard the astronomer, where he lectured with much
> applause in the lecture-room,
> How soon, unaccountably, I became tired and sick;
> Till rising and gliding out, I wandered off by myself,
> In the mystical moist night-air, and from time to time,
> Look'd up in perfect silence at the stars.[40]

For the Romantics, the emotions, fed by the senses, and the unconscious, which surfaced in dreams and madness, were more reliable guides to

FIGURE 11. *An Experiment on a Bird in the Air Pump,* by Joseph Wright, ca. 1767–68. National Gallery, London.

Truth than the rational mind, confined as it was to the processes of logic. These approaches had little in common with science, and each group became increasingly intolerant of the claims of the other. The dichotomy is reflected not only in the poets but in many nineteenth-century prose writers, such as Dickens and Carlyle, who saw spiritual values being increasingly eroded by the march of industrialism and the economically successful alliance between science and technology. These writers frequently depicted the scientist as cold and unresponsive, imprisoned in a wholly intellectual pattern of behavior, to the detriment of natural feeling and human relationships. "Men are grown mechanical in head and in heart, as well as in hand," lamented Carlyle,[41] and Dickens makes a similar point in his "Full Report of the First Meeting of the Mudfog Association for the Advancement of Everything" (1837), a satire on the British Association for the Advancement of Science, which met for the first time in 1831.[42] The "great scientific stars, the brilliant and extraordinary luminaries" gathered there, are characterized by complete insensitivity to the suffering of a pet dog, which they proceed to vivisect. The "scientific reporter" records: "You cannot imagine the feverish state of irritation we are in, lest the interests of science should be sacrificed to the prejudices of a brute creature, who is not endowed with sufficient sense to foresee the incalculable benefits which the whole human race may derive from so very slight a concession on his part" (400).

Sometimes, although rarely, the unfeeling scientist in literature is pitied and permitted to reform. Browning's long dramatic poem *Paracelsus* (1835), based on the life of the Swiss alchemist Theophrastus Bombast von Hohenheim, depicts a scientist who in his proud and single-minded pursuit of knowledge sacrifices everything else in life, only to realize at the moment of death that he has omitted the one important factor, love. Similarly, in Dickens's Christmas story "The Haunted Man" (1848) the introspective chemist Redlaw, a latter-day Frankenstein who is full of regrets for the past and emotionally unfulfilled in the present, is permitted to be reclaimed by the simple, loving Milly Swidger, counterpart of Frankenstein's Elizabeth, a Dickensian angel-in-the-house.

But more often there is no reprieve for the scientist, whether alchemist or chemist, who has devoted his life to knowledge, to the exclusion of the emotions. In such accounts, mathematicians feature prominently as exemplars of the dehumanization process.[43] This is comically expressed in W. J. M. Rankine's poem "The Mathematician in Love" (1874). The mathematician is mocked for his inability to relate emotionally to the young lady, and his obsession with formulas is duly punished in the living world, where emotions rather than abstractions are the accepted currency.

> A mathematician fell madly in love
>   With a lady, young, handsome and charming:
> By angles and ratios harmonic he strove
> Her curves and proportions all faultless to prove.
>   As he scrawled hieroglyphics alarming
>
> . . . .
> "Let x denote beauty,—y, manners well-bred,—
>   "z, Fortune,—(this last is essential),—
> "Let L stand for love"—our philosopher said,—
> "Then L is a function of x, y, and z,
>   "Of the kind which is known as potential."
> "Now integrate L, with respect to dt,
>   "(t standing for time and persuasion);
> "Then, between proper limits, 'tis easy to see,
> "The definite integral *Marriage* must be:—
>   "(A very concise demonstration)."
>
> . . . .
> But the lady ran off with a dashing dragoon,
>   And left him amazed and afflicted.[44]

More usually, however, the treatment is tragic rather than comic. Two of the most successful stories of Fitz-James O'Brien explore the psychology of the obsessed scientist. In "The Diamond Lens" (1858) Linley, a microscopist, is haunted by the desire to construct a perfect microscope, one that will provide hitherto unimagined magnification. To acquire the diamond necessary for such a lens, he kills a fellow lodger, and then he labors in solitude for months to produce the microscope. His nemesis comes in an interesting psychological form. With his superlens he discovers in a drop of water an infinitesimal world inhabited by a beautiful girl, whom he calls Animula and with whom he falls in love. Of necessity this love is hopeless, since the lovers are separated by that very barrier of magnitude the scientist had thought to transcend with his microscope. He is released from his obsession only by the extinction of the girl when the water droplet evaporates, whereupon the student is declared insane. The story combines, in effect, two traditions: that of the introverted, isolated scientist who stops at nothing, not even murder, to pursue his research and that of the Faustian overreacher, whose nemesis, like Frankenstein's, arises from his very success, in this case his achievement in constructing the microscope. The scientist's infatuation with the girl symbolizes both his obsession with his experiment and his arrogant, vain attempt to transcend natural barriers.

In "The Golden Ingot" (1858) O'Brien depicts another scientist, Blacklock, obsessed with the alchemists' search for a means of transforming base metals into gold. When one day he finds a gold ingot in his crucible, he thinks he has at last succeeded, until he discovers that his devoted daughter, intending to bring him happiness, has pawned everything to purchase the ingot and has secreted it there. Thereupon he dies of a stroke.

Many of the best-known scientists of this type are depicted as destroying not only themselves but those close to them as well. Balthazar Claës of Balzac's *La Recherche de l'Absolu* (1834) is one of the most complex of these studies. Although Balzac's major interest is the psychological, almost clinical, study of a genius and the effect on his family,[45] the underlying moral concerns the Romantic belief that preoccupation with science atrophies the normal emotions that sustain personal relations and social responsibilities. Claës's wife Josephine pleads the case for the emotions when she tells him, "Science has eaten away your heart" (84), and contrasts her own selfless devotion with his uncaring obsession with his chemistry. His response, a piece of unwitting self-parody, is to redefine feelings in the current chemical term, "affinities."[46]

Claës is portrayed, in fact, as a latter-day alchemist (the local townspeople mockingly call him "Claës the Alchemist"), with all the stereotypical trappings—his physical appearance, his alchemist's laboratory, and an assistant who is popularly supposed to be akin to the devil. Balzac also includes a Mephistopheles figure in the Polish stranger, through whose persuasive agency Claës effectively sells his soul to the Devil as he becomes obsessed with finding a method to produce diamonds by transmutation, ruining his family in the process. Balzac's analysis is more complex, however, than this suggests. The word "Absolute" suggests a transcendent reality beyond the analytical procedures and chemical terms of the particular experiment. Thus although Claës is presented as deficient in feelings towards his family, there is also an aura of grandeur attaching to him, not only in his acknowledged genius but in his devotion to an ideal, his self-sacrifice, and the insults he endures. Balzac's characterization in fact reflects the same complexity of response as Blake's rendering of Newton and the Romantic ambivalence towards the scientist "voyaging through strange seas of thought alone" while attempting to confine the universe within a reductive, analytical system. Like Browning's character, Paracelsus, Claës discovers the secret of the Absolute only at the moment of death, for it is a metaphysical rather than a physical truth.

Claës is presented as a tragic figure whose essential nobility and human potential are wasted; more often, however, the Romantic image of the scientist is evil rather than pitiable. Charlotte Dacré's otherwise unremarkable novel *Zofloya: the Moor* (1806) not only uses the stereotype of the demonic Moorish alchemist dabbling in the forbidden black arts (the Arabic association still lingers) but discloses, on the last, climactic page, that the alchemist is not merely in league with Satan; he *is* Satan![47]

Two of the stories in E.T.A. Hoffmann's *Nachtstücke* (1817),"Ignaz Denner" and "Der Sandmann," examine the destruction of innocent if gullible victims by evil scientists. "Der Sandmann" in particular is worth examining in some detail.[48] The title character, Dr. Coppelius, overtly a lawyer, is a closet alchemist who apparently exerts an extraordinary influence over the protagonist's father. As the child Nathanael watches from his hiding place with fascinated horror, Coppelius and the boy's father, in a setting heavily suggestive of an alchemist's laboratory, engage in an experiment intended to produce an automaton (the representative par excellence of mechanistic science). On discovering the child, Coppelius threatens to extract his eyes, and the boy faints with terror.

A year later, Coppelius is implicated in the death of Nathanael's father from an explosion in their laboratory. Thus, from the beginning he is associated with the sinister and the secretive. His experiment is also mysteriously linked with the removal of natural eyes to produce artificial vision. This latter theme is amplified when Coppelius subsequently returns, disguised as Coppola, an Italian hawker of scientific glasses—barometers and telescopes. These distort the vision of those who look through them, bewitching them with an illusory image, so that, symbolically, they have indeed lost their natural sight. It is by gazing through such a telescope that Nathanael first sees the beautiful automaton Olimpia, technically flawless but lacking emotions and spontaneity. In the descriptions of Olimpia, Hoffmann vividly expresses the Romantic abhorrence of mechanism. Nathanael, infatuated with Olimpia, dances with her at a ball given in her honor but is perturbed to discover how stiffly she holds herself and how mechanically she dances. She plays and sings like a clockwork model, and when Nathanael bends to kiss her, her lips are ice-cold.[49] Her mechanical nature is finally demonstrated when he hurls her to the floor, causing her dismemberment. But Nathanael is also destroyed. Under the spell of Coppola's telescope, he flings himself to his death from a tower, itself the symbol of his isolation from both society and nature. Hoffmann thereby reinforces his moral that scientific obsession, whether in the form of the automaton or the fabricated glass that distorts nature's truth, erodes human emotions and finally destroys life itself.

The American novelist Nathaniel Hawthorne also wrote from a strongly Romantic position, insofar as he distrusted rationality and regarded the emotions as a truer guide to morality and to spiritual and mental health. In this he may have been reflecting the frontier values of nineteenth-century America, where, unlike urbanized Europeans, Americans experienced a world that was not orderly and rational but violent and unpredictable. In Hawthorne's novels and stories the excessive cultivation of the intellect is seen as destructive not only of the one so obsessed but often of others whose lives are entangled with his. In a *Notebook* entry of 1844 Hawthorne wrote:

> The Unpardonable Sin might consist in a want of love and reverence for the Human Soul; in consequence of which, the investigator pried into its dark depths, not with a hope or purpose of making it better, but from a cold philosophical curiosity.—content that it should be wicked in whatever kind or degree, and only desiring to study it out. Would not this, in other words, be the separation of the intellect from the heart?[50]

This idea clearly underlies much of his fiction, for Hawthorne was more interested in psychology and in the many fringe sciences, from phrenology to mesmerism, which flourished in mid-nineteenth-century America, than he was in physics or chemistry. His scientists are therefore symbolic rather than realistic, and most are cast in the alchemist mold.[51] Aylmer, the protagonist of his short story "The Birthmark" (1845), is a chemist whose love for his beautiful wife is subsumed by his obsession with her birthmark, which he sees as abhorrent unless it can function as a scientific challenge. Her subsequent death, when he finally persuades her to submit to the removal of the mark, is both literal and symbolic. In the tradition of Faustus, this sacrifice of life and the emotions for an abstract ideal is seen as the end result of his arrogant desire to explore what "seemed to open paths into the region of *miracle*" (my italics). Science has become a religion for Aylmer, who has "faith in man's ultimate control over Nature." His inability to accept the "symbol of imperfection," the birthmark, with its implications of original sin and the human condition, represents his determination to perfect Nature through science. Hawthorne makes here a pertinent comment on the meliorist claims of nineteenth-century scientists, who implicitly assumed they could eliminate sin along with disease and offer a utopia cleansed of physical and moral blemishes.

In "Rappacini's Daughter" (1844) Hawthorne presents another cold and inhuman scientist. Rappacini is "as true a man of science as ever distilled his own heart in an alembic."[52] Like Aylmer, his physical attributes have atrophied in favor of his intellectual obsession, and again this condition is associated with heartlessness. He is "a tall, emaciated, sallow, and sickly-looking man, dressed in a scholar's garb of black. He was beyond the middle term of life, with gray hair, a thin gray beard, and a face singularly marked with intellect and cultivation, but which could never, even in his more youthful days, have expressed much warmth of heart" (258).

Rappacini has cultivated a garden of exquisite but poisonous flowers, not for the sake of their beauty but in order to study their fatal effect on other forms of life. By a process of gradual assimilation, his beautiful daughter Beatrice has become immune to the fatal poison emanating from the flowers, but the cost of this is that she too exudes the poison, so that whatever she breathes upon soon dies. Just as he observes the effect of the deadly poison on insects nearing the flowers, the vampirelike Rappacini encourages suitors for his daughter in order to study the speed with which the poison will act on them. The characterization of Rappacini clearly testifies to Hawthorne's perennial mistrust of the scientific mind and his conviction that its exclusive cultivation leads to death. His colleague Baglioni warns: "I know that look of his: it is the same that coldly illuminates his face as he bends over a bird, a mouse or a butterfly which in pursuance of some experiment he has killed by the perfume of a flower—a look as deep as Nature itself, but without Nature's warmth of love."[53]

Characteristically, in Hawthorne's work as in much Romantic writing, the rationalist head is contrasted with the potentially noble but ineffectual heart. This is rendered almost diagrammatically in Chillingworth, the learned physician-cum-alchemist of Hawthorne's best-known novel, *The Scarlet Letter* (1837). Like Faust, Chillingworth has studied at a German university and relies strongly on herbs and cures learnt from the Indians and concocted in his alchemist's laboratory.[54] He is also said to be a magician, and, as in the medieval tradition of witchcraft, Chillingworth, cold and unfeeling as his name suggests, becomes a "deformed old figure . . . stooping away along the earth" (193) as an index of his spiritual corruption.

Chillingworth is driven by a desire to discover the identity of his young wife's lover and take his revenge. Hawthorne's real interest centers on the psychological causes for Chillingworth's behavior as he studies and assumes control over his victim in the manner of the scientist observing a passive specimen. In Hawthorne's interpretation, Chillingworth's great sin

lies in exploiting his victim's psychosomatic dependence upon him, studying him coldly and without sympathy, in order to destroy him. His victim
Dimmesdale, the adulterous minister, speaks for Hawthorne when he says,
"That old man's revenge has been blacker than my sin. He has violated, in
cold blood, the sanctity of a human heart" (212).

In *Septimius Felton, or the Elixir of Life* (1871) Hawthorne created yet
another Faustian scientist, whose little store of knowledge only proves to
him the impossibility of acquiring the vast amount he desires. In his solitary,
obsessive search for all knowledge, he becomes merely "a cold and discon-
nected spectator of all mankind's warm and sympathetic life."[55]

Hawthorne's scientists clearly owe their basic conception to the Faust
story, and like their prototype, most began as idealists. Aylmer's life had
formerly been devoted to lofty scientific ideals, Chillingworth was once a
dedicated physician intent on helping others, and Septimius Felton was
famed for his impeccable behavior. Like Faust, however, they all succumb to
hubris, desiring to know too much, to transcend the limitations of the human condition. But Hawthorne imparts a specifically Romantic cast to the
Faust myth by grafting on the additional psychological implication of the
effect on the human personality of the head, symbolized most often as
the pursuit of scientific knowledge, dominating the heart, representing compassion. In these parables of the intellect, Hawthorne warned that a science
that progressed more rapidly than society's moral development would create its own ethical standards before society was even aware of the moral
issues being raised.

While Hawthorne's attack on scientific hubris and its consequences is
couched in psychological, even moral, terms, his English contemporary
Edward Bulwer, Lord Lytton, was a vitalist who mounted his attack on
scientific materialism in terms that mimicked those of the very science he
was condemning. In particular, Lytton was fascinated by the then unexplained phenomenon of electricity, which he saw as one manifestation of a
supernatural, all-pervading power or vital principle. His early novel *Zanoni*
(1842) is an allegorical attack on the mechanistic interpretation of life in
which Mejnour, the representative of true wisdom, "Contemplation of the
actual—SCIENCE," attacks contemporary experimental science and defends
the occult practices of the Rosicrucians. The universal power to which he is
privy and which is also, it seems, the basis of medicine is described as "an
indivisible fluid resembling electricity, a fluid that connected thought to
thought with the rapidity and precision of the modern telegraph."[56] Despite
his opposition to the science of his time, Lytton attempted to borrow credibility by claiming familiarity with its ideas and practitioners. Even the passages attacking orthodox science are bolstered with copious references to
contemporary scientists.[57] The net effect in the novel is a curious pastiche of

the occult, physical science, theories of animal magnetism, and Schopenhauer's impersonal, cosmic Will.

In *A Strange Story* (1861) and later in his best-known novel, *The Coming Race* (1871), Lytton returned to the theme of a vital force more powerful than any technology could produce. Lytton called *The Coming Race* a fictionalized account of "the Darwinian proposition that a coming race is destined to supplant our races,"[58] and he suggests that the underground society "Vril-ya," whose members control a universal power, "vril," a combination of galvanism and magnetism, will overthrow humanity.[59] Lytton's attitude in this novel is strangely ambivalent, for while he still deplores the ruthlessness of this society, with its practice of eugenics, there is a sense of calm inevitability about the final evolutionary victory of the Vril-ya.

The Romantic writers' reaction against rationalism and reductionism and their development of an alternative explanation of the natural world spanned nearly a century; therefore their emphasis inevitably varied at different times and places. There are, however, certain constant factors that can be taken as categorizing the Romantic attitude towards natural science and scientists. These include a deep suspicion of abstraction alongside a rejection of materialism; an assertion of the importance of secondary characteristics and values as well as the measurable quantities of Newtonian physics; the rejection of mechanism in favor of organism, which in the last decades of the nineteenth century became formalized as the doctrine of vitalism; and a preference for spontaneity over causality as an explanation for phenomena. In much Romantic writing scientists are characterized by their failure to espouse these principles, a failure that is depicted as so directly affecting their relationships with people and Nature that it functions as an index of moral and spiritual health.

The Romantic image of the scientist as cold, inhuman, and unable to relate to others has been one of the most influential in twentieth-century stereotyping of the scientist in both literature and film, as will be seen in chapters 13 and 14. A particular, even archetypal, version of this Romantic image is the character of Frankenstein.

CHAPTER SEVEN

FRANKENSTEIN AND
THE MONSTER

> Life and death appeared to me ideal bounds which I should first
> break through, and pour a torrent of light into our dark world. A
> new species would bless me as its creator and source; many
> happy and excellent natures would owe their being to me. . . . I
> thought that if I could bestow animation upon lifeless matter, I
> might, in process of time . . . renew life where death had
> apparently devoted the body to corruption.
> —Mary Shelley

Although *Frankenstein* (written in 1816, when its author, Mary Shelley, was
eighteen, and published in 1818) is clearly a Romantic work and indeed
predates many of the works considered in the preceding chapter, it merits
separate discussion because of the extraordinary influence it has had on
subsequent presentations of the scientist. Frankenstein has become an archetype in his own right, universally referred to and providing the dominant image of the scientist in twentieth-century fiction and film. Not only
has his name become synonymous with any experiment out of control but
his relation with the Monster he creates has become, in the popular mind at
least, complete identification: Frankenstein *is* the Monster. The archetypal
power of the Frankenstein story can be attributed to the fact that in its
essentials it was a product of the subconscious rather than the conscious
mind of its author and thus, in Jungian terms, draws upon the whole collec-
tive unconscious of the race.

The circumstances of the composition of *Frankenstein*, as described by
Mary Shelley in her introduction to the 1831 edition, are almost as well
known as the story itself and have themselves inspired other fictional accounts, including a film and an opera.[1] Yet it is worth stressing here that
according to her account, the story was produced by the concurrence of two
specific factors: the need to produce a horror story and the account of an
alleged scientific experiment. Mary and Percy Shelley, their baby son, and
Mary's stepsister Claire Clairmont were spending the summer of 1816 near
Geneva, where they were neighbors of the poet Lord Byron and his personal

physician Polidori. Kept indoors by a stretch of bad weather, Byron, Polidori, and Percy and Mary Shelley each agreed to write a ghost story as entertainment. Mary records that she found great difficulty in thinking of a suitable plot until the evening when the others were discussing the latest experiments allegedly conducted by Erasmus Darwin. The latter was supposed to have "preserved a piece of vermicelli in a glass case till by some extraordinary means it began to move with voluntary motion. Not thus, after all, would life be given. Perhaps a corpse would be reanimated; galvanism had given token of such things. Perhaps the component parts of a creature might be manufactured, brought together, and endued with vital warmth." After hearing this philosophical discussion, Mary dreamed the central scene of her novel. The esteemed Doctor Darwin has been translated into "the pale student of unhallowed arts, kneeling beside the thing he had put together." This suggests that the very attempt to create life was already associated, at least in Mary's subconscious mind as accessed by her dream, with the demonic and the horrific. The problem of finding a subject for her story was instantly solved: "What terrified me will terrify others; and I need only describe the spectre which had haunted my midnight pillow. . . . making a transcript of the grim terrors of my waking dream."[2]

It is not difficult to supply reasons why the account of Darwin's alleged experiments should have had such a deep effect on Mary Shelley. As the youngest and least assured person present, and clearly intellectually overawed by the discussion (she tells us that she was "a devout but nearly silent listener"), Mary, who had only the preceding year lost her first child, born prematurely, and who had recently undergone a second, difficult confinement, would have felt emotionally disturbed, even violated, by a discussion that not only abolished the role of the female in the creation of life but trivialized the process by reducing it to "a piece of vermicelli in a glass case." Unable to argue at a rational level with the intellectual giants Byron and Shelley, she doubtless suppressed her disquiet, which emerged violently in her subsequent dream. What is more interesting for the purpose of this exploration of images is her immediate identification of the highly visual nightmare image of the attempt to create life with her earlier aim "to think of a story . . . which would speak to the mysterious fears of our nature" (8).

The power and immediacy of *Frankenstein* owe much to its highly visual sequences (hence the power of the film versions that exploit these), but its more subtle and pervasive implications derive from the older myths relating to the desire for knowledge and creativity on which it draws—Pandora's box, both the Prometheus myths, the Genesis stories of the Creation and the Fall, and the Faust legend, particularly Goethe's treatment of it.[3] However, Shelley's treatment also involves a more complex assessment of the issues presented by Goethe insofar as Frankenstein embraces both the scientific rationalism and reductionism condemned by the Romantics and the ultimate Romantic quest for knowledge of the absolutes of life and death. She thereby suggests that these apparently contrary positions are, in the final analysis, merely variations of the same basic type—the overreacher whose aspirations lead inevitably to the destruction of himself and others yet who is instinctively admired as the heroic genius.

Although the scientific background of the novel, with its Gothic trappings, may now seem fantastic, Mary Shelley was careful to make it consistent with current scientific theories, including ideas about electricity that were still somewhat magical. Percy Shelley had long been interested in electricity and galvanism. He had personally experimented with making a large-scale battery and had repeated Benjamin Franklin's experiment with a kite in an electrical storm,[4] so that the methods whereby Frankenstein brings his Monster to life—an electric discharge—would have been regarded by many contemporaries as at least feasible.[5] In his preface to the novel, Percy Shelley explicitly insisted that "the event on which this fiction is founded has been supposed, by Dr. Darwin and some of the physiological writers of Germany, as not of impossible occurrence." This adherence to what was scientifically credible was essential to Mary's purpose of exploring both the premises of scientific materialism and the values of the Enlightenment.

Frankenstein is not only the Romantic overreacher determined to transcend human limitations; he is also the heir of Baconian optimism and Enlightenment confidence that everything can ultimately be known and that such knowledge will inevitably be for the good. "I doubted not that I should ultimately succeed. . . . A new species would bless me as its creator and source; many happy and excellent natures would owe their being to me" (53–54).[6] Frankenstein also accepts uncritically the reductionist premise of the eighteenth-century mechanists, that an organism is no more than the sum of its parts. As heir to a mechanist view of science, he has no sense of the extraordinary irony involved when he sets out to create a "being like myself" from dead and inanimate components, ignoring the possible need for any living or spiritual elements. Even in retrospect he seems to see no anomaly in this, for he tells Walton, not without pride: "In my education my father had taken the greatest precautions that my mind should be impressed with no supernatural horrors. I do not ever remember to have trembled at a tale of superstition, or to have feared the apparition of a spirit" (51).

But the being he creates is not merely a mechanism, the sum of its inanimate parts; it is indeed a being like himself with free will and a soul, which are not subject to Frankenstein's control. As such, it is a recreation of Frankenstein's own unconscious desires, both good and evil, which have been suppressed by the discipline of his research program and by cultural censorship.[7] The Monster is thus both an alter ego and a substitute for the natural child to whom he has denied existence by deferring his marriage with Elizabeth. This Doppelgänger relationship symbolizes the essential duality of man, the complex of rational and emotional selves (Hawthorne's notion of the head versus the heart), mutually alienated but finally inseparable.[8] In the image of the larger than human Monster, Shelley reaffirms the Romantic position that the unconscious is an intrinsic and more powerful part of the human experience than the rational mind and that, if suppressed, it will ultimately emerge to destroy the latter.

Shelley further aligns herself with the Romantic protest against scientific rationalism in emphasizing the fact that the obsessive desire to acquire scientific knowledge is achieved at the cost of emotional well-being. In order to pursue his experiments (conducted, significantly, among graves and charnel houses, the symbols and venues of death) Frankenstein denies the affections of the living, isolating himself from the loving family and friends who would both impede his studies and save him from his fatal obsession. Typically, he becomes insensitive to natural beauty, and his health and emotional state deteriorate. He tells Walton: "My cheek had grown pale with study, and my person had become emaciated with confinement. . . . It was a most beautiful season, . . . but my eyes were insensible to the charms of nature. And the same feelings which made me neglect the scenes about me, caused me also to forget those friends who were so many miles absent" (54–55).[9] These aspects of Frankenstein's character were to form the basis for many representations of the emotionally crippled scientist isolating himself from life, nature, and human affections. Some of these have already been considered in the preceding chapter, but they have become, if anything, more numerous in twentieth-century representations (see chapter 13 below).

Shelley's critique of science and its assertion of the right to pursue knowledge, wherever it might lead, is both more complex and more radical than the Romantic condemnation of intellectualism per se, for Frankenstein is also portrayed as partaking of many of the very qualities that characterized Romanticism itself and are still associated with the image of the noble scientist seeking truth. Frankenstein is simultaneously both the scientific rationalist and the passionate idealist. Dissatisfied with the mere accumulation of facts, he desires to know the unknowable, to press beyond the boundaries of human knowledge, "to penetrate the secrets of Nature." There was no secret more powerful to the imagination than the creation of life, the equivalent of the alchemists' quest for the elixir of life, which in Frankenstein's research program, involving the materials of death, also includes the characteristically Romantic attempt to reconcile opposites.

Again, although Frankenstein is explicitly exonerated from any desire for mere wealth and justifies his research on the grounds that it will benefit the whole human race, Shelley makes it clear that his search is by no means disinterested: "Wealth was an inferior object; but what glory would attend the discovery, if I could banish disease from the human race and render man invulnerable to any but a violent death" (40). It becomes apparent that "glory" is closely akin to power, and in relating the story of his life to Walton, Frankenstein unwittingly reveals how strongly he is attracted to power, regardless of the devastation it leaves in its wake.[10] He tells how he stood enraptured by the violence of the electrical storm, undismayed by the consequent destruction of the living oak tree. And it is no coincidence that his inspiration at the University of Ingolstadt is not the empiricist M. Krempe but the charismatic M. Waldman, who, in language reverberating with biblical echoes, claims for modern chemistry supremacy over other branches of knowledge because of the power it offers. "[The chemists] have indeed performed miracles. They penetrate into the recesses of nature and show how she works in her hiding places. They ascend into the heavens; they have discovered how the blood circulates and the nature of the air we breathe. They have acquired new and almost unlimited powers; they can command the thunders of heaven, mimic the earthquake, and even mock the invisible world with its own shadows" (47–48).

Thus, what captivates Frankenstein is less the lure of knowledge for its own sake than the promise of the power it confers. The reference to Faust's bargain with Mephistopheles is clear. Frankenstein says, "I felt as if my soul were grappling with a palpable enemy"; and he formulates his obsessive purpose in terms of a spiritual conquest. "More, far more will I achieve. . . . I will pioneer a new way, explore unknown powers, and unfold to the world the deepest mysteries of creation" (48).[11] "Life and death appeared to me ideal bounds which I should first break through, and pour a torrent of light into our dark world" (54). That is, he sees himself as reenacting the role of the Creator[12] and, in accordance with the Romantic quest, wrenching opposites into unity at his will.

Another interesting facet of Shelley's portrayal is that Frankenstein's idealism is presented as one of his most dangerous qualities. In this she was almost certainly referring to the contemporary political situation in France, where the original ideals of the French Revolution (warmly welcomed at first by the Romantic writers) had plunged the country into the Reign of Terror and led, in the long term, to the Napoleonic Wars. Indeed, there are striking similarities between the descriptions of the Monster and the iconography associated with Napoleon Bonaparte.[13]

It was largely as a result of her ambivalent characterization of Frankenstein as both scientist and Romantic idealist that Mary Shelley was able to suggest new and complex facets of the motivation of the scientist and his relation to his work. In so doing, she preempted many of the philosophical and psychological considerations that have subsequently been recognized as inextricably linked to scientific research. First, since almost everything that pertains to Frankenstein is equally relevant to the career of the artist, Shelley indicates the close nexus between scientific research and the creative imagination and suggests that they entail both opportunity and risk. Like art, science represents the highest creative endeavor of humanity and has the power to produce either benefits or disasters. Second, her chief purpose, as outlined in her preface, is to explore the ethical consequences of the *success* of Frankenstein's experiment. In scientific terms, the creation of the Monster is a brilliant achievement; yet Frankenstein's horror begins at the precise moment when the creature opens its eye, the moment when for the first time Frankenstein himself is no longer in control of his experiment. His creation is now autonomous and cannot be uncreated any more than the results of scientific research can be unlearned, or the contents of Pandora's box recaptured. Hence, paradoxically, it is at the moment of his anticipated triumph that Frankenstein qua scientist first realizes his inadequacy (see fig. 12).

As soon as he perceives the power and horror of the living Monster, Frankenstein rushes from the room, and when the creature seeks him out, he runs away and hides from him. Subsequently he falls into a "nervous fever" for several months; later he blames his unconscious state for his failure to prevent the Monster's escape. He thus effectively puns on the idea that he was unconscious (i.e., unaware) of the consequences of his work. However, it is clearly implied in the text that this unawareness was at least partly willed and postdated opportunism. The hideousness of the Monster and the potential for power and destruction contingent on his size were apparent throughout the construction program. Shelley thus suggests that claims of ignorance on the part of scientists for their failure to foresee the consequences of their work are too glib to be credible and cannot be socially acceptable. By his isolation, Frankenstein had deliberately excluded those, such as his father and Clerval, who might have given him the ethical advice he did not wish to hear.

Shelley also explores the relation between Frankenstein's pursuit of scientific success, his failure as a human being, and his social guilt. The inevitable neglect of human ties involved in the scientist's dedication to his research results not only in his own isolation and loneliness but in a moral and emotional loss to society. Whereas many other Romantic treatments of the scientist's isolation assumed that this was a voluntary state that could be reversed at will (Paracelsus, Redlaw, Aylmer, and Claës are indicted for their refusal to reform), Shelley suggests that there is a necessary loneliness and guilt contingent on scientific research. Frankenstein begins by frequenting remote and lonely places. At first this isolation is dictated by the requirements of his research, since he collects his materials from graveyards and charnel houses; but subsequently his separation from society becomes a necessity imposed by the result of his experiment—the existence of the Monster. In relating his tale to Walton, another scientist pursuing an obsession in contravention of the natural ties of affection, Frankenstein digresses

FIGURE 12.
An illustration from
Frankenstein, the
1831 Standard
Novels edition, of
the critical moment
when the Monster
comes to life, by
T. Holst.

to moralize explicitly: "If the study to which you apply yourself has a tendency to weaken your affections, and to destroy your taste for those simple
pleasures in which no alloy can possibly mix, then that study is certainly
unlawful, that is to say, not befitting the human mind" (56).

Similarly, the Monster, deprived of any affection from his creator and
driven away from the company of those he wishes to befriend, becomes
through this enforced isolation from society a ruthless murderer. Both,
therefore, through isolation from society lose their sense of social morality.
Significantly, Mary dedicated *Frankenstein* to her father, William Godwin,
who, like Victor Frankenstein's father, rejected religious sanctions for behavior and sought to ground morality wholly on a social and psychological
basis.[14]

Shelley also probes the intellectual pride of the scientist, masquerading as a desire to serve mankind. Like many researchers, Frankenstein justifies his experimental program as a moral endeavor designed to "pour a
torrent of light into our dark world. A new species would bless me as its

creator and source; many happy and excellent natures would owe their being to me. . . . I might, in process of time, . . . renew life where death had apparently devoted the body to corruption" (54). She thus points to the opportunities for self-delusion that accompany the pursuit of glory and power.

Philosophically, Shelley also examines the paradox involved in the pursuit of freedom through knowledge. The intrinsic dilemma of *Frankenstein* resides in the irony that the mind, which can conceive of freedom from limitations, is also the source of man's most acute agony, because it perceives both its own restrictions (the more Frankenstein learns, the more aware he is of his own ignorance) and the bondage that it has itself constructed. This perception, symbolized in Frankenstein's involuntary subjection to the very creature that represents his triumph over the former boundaries of knowledge, was to be further explored by H. G. Wells in "The Time Machine," *The Island of Doctor Moreau,* and *The Invisible Man.*

There is a related irony attaching to Frankenstein's desire to act the part of God. In usurping this role, Frankenstein acquires all the burdens of the creator and none of the glory. Like the God of the Genesis story, he plans a paradise from which evil and death will be absent, only to have his creature turn against him in defiance and introduce sin into the world. Indeed, like Milton's Adam, the Monster blames his creator for this sin.[15] But unlike the Christian God, Frankenstein fails to engineer a redemptive scheme for his creature, even though the Monster begs him: "Make me happy and I shall again be virtuous" (100). In a parody of the God of the deists, Frankenstein is isolated from his creation; unable to bring himself to accept it in its sinful state, he leaves it to fend for itself.[16]

The presence in the novel of Walton, the polar explorer and hence another type of scientist, is important in suggesting the universality of the statements concerning Frankenstein. There are many significant parallels between the two men, both in their situations and in their attitudes. Both have left the security and affection of family in the pursuit of a goal that will inevitably cost the lives of innocent people. Both are motivated by the desire for glory and personal achievement but attempt to justify their search in terms of the benefits it will confer on a grateful posterity. Both seek to yoke together natural opposites. Whereas Frankenstein seeks to eliminate the boundaries between life and death, Walton is bent on discovering the tropical paradise in the polar wastes that one contemporary theory proposed. The poles, being the furthest points from the inhabited world, also represented boundaries and possessed an almost mystic significance symbolized in their supposed link with the tropical lands of the equator. To reach them symbolized completion, the final definition of the world; hence the nineteenth- and early-twentieth-century fascination with polar exploration and Walton's personal obsession with achieving this ultimate transcendence of geographical limitations. The charismatic effect on Frankenstein of Wald-man's words about the acquisition of "new and almost unlimited powers" is mirrored in Walton's admiration for Frankenstein and the compulsive eagerness with which he listens to his story.

Ironically, Mary Shelley also provided an image for the concept of women in relation to science, an image that was to continue well into the twentieth century, if indeed it has ever been significantly revised. Her own subconscious protest against the idea that women were unnecessary in the creation of life has already been discussed, but it was also to issue in the critically neglected figure of Frankenstein's fiancée Elizabeth. In the terms of his research proposal, she too is rendered redundant to the life process, as Frankenstein, while professing to be in love with her, keeps deferring his marriage and hence the creation of a child by natural means. Officially this delay is only until he has perfected the unnatural creation of his monstrous "child"; however, the Monster's subsequent murder of Elizabeth enacts Frankenstein's own repressed desire to rid himself both of his social responsibilities and of the much simpler natural method of procreation.[17] It is significant that, immediately after the animation of his Monster, Frankenstein dreams of Elizabeth transformed by his embrace into the corpse of his dead mother.

This aspect of the story was to provide a new and pertinent twist to the powerful Gothic image depicted in Henry Fuseli's painting *The Nightmare* (1782): the vulnerable female stretched passively on a bed, her head hanging down, menaced by the surrealist intruders (see fig. 13). In *Frankenstein* the menacing intruder is no supernatural visitor that can be banished by the rational, waking mind, but the material creation of science. Frankenstein awakens to find his nightmare actualized. The sexual connotations of this neo-Gothic image, which filmmakers were quick to exploit, also dominated the pulp science fiction of the twentieth century and were clearly regarded as an important aspect of their attraction, since variations of this icon formed a staple part of the film posters of *Frankenstein* and the lurid cover designs of the pulp magazines.[18] By extension, the suggested rape and subsequent death of the beautiful and vivacious Elizabeth by the Monster created by science can also be seen as a figurative enactment of Bacon's perception of science as penetrating the secrets of a symbolically female Nature, laid inert on the rack.

By bringing together in Frankenstein the apparently opposite qualities of the scientist and the Romantic visionary, Mary Shelley not only enriched immeasurably her depiction of the scientist but extended the basic Romantic protest against materialism and rationalism. She showed Frankenstein, apparently so rational, so desirous of secularizing the world and removing its mysteries, to be, at crucial points, highly irrational, suppressing those considerations that might conflict with his obsession. George Levine points out

FIGURE 13.
Nightmare, by
Henry Fuseli, 1782.
Goethe Museum,
Frankfurt.

that *Frankenstein* "as a modern metaphor implies the conception of the divided self, the creator and his world at odds. The civilised man or woman contains within the self a monstrous, destructive, and self-destructive energy."[19] The novel thus becomes a modern scientific formulation of the archetypal myths of *psychomachia*, or the conflict within the soul. In this wholly secular version, science and technology are a concretization of inner desires, masquerading as rational but, like the Monster, equally capable of springing from the dark, unacknowledged depths of their creator's subconscious.

This perception suggests some important qualifications of the Enlightenment belief that the pursuit of knowledge is, by definition, rational and good and should not be restricted by any sociomoral considerations. Few nineteenth-century readers, however, were able to follow these implications. It has taken such twentieth-century Monsters as psychoanalysis, nuclear power, in vitro fertilization, and genetic engineering, bursting upon an ethically unprepared world with their dual potential for good and evil, to illuminate fully the depths of meaning in *Frankenstein*. Not surprisingly, playwrights and filmmakers have returned with great frequency to the story, modifying it to suit the prevailing tastes, values, and scientific debates of their time. It is interesting, however, that no screen version has retained Shelley's pessimistic ending.

The first physical presentation of Frankenstein was H. M. Milner's play of 1826, *Frankenstein; or, the Man and the Monster,* which became the subject of one of the earliest films, the Edison Company's *Frankenstein* (1910). This film eliminated most of the repulsive physical situations and concentrated instead on the psychological aspects of the story, emphasizing that the creation of the Monster was possible only because Frankenstein allowed his normally healthy mind to be overcome by evil and unnatural thoughts. Edison's ending, in particular, was far more positive and romantic, echoing contemporary optimism about science: the Monster finally fades away, leaving only his reflection in a mirror, and even this is subsequently dissolved into Frankenstein's own image by the power of Elizabeth's love. Frankenstein has been restored to mental health, and hence the Monster can no longer exist.

Carlos Clerens, a historian of horror films, rates the 1931 Universal film classic, *Frankenstein,* which introduced Boris Karloff as the Monster, as "the most famous horror movie of all time";[20] yet, compared with Shelley's novel, the film is hardly horrific at all. The heavily underlined moral, stated at the beginning, that "it is the story of Frankenstein, a man of science who sought to create a man after his own image without reckoning upon God," restores a religious dimension of supernatural order and justice to Mary Shelley's entirely secular and unredeemed situation. In this version, Henry Frankenstein (who, following Peggy Webling's 1930 play, on which the film is based, has changed given names with Clerval) is presented as the innocent victim of a mistake whereby his careless assistant has brought him the brain of a murderer instead of one from a noble person for insertion into his creature. The evil character of the Monster is therefore an experimental error rather than the result of Frankenstein's hubris. The implication is that the creation of the Monster per se posed no real or abiding problem and that with due precautions, a better result could be obtained the next time. This treatment of the story, including the otherwise anomalous introductory moral, was consistent with the adulation of scientists in the United States during the 1930s.

Although the film ended with the Monster being burnt to death and the celebration of Frankenstein's wedding to Elizabeth, the box office success indicated a sequel. The final scenes of the 1931 film were cut from all prints in circulation, and *Bride of Frankenstein* (1935) opened with Mary Shelley telling Shelley and Byron the sequel to "her" novel. In this film, Frankenstein becomes the pawn of another scientist, the mad and evil Dr. Pretorius, who, having constructed various homunculi, now wishes to produce something larger. He forces Frankenstein to create the mate for which the Monster of the novel had begged. The female Monster (in an extension of the Doppelgänger effect in the novel, she is played by the same actress who plays Mary Shelley, Elsa Lanchester) is striking but not hideous, and she immediately rejects the Monster, who in despair electrocutes her, Dr. Pretorius, and himself. In this film Frankenstein has become entirely absolved of guilt, and the function of the evil scientist bent on creating life has passed to the alchemistlike Pretorius.

Bride of Frankenstein was followed by a long succession of Frankenstein derivatives whose titles are sufficiently indicative of their content and of the way in which Frankenstein has been integrated into Western culture as an ever-contemporary byword, almost as a real person.[21] At different periods the emphasis falls variously on horror, space travel, sexuality, or comedy associated with the figure of the scientist. One of the most interesting films in terms of the application of the Frankenstein story to a contemporary scientific debate is *Frankenstein 1970* (1958), in which Boris Karloff returned to the screen as the disfigured Victor Frankenstein, victim of Nazi torture. By means of an atomic reactor, he revives the Monster from his ancestor's 1757 experiment, but they both die a horrible death from radioactivity when the reactor blows up. Only then is the Monster's face revealed. It is the face of a youthful Victor Frankenstein, symbolizing in startling visual imagery the identification of creator and creature, in this case the atomic scientist and his dangerous and faulty creation, atomic power. More recently, Ken Russell's film *Gothic* (1986) attempted to recreate the scene of the composition of Frankenstein, while David Wickes's *Frankenstein: The Real Story* (1993) stresses the twentieth-century relevance of the novel. The undiminished fascination of Frankenstein as perennial myth seems assured.

CHAPTER EIGHT

VICTORIAN SCIENTISTS: DOUBT AND STRUGGLE

"The stars," she whispers, "blindly run;
  A web is wov'n across the sky;
  From out waste places comes a cry,
And murmurs from the dying sun;

"And all the phantom, Nature, stands
  With all the music in her tone
  A hollow echo of my own,—
A hollow form with empty hands."
—Tennyson

At no other time in history has a scientific theory evoked so much public controversy and private anguish as occurred in Britain during the latter half of the nineteenth century. Although this was associated directly with the publication in 1859 of Charles Darwin's book *On the Origin of Species by Means of Natural Selection*, the unease had been building for at least a decade, fueled at one level by a fear of social change and possible revolution (such as occurred in almost every other European country during the 1840s), at another level by insecurity resulting from the eroding of traditional religious faith, and not least by a realization derived from popular accounts of astronomy and geology that humanity occupied only a tiny speck in an alien universe that was inconceivably vast in time and space.[1]

The extent of the popular controversy over Darwinism and the degree to which loss of religious faith was attributed to a knowledge of science are clear indicators of radical change in the public awareness of science and of scientists during Queen Victoria's reign. Whereas in the 1830s the only scientific research was that carried out by wealthy amateurs or doctors whose practice allowed them sufficient time for medical research, by the end of the century there was a popular groundswell of enthusiasm for a scientific education, and science as a paid profession was becoming a possibility. This change did not emanate from the universities. For most of the nineteenth century little science was taught at either Oxford or Cambridge, and almost no scientific research was performed there. The curriculum at these prestigious institutions remained essentially that of the grammar school, with Oxford emphasizing the classics and Cambridge pure mathematics. Experimental science was not considered worthy of study or appropriate for the awarding of degrees. There was little change in this reactionary attitude until the 1870s, when Michael Foster introduced a course in general biology at Trinity College, Cambridge, and James Clerk Maxwell was appointed Cavendish Professor of Experimental Physics at Cambridge and charged with setting up the Cavendish Laboratory.[2]

Despite the continuing indifference to science by the British government and the two principal British universities, the proliferation during this period of mechanics' institutes offering scientific studies testifies to the determination by members of the working class to gain the kind of education they themselves considered relevant rather than the classics-based education offered by local grammar schools. This push for the rethinking of the content of education was increasingly supported by those of liberal and progressive views, exemplified by Henry Brougham's Society for the Diffusion of Useful Knowledge. And in contrast to most of the specialist scientific societies, which, like the Royal Society, had been originally established for the benefit of wealthy amateurs, the British Association for the Advancement of Science was set up in 1831 as a democratic organization specifically intended to counter this kind of elitism.[3] The British Association organized popular public lectures on science in towns all over England and encouraged some of the nation's foremost scientists to take on the responsibility of communicating science to the general public (see fig. 14).

In addition, Victorian Britain produced a formidable array of eminent scientists, including Sir Frederick William Herschel, Sir Humphrey Davy, John Dalton, Michael Faraday, Robert Chambers, Charles Babbage, Sir Charles Lyell, James Joule, Charles Darwin, Alfred Russel Wallace, Thomas Henry Huxley, John Tyndall, and James Clerk Maxwell. It was inevitable that these men, many of whom were skilled communicators of science, would capture the public imagination and have a considerable effect on the status of science in an increasingly literate society.

Notwithstanding this popular following, all was not harmony within the scientific edifice itself. In the latter half of the nineteenth century, the hitherto close liaison between scientists and philosophers represented in the term *natural philosophers* began to break down. Disputes between empirical scientists and theoreticians became increasingly acrimonious, the experimentalists denouncing the theoreticians as vague and untrustworthy and the latter retaliating with charges of narrow specialization. In practice, science moved further and further away from its origins as the whole body of knowledge, becoming more specialized, more occupied with specific problems, and less concerned with questions of Truth or universal values. And

FIGURE 14. *Scientific Researches!* by James Gillray, 1802. This caricature of a lecture at the Royal Institution reflects the humor of the time. It shows Dr. Thomas Garnett lecturing to a fashionable audience, with Humphrey Davy, standing by with a pair of bellows, as assistant lecturer. On the lecture bench are bottles of oxygen and hydrogen and a collapsed bladder. Count Rumford stands on one side, and the audience includes Isaac Disraeli, Lord Stanhope, Earl Pomfret, and Sir H. Englefield. Reproduced by kind permission of a private collector, to whom the author extends her thanks.

although some philosophers regarded the new, specialized disciplines as trivial,[4] by the mid-nineteenth century the Romantic search for universal Truth had disappeared almost entirely from science, being replaced by a thoroughgoing scientific materialism.

This trend towards specialization was reflected in the Victorian novel, where characters were no longer described merely as "scientists" in the general sense, but as geologists, astronomers, biologists, and mathematicians. Causality, too, in the form of social and psychological determinism, became an integral part not only of plot but of characterization as the new science of psychology began to arouse wide public interest and Social Darwinists attempted to account for socioeconomic disparities in biological terms.[5]

Whereas physics and chemistry had been the sciences featuring most prominently in literature before the 1850s, they were replaced in the latter half of the century by astronomy and biology. These sciences became the focus of popular debate and the cause of widespread anxiety as there emerged a general realization of the vastness of the universe. Astronomical scales of time and distance necessarily reduce human concerns and values to the point of insignificance, rendering the human scale of events trivial or irrelevant. In 1831 Thomas Henderson had shown that the nearest star to the sun was twenty-four trillion miles distant, and suddenly the order and security the compact Newtonian model of the solar system had seemed to offer were extinguished by unimaginable spaces. The fear engendered by such spatial magnitude was made more terrible by the hitherto unthinkable time scales being revealed by geology. In the seventeenth century Bishop Ussher, using Old Testament genealogies, had set the date of the Creation at 4004 B.C., but this figure, based on a human-related unit of measurement, the length of a generation, was increasing rapidly to orders of magnitude that rendered the human life span wholly insignificant. After the publication of Sir Charles Lyell's three-volume *Principles of Geology* (1830–33), in which the word *evolution* was first used in its modern sense, and his later *Geological Evidences of the Antiquity of Man* (1863), it became increasingly difficult to reconcile the idea of such a vast universe, in which man was effectively a mere speck of matter marooned on a tiny planet, with the Christian belief in a personal, caring God.

More difficult still for many Victorians to accept was Darwin's formulation of evolutionary theory, which appeared to strip man of any special claim on the deity by relegating him to the same status as other creatures. It would be hard to overestimate the impact of evolutionary ideas, even before Darwin's *On the Origin of Species* exploded into print in 1859, selling out the first edition in one day. With Darwinism, the cutting edge of scientific materialism moved from physics to biology. And while religious orthodoxy had been able to assimilate Newton by declaring his cosmic scheme a triumph for God, who thereby emerged as a first-class Watchmaker, it was much less able to digest Darwin. It now seemed that not only the inanimate world but the realm of biology and even, after Darwin's yet more iconoclastic *Descent of Man* (1871), man himself could be accounted for in wholly material terms of cause and effect. The German chemist Jacob Moleschott spoke for many scientists of the period when he asserted that the secret of life was chemistry and that people were no more than the sum of their heredity, food, and environment.[6]

Even more shattering for Darwin's contemporaries than a mechanistic biology was the absence of purpose in the Darwinian scheme. The ideas of evolution put forward in earlier works, such as Erasmus Darwin's long poem *Zoonomia* (1794–96) and Robert Chambers's *Vestiges of the Natural History of Creation* (1844), had posited a benevolent purpose on the part of the Creator in producing life forms exquisitely adapted to their circumstances. But with the Darwinian scheme the concept of directed evolution, and hence of purpose or design on the part of Creator, were swept away. The bases of the new theory were chance, waste, and suffering, all of which appeared irreconcilable with the Christian belief in a loving God. To many it now seemed that concepts such as free will, sin, redemption, and the other foundation stones of religion were at best irrelevant, at worst a longstanding deception. The mental and emotional turmoil that this caused among thinking people of the late nineteenth century was the theme of countless novels, poems, and plays, of which only a small number are now read, and while the *Angst* was not exclusively associated with science,[7] it was frequently examined in terms of the dilemma facing the Christian scientist.

The nineteenth-century interest in realism, the literary counterpart of scientific materialism,[8] was at first an impediment to the development of the scientist as a realistic character, since in order for such a figure to be portrayed with sufficient credibility it was necessary for an author to have considerable understanding of the concerns, motivation, and language of scientists, yet most novelists were quite ignorant of what scientists did or said when they were not present at social gatherings or giving public lectures. It is scarcely surprising therefore that doctors, who were known to everyone, preceded scientists into the realistic novel. The Victorian period saw the appearance of some fine realistic portrayals of doctors, and the first realistic scientists in fiction were doctors who also engaged in medical research. Although doctors in general practice form no part of this study, those who were also medical researchers will be considered here in that latter capacity, since they represent an important stage in the emergence of the professional scientist in literature. Several of the writers whose work is discussed in this chapter had some personal acquaintance with actual scientists, and certainly Charles Kingsley, Elizabeth Gaskell, George Eliot, and Thomas Hardy took the trouble to acquire a working knowledge of the science they wrote about. Their accounts therefore carry more credibility at a factual level than many of the earlier studies discussed above. This in itself is an indication both of the seriousness with which science was now approached in literature and of the interest in scientific issues on the part of the reading public.

Because of the increased differentiation of sciences and the complexity of the issues involved, Victorian writers developed a variety of realistic treatments of the scientist rather than a single prominent character type such as we have discussed from earlier periods. In the sections that follow, we shall consider scientist characters under the following groupings, which indicate the major concerns raised in relation to science during Victoria's long reign: idealized natural scientists, professional scientists, scientists and religion, immensities of space and time, the exploiters.

Idealized Natural Scientists

Unlike the unflattering eighteenth-century caricatures of doctors, the realistic portrayals of nineteenth-century medical researchers are, almost without exception, complimentary to the point of eulogy. This was due in part to the recognition of the widespread public benefits deriving from Edward Jenner's introduction of vaccination against smallpox and Joseph Lister's pioneering work in antisepsis. Doctors in Victorian literature are seen as heroic figures of unfailing integrity, battling against the diseases of the poor for little financial reward, and the early depictions of natural scientists are endowed with a similar catalog of virtues, moral and physical as well as intellectual.

One of the first such portraits is the heroic Tom Thurnall, doctor-scientist of Charles Kingsley's novel *Two Years Ago* (1857). Kingsley, an enthusiastic amateur zoologist as well as a novelist and clergyman, firmly believed in the ideal of *mens sana in corpore sano*.[9] Tom, modeled on the author's brother George Henry Kingsley, has impeccable qualifications, having studied in the laboratories and hospitals of Glasgow and Paris and having been a ship's surgeon and army doctor. Ever ready with his microscope and collecting equipment, he is no mere theoretician but a genuine, born-again muscular scientist, extolling the virtues of experience as the basis of scientific method: "We doctors, you see, get into the way of looking at things as men of science; and the ground of science is experience."[10]

Like his author, Tom is a passionate exponent of sanitary reform.[11] When an outbreak of cholera threatens a village, he leaps into the fray. For him it is a holy war against disease. As he tells the curate: "You hate sin, you know. Well, I hate disease. Moral evil is your devil, and physical evil is mine" (212). Kingsley introduces other contemporary medical issues as well. Scrupulously honest, Tom goes against the normal practice in refusing to prescribe medicines he thinks unnecessary and invariably charges the minimum for those he does prescribe. Kingsley is also careful to indicate that Tom's interest in collecting zoological specimens along the Devonshire coast, which makes him the first medical research scientist in literature,[12] arises not only from the specimens per se, although there is no doubt that his fascination is genuine, but also from their value in medical practice, for Kingsley, like Bacon, justifies science in terms of its social usefulness. "This little zoophyte lives by the same laws as you and I; . . . he and the sea-weeds, and so forth, teach us doctors certain little rules concerning life and death" (150). In Tom Thurnall, the scientist-doctor, Kingsley contrived to combine the qualities he most admired in the scientist, the soldier, and the social reformer by rendering the doctor's war against disease and unsanitary conditions almost indistinguishable from a sociomoral war against greed and stupidity.

It was relatively easy to idealize the doctor's pursuit of scientific knowledge in altruistic terms; his search was not for himself but for the benefit of a suffering humanity. It was during the nineteenth century that doctors acquired the elevated social and moral status that they were to maintain for more than a century.[13] It was much more difficult to justify the pure scientist, about whom there still lingered the suspicion of intellectual and perhaps, therefore, spiritual pride, the taint of Faustus. Considerable public relations work had to be done before a pure scientist could enter the realistic novel with his moral reputation intact.

Once again, one of the first champions in the field was Charles Kingsley. His prolific correspondence with such scientists as Huxley, Darwin, and Philip Gosse was mutually fruitful and he was honored by being elected a fellow of both the Linnean Society (1857) and the Geological Society (1863). Despite the stereotyped view of the Victorian age as a battleground between science and religion, there is no suggestion in Kingsley's writings that science need be inimical to the full development of the emotional life or to religion. On the contrary, Kingsley, like the seventeenth- and eighteenth-century exponents of natural theology, believed that a better knowledge of the marvels of nature would engender a deeper love of the Creator. "I am sure that science and the creeds will shake hands at last . . . and that by God's grace I may help them to do so."[14] His fairy tale *The Water Babies* (1863) was written in order to introduce children to the concept of evolution in a Christian context.[15] It is therefore not surprising that the several naturalists who appear in his works, far from being reclusive intellectuals, are all of the vigorous, "muscular Christian" stamp favored by Kingsley.[16] In *Glaucus; or, the Wonders of the Shore* (1855) Kingsley even extols the natural scientist as an ideal type, a kind of latter-day crusader, embodying bravery, patience, modesty, reverence, and chivalry: "It is these qualities, . . . which make our scientific men, as a class, the wholesomest and pleasantest of companions abroad" (see fig. 15).[17] Such praise from a well-known and respected clergyman did not pass unheeded. Despite the subsequent clash with the religious establishment, an aura of selfless devotion to Truth began to flicker around the scientific cranium.

An unusually late example of a scientist idealized without qualification is to be found in *Melampus* (1883), a long poem by George Meredith, who alone among the major Victorian poets seems to have been able to integrate the Darwinian picture of nature with Romantic values. His ideal scientist-poet Melampus observes and learns from nature the secrets that will allow man to overcome his alienation from the natural world. Essentially he is an intermediary through whom the healing power of nature can flow to others, curing the psychosomatic diseases contingent upon their alienation from nature. In *Melampus*, the search for knowledge has no taint of evil associated with it. Meredith indeed goes out of his way to counter

FIGURE 15. *A Victorian Family Beachcombing*, from *La Musée des Enfants*. Mary Evans Picture Library.

such a suggestion; the serpent of Eden is here recast as the benign snake of Hippocrates, which confers upon Melampus the gift of understanding the voices of nature and the pattern of causality, one of the main points at issue between nineteenth-century science and faith. However, despite Meredith's concern to incorporate realistic details, Melampus remains a symbolic rather than a realistic figure, embodying the author's pious hope that the discoveries and values of science are not totally alien to those of Romantic poetry. Moreover, there is a real sense that Meredith is cheating. Either he has carefully tailored his science to fit the aesthetic values of art and poetry or he has failed to understand the full significance of Darwin's theory. He glosses over the terror of nature, partly by ignoring the inanimate universe and partly by taming the internecine struggle for existence that had so appalled Tennyson into a mild and remediable injury for which nature herself provides a panacea if we but search for it in the right spirit. The failure of Melampus to convince is an index of how difficult it was to fit the idealized scientist into realistic literature.

Professional Scientists

The more innovative portrayals of the Victorian period are those non-idealized figures resulting from a genuine effort by the author to understand the struggles and problems of actual scientists. Elizabeth Gaskell's novel *Wives and Daughters* (1866) is a particularly interesting piece of social history, depicting three scientific characters who span a broad social spectrum. Lord Hollingford, eldest son of an earl, is a wealthy amateur scientist in the eighteenth-century tradition. Intensely shy and lacking in small talk, he relates easily only to those with whom he can discuss science, namely, his scientific friends from London and the local doctor Gibson, who also finds time to contribute to scientific medical journals. Gaskell is thus the first to show the bond of scientific interest as a social leveler, for there were few other grounds on which the lord and the local doctor could have met as equals. This scientific egalitarianism, which could override class distinctions, was certainly one of the attractions of the scientific societies.

The third scientist in the novel, Roger Hamley, is the most interesting, since he represents the first example in literature of a professional scientist in the modern sense. Roger is regarded by his family as a mere plodder and contrasted unfavorably with his elder brother Osborne, who seems ready to distinguish himself in the expected classical education at Cambridge. His mother considered that Roger "was so little likely to distinguish himself in intellectual pursuits; anything practical—such as a civil engineer—would be more the kind of life for him."[18] (We may note the nineteenth-century distinction between *intellectual,* meaning classical, and *practical,* meaning scientific.) However, the unlikely Roger emerges as a new kind of hero. Modeled on Charles Darwin, whom Gaskell knew well, his scientific enthusiasm and careful observation outweigh his social awkwardness.[19] He becomes Senior Wrangler, wins acclaim for his papers on comparative anatomy and osteology, and, like Darwin, embarks on an exploratory voyage as a naturalist, laying the foundations for a distinguished career as "professor of some great scientific institution" (707).

In her portrayal of the two Hamley brothers, Gaskell contrasts not only two distinct personalities but also the results of two different approaches to education. The classical-humanist criteria favored Osborne with his wit and his ability to write poetry and savor intense impressions, and discriminated against Roger's shyness and slow thoroughness. A different set of parameters however—and Gaskell implies that these parameters are increasingly those of the real world—encourages the abilities of Roger, who, it is clearly suggested at the end of the novel, is the man of the future. Thus, in *Wives and Daughters* there is, in effect, a dramatization of the protracted dispute between Huxley and Matthew Arnold over the relative importance of the classics and science in the educational curriculum, the latter emerging triumphant.[20]

Gaskell also suggests one of the reasons for the slow emergence of the professional scientist in England, namely, the financial precariousness of the profession in this period. Despite his outstanding university results and his reputation in the scientific journals, Roger is still dependent on wealthy amateurs for an income. Lord Hollingford remarks, "Science is not a remunerative profession, if profession it can be called" (676).

The truth of this statement is developed in George Eliot's greatest novel, *Middlemarch: A Study of Provincial Life* (1871–72), which depicts the struggles, financial and personal, contingent upon medical research.[21] *Middlemarch* is set some seven years later than *Wives and Daughters*, in a country town that, compared with the traditional village community of Hollingford, is undergoing vastly accelerated social change, including the advent of railways, the repercussions of the first reform bill, and the beginnings of medical reform. Against this background, Eliot's detailed study of Tertius Lydgate, the "new, young surgeon" who aspires both to reform general practice and to pursue first-class research in tissue structure, constitutes the preeminent psychological and social portrait of a scientist in nineteenth-century literature, the realistic counterpart of Mary Shelley's *Frankenstein*. The complexity of Lydgate as a character stems, first, from the density of scientific background Eliot amassed during the writing of *Middlemarch* and, second, from the fact that she perceived the society of Middlemarch and in particular Lydgate's role in it not merely in literary terms but as a kind of scientific experiment through which to examine the social implications of Darwinian theory. Thus Lydgate, while presented with great realism, comes to have a symbolic significance as well.

As Anna Kitchel's *Quarry for Middlemarch*[22] and Eliot's own *Journal*[23] conclusively demonstrate, Eliot read exhaustively not only all the relevant literature of medical and scientific research in the years preceding 1829—the year in which Lydgate is said to have arrived in Middlemarch—but also the background controversy surrounding attempts at medical and sanitary reform in England, as it was reported and argued in the columns of the *Lancet*.[24] Thus, unlike earlier novelists, she does not merely *tell* us that Lydgate is a medical researcher; she places him in the milieu of convincing medical research so that his actions and assumptions, his motives and behavior, result from and interact with this background. Eliot also shows the resistance Lydgate receives from colleagues and the petty jealousies of social intercourse, thereby integrating the procedures of research into the wider social picture.

At the time Eliot was writing *Middlemarch*, medicine had only recently been accorded scientific status.[25] Owing to the way in which doctors were trained in England, experimental research in medicine during the first half of the century was almost nonexistent.[26] "About 1829," Eliot remarks in *Middlemarch*, "the dark territories of Pathology were a fine America for a spirited young adventurer" (1:128). Having studied anatomy and pathology in Edinburgh and Paris, Lydgate is far ahead of his provincial colleagues in his knowledge of the causes and treatment of fevers, including cholera and typhoid,[27] in his use of the stethoscope[28] and the new achromatic micro-scope, and in his insistence on conducting postmortem examinations when the cause of death is not adequately known. We are also shown several instances of his superiority in diagnosis, which, understandably, affront the medical fraternity of Middlemarch. Thus Lydgate realizes that Fred Vincy is in the pink-skinned stage of typhoid fever, whereas the well-established Dr. Wrench, believing his patient to be suffering merely from "a slight derange-ment," has been prescribing drugs harmful for Fred's actual condition. Sub-sequently, Lydgate rediagnoses Nancy Nash's "tumour" as cramp, and his knowledge of the new methods of treating pneumonia (by permitting the patient's own resistance to combat the disease[29] rather than bleeding and thus further weakening him) allows Trumbull to recover. Perhaps the most impressive indication of Eliot's scrupulous research is seen in Lydgate's treatment of Raffles's *delirium tremens*, which follows the method proposed only the preceding year by the American John Ware.[30] He has, moreover, higher moral standards than most of his colleagues, refusing to dispense his own medicines (and so avoiding the temptation to overprescribe), which further aggrieves his opportunistic colleagues.[31]

These incidents indicate not only a sense of authenticity about the character of Lydgate but also the new attitudes emerging within the disci-pline of science. The gradual change from the mechanism that had charac-terized medicine since the time of Harvey to the idea of process and change implicit in evolutionary theory is mirrored in Lydgate's own attitudes.[32] Lydgate began as a mechanist, fascinated by "his first vivid notion of finely-adjusted mechanism in the human frame" (1:124–25). Then in Paris he came under the influence of the great histologist Marie-François-Xavier Bichat, who rejected the reductionist philosophy, asserting that living bodies "are not associations of organs which can be understood by studying them first apart, and then as it were federally" (1:128).[33] Finally Lydgate comes to believe that subjective elements are a necessary part of scientific procedure.

Lydgate's contact with the work of Bichat, who had shown that all the apparently diverse organs were composed of comparatively few kinds of tissues, has filled him with a neo-Romantic desire to discover the fundamen-tal or primitive tissue, of which Bichat's tissues are but modifications. In other words, Lydgate hopes to discover a single common basis for all living structures, a preoccupation parallel to the desires of the alchemists and of Newton, to discover a comprehensive unifying system. What is more signif-icant, his quest is also parallel to that of Frankenstein, since he too wishes to find the basic secret of life. Although they approach their goal from opposite ends—Frankenstein attempts to turn dead matter into a living organism, while Lydgate strives to reduce living tissue to a basic material—they both assume a wholly materialist view of life. For George Eliot, who had already dealt with a similar issue in her frankly Gothic story "The Lifted Veil," the issue was of immediate, contemporary relevance, since this was the pivotal point of Darwinism, its essentially materialist philosophy.

As with Frankenstein, the obverse of Lydgate's intellectual passion is a certain intellectual arrogance and scorn for the second-rate. Lydgate takes no trouble to conceal his contempt for the backward medical practices of Middlemarch, thereby arousing the hostility of both his colleagues and his clientele. "Lydgate's conceit was of the arrogant sort, . . . massive in its claims and benevolently contemptuous" (1:130). In Lydgate, this intellectual pride is exacerbated by a residual social snobbery contingent upon his aristocratic family connections, an attitude that ill accords with his progressive notions in science. Eliot is acutely aware that advanced ideas in one area may fail to modify conventional thinking in another so that even a distinguished mind may be "a little spotted with commonness." "That distinction of mind which belonged to his intellectual ardour, did not penetrate his feeling and judgment about furniture, or women, or the desirability of its being known (without his telling) that he was better born than other country surgeons. . . . neither biology nor schemes of reform would lift him above the vulgarity of feeling that there would be an incompatibility in his furniture not being of the best" (1:130–31). It is when Lydgate sets himself up as the great man that he becomes most ordinary. Despite his superior abilities and motivation, Lydgate, like the protagonist of classical tragedy, is brought low by hubris, including not only intellectual arrogance (to which scientists might be considered especially prone) but also a more mundane snobbery concerning social status and possessions.

Outside his professional interests, then, Lydgate is as conventional and small-minded as the Middlemarch society he so keenly despises. In contrast to his meticulous procedures in science, he leaps, in his private life, from superficial observations to the erroneous conclusions that lead to his disastrous marriage with an incompatible partner and thence, inevitably, to debts and social disgrace. Lydgate thus represents an important development in the representation of the scientist from the Romantic depiction, where the promptings of the heart were depicted as morally superior to those of the mind and submission to the latter invariably issued in evil. It is when Lydgate *fails* to use his mental faculties to the full and succumbs instead to his emotions that he falls. In Lydgate's case this is particularly ironic, because it involves a violation of those very habits of rigorous enquiry that have made him so promising a student of Bichat and Louis. When the latter's new book on fevers arrives, Eliot remarks caustically that Lydgate "read far into the smallest hour, bringing a much more testing vision of details and relations into this pathological study than he had ever thought it necessary to apply to the complexities of love and marriage, these being subjects on which he felt himself amply informed by literature, and that traditional wisdom which is handed down in the genial conversation of men" (1:143).

Eliot also revises the image of the scientist as isolationist. There is considerable irony in the fact that Lydgate, who hopes to extend Bichat's work on tissues and eventually to discover the common basis of all living structures, should endeavor to dissociate himself from society. Bichat had used the old French word *tissu* (derived from the Latin *textere*, to weave) for the structures he discovered, and the study of tissues was subsequently named histology (from the Greek *histos*, web).[34] Yet Lydgate, aware of the intrigues and professional detraction (as well as the expense) contingent on living in London, has come to provincial Middlemarch in order, as he naively thinks, to avoid those very interconnections he so earnestly seeks in the laboratory. He soon discovers that causality operates in society as inevitably as in science. His medical colleagues resent his new ideas and more particularly the arrogance with which he introduces them.[35] "Middlemarch in fact, counted on swallowing Lydgate and assimilating him very comfortably" (1:134). He learns that visiting frequently at the Vincys' house and flirting with their pretty daughter carries social obligations of marriage; that the extravagant lifestyle favored by Rosamond leads to debts and hence to further entanglements; that even an honorary position as director of Bulstrode's new hospital has adverse consequences.[36]

Hence each of Lydgate's assumptions—about society, women, and money—not only plays a part in his downfall but runs directly counter to both his scientific training in logic and his search for the fundamental unit of life. If Lydgate fails to find the primary tissue in his laboratory, he finds it, symbolically and without ever recognizing it, in the fabric of the community; he, like everyone else in the novel, is an integral part of the tissue of Middlemarch, whatever superficial modifications he may have acquired, and isolation is as impossible in society as in anatomy and histology.[37] His nemesis for failing to realize this is that he is evicted from Middlemarch and ends as a society doctor "alternating, according to the season, between London and a Continental bathing place; having written a treatise on Gout, a disease which has a good deal of wealth on its side," much to the satisfaction of his wife and to his own bitter disgust (2:360).

Eliot also introduces another factor that was to become a perennial difficulty for scientists in the increasingly competitive professional environment of research, namely, the race to establish a discovery before others do so and the difficulty of accomplishing this without private finance.[38] This is a new element in the presentation of science, being opposed to both Bacon's ideal of the open communication of research and the tradition of the scientist of private means funding his own research. It represents a situation between the time when research was possible only for gentlemen of private means or those with wealthy patrons and the era of publicly funded science, which was not to occur until the end of the century. Lydgate is acutely conscious that "when one has notions in science, every moment is an opportunity" (1:307) and that "Raspail and others are on the same track" (2:23). It was, of course, Raspail who, in 1833, the year after the novel ends, discovered that very "substance membreuse des organes Animaux" that Lydgate had been seeking.[39]

There is yet another level of scientific interest in the presentation of Lydgate. Not only does Eliot show the inextricable threads of consequence emanating both from Lydgate's own flaws of character and from his interaction with a given society at a given place and time but she explains them in scientific terms. If Lydgate is unaware of the tightly woven threads of the communal fabric, Eliot is not. With post-Darwinian vision, she interprets it as the social counterpart of the process of natural selection. Lydgate, with his new intellectual modifications, cannot succeed in Middlemarch because he cannot adapt to the environment of that time; nor can Middlemarch tolerate this strange, incompatible organism. When the novel was published, forty years after the time of its setting, increased concern with public health and sanitation, spurred on by several outbreaks of cholera and typhoid in the intervening years, had sharpened public appreciation of Lydgate's schemes for preventive medicine and the need for scientific research in medicine. Had Lydgate lived forty years later, he would have been lionized.

This time lapse also permits a further irony, for as W. J. Harvey has pointed out, Lydgate's failure to find the "primitive tissue" results in part from his asking the wrong question.[40] What is missing from his theoretical consideration of the problem is any reference to cell theory, which had yet to be enunciated. Robert Brown, whose earlier work Lydgate has read without being aware of its importance (he exchanges Brown's book for a preserved sea creature), did his seminal work on the cell nucleus only in 1831, by which time Lydgate was already bedeviled by marital and financial problems; and the crucial work by Schwann and Schneider (the "plodding Germans" whom Lydgate fears as competitors) was completed only in 1838–39. Thus Eliot shows that Lydgate's research, like his progressive ideas of medical practice, is premature.[41] Time and place are almost as essential for the evolution of science itself as for biological evolution.

The effect of this is to repudiate, or at least modify, the Romantic idea of the scientific man of genius who arises in unlikely circumstances and transcends an unfavorable environment to produce a great discovery. Eliot's study, however, in accordance with the Victorian emphasis on society and her own criterion of responsibility as the preeminent virtue, shows the scientist as unable to rise above the limitations of self and society. Her image of the struggling scientist fighting against financial, personal, and social impediments was to be a popular one in the early decades of the twentieth century, although these later presentations, influenced by the heroic status accorded the Curies, who came to typify this struggle, are usually idealized far beyond the character of Lydgate.

Scientists and Religion

Newtonian cosmology, though indicating in theory the vastness of the solar system, had nevertheless left man in a privileged position as the intellectual if not the physical center of the universe. His status as a being only slightly lower than the angels was confirmed by the very fact of his being able to comprehend the divine plan. But by the mid-nineteenth century such complacency was no longer possible. Man had been demoted to the status of a local accident in an immense, chaotic system that might at any moment obliterate him by a cataclysm such as the geological record displayed with great frequency. If whole species could be rendered extinct, what faith could there be in a loving Creator of specific creatures, in the uniqueness of man, or in the value of the individual life?

Probably the most eloquent presentation of the dilemma facing the Victorian intellectual is to be found in Tennyson's long poem *In Memoriam A.H.H.* (1850), which represents the author's attempt to come to terms with the death, at twenty-three, of his friend Arthur Hallam and to find some meaning in the loss of so promising a life. Although *In Memoriam* was published nine years before Darwin's *Origin of Species* and had already been some eighteen years in the writing, the view of nature expressed in it epitomizes the Darwinian problem. Nature is fundamentally chaotic, governed only by chance, or if there is any perceptible tendency, it is in the direction of destruction.

> "The stars," she whispers, "blindly run;
>   A web is wov'n across the sky;
>   From out waste places comes a cry,
> And murmurs from the dying sun;
>
> "And all the phantom, Nature, stands
>   With all the music in her tone
>   A hollow echo of my own,—
> A hollow form with empty hands."
>
> . . . .
>
> So careful of the type she seems,
> So careless of the single life;

"So careful of the type"? But no.
From scarped hill and quarried stone
She cries, "A thousand types are gone;
I care for nothing, all shall go."[42]

With the appearance of Darwin's *Origin of Species* a storm of controversy broke out and swept the nation. Eminent scientists and equally eminent churchmen, statesmen, and other public figures engaged in heated public debates that were eagerly attended. The best-known debate in the nineteenth century was that between Thomas Henry Huxley, popularly styled "Darwin's bulldog," and the bishop of Oxford, but similar confrontations took place even in the British Parliament, often at the level of an exchange of insults rather than an exploration of the issues involved. The large number of cartoons on the subject of evolution that appeared in *Punch* and the spate of minor novels dealing with the anguish of those who had lost their religious faith through a reading of Darwinism indicate the extent to which the head-on collision between science and religion impacted on all sections of society (see fig. 16). In such a climate of contention between science and religion, it is not surprising to find a revival of the medieval stereotype of the godless scientist, setting himself up against the authority of the church.

Benjamin Disraeli's response to Darwinism, like that of most non-scientists before about 1870,[43] was conservative and satirical, and the representatives of evolutionary theory in his novels are mere caricatures reflecting this bias. In *Tancred or the New Crusade* (1847) Disraeli reaffirmed the doctrine of special creation by ridiculing the new theory. His model gentleman-scientist Tancred remains steadfastly contemptuous in the face of a foolish society lady's gushing exegesis of an evolutionary work, *The Revelations of Chaos* (almost certainly intended to refer to Robert Chambers's *Vestiges of the Natural History of Creation* [1844]).[44] "What is most interesting is the way in which man has been developed. . . . First there was nothing, then there was something; then, I forget the next, I think there were shells, then fishes; then we came. . . . And the next change there will be something very superior to us, something with wings. Ah! that's it: we were fishes, and I believe we shall be crows. . . . Everything is proved: by geology, you know."[45]

A more realistic and sympathetic study of a scientist struggling to maintain his fundamentalist religious faith in the face of the Darwinian onslaught is Edmund Gosse's depiction of his own father, Philip Gosse, a keen and highly respected amateur biologist who devised and popularized the aquarium as a means of studying living organisms in their own environment. Although his autobiographical novel *Father and Son* was not published until 1907, it records the struggles of the mid-nineteenth century. Philip

FIGURE 16.
*Charles Robert
Darwin.* Cartoon by
Linley Sambourne
from *Punch.* Punch
Publications, Ltd.

Gosse attempted to maintain both his intellectual integrity and his fundamentalist faith in creationism. His son records that when he first heard of
Darwin's theory, "every instinct of his intelligence went out at first to greet
the new light. It had hardly done so, when a recollection of the opening
chapter of 'Genesis' checked it at the outset."[46] After long reflection, Gosse
prepared a theory of his own, outlined in his book *Omphalos* (1857),[47] "which
he fondly hoped would take the wind out of Lyell's sails and justify geology
to the godly readers of 'Genesis.'" According to Philip Gosse's thesis, God
created the earth complete with the geological strata and their fossils, suggestive of a much longer period of time than the four thousand years calculated from biblical genealogy, solely in order to test our faith. By such an
argument he thought "to bring all the turmoil of scientific speculation to a
close, fling geology into the arms of Scripture" (105).

Edmund Gosse describes the trauma his father experienced when this
"system of intellectual therapeutics," so far from eliciting universal grati-

tude from reconciled opponents, received only scorn from "atheists and Christians alike [who] looked at it, and laughed, and threw it away." Even Charles Kingsley, from whom Gosse had expected the warmest acclaim, wrote that he could not "give up the painful and slow conclusion of five and twenty years' study of geology, and believe that God had written on the rocks one enormous and superfluous lie" (105).[48]

Gosse diagnosed his father's mental turmoil as arising, in the last analysis, from a lack of humility, from an "obstinate persuasion that he alone knew the mind of God, that he alone could interpret the designs of the Creator" (73). This comment is particularly interesting insofar as it represents a complete inversion of the conventional judgment that scientists who questioned theology were guilty of hubris, while those who clung to religious dogma were walking humbly before their God. Edmund Gosse has, in effect, flipped the point of reference: geology is now the new orthodoxy, and the individual who questions or rejects it is cast as arrogant. This shift in evaluation is symptomatic of the compromise that had been reached by the turn of the century. The rival factions of science and religion made their public peace by an implicit treaty of partition whereby each side agreed not to trespass on the other's designated territory. However, there have continued to be infringements of this unspoken agreement, as in the notorious "monkey case" in the United States and in the attempts by militant creationists to have evolution theory expurgated from school and university syllabuses. These incidents have provided material for a number of twentieth-century plays, novels, and films, but the issues are still nineteenth-century in essence.

Immensities of Space and Time

For the scientist characters of Thomas Hardy's novels there is no longer any conflict between science and religion, for the latter has ceased to have meaning. Correspondingly, there is no comfort or psychological bulwark against the immensities of time and space that science was formulating. Instead there is the terrifying vision of a vast, impersonal, uncaring universe as revealed by astronomy and evolutionary theory, in which any appearance of benign purpose is merely a façade, a trap for the unwary; Nature will soon return to her normal modus operandi, chance and destruction. This attitude influences Hardy's presentation of the pursuit of science in an interesting way, for while he himself rejects any alternative belief, he nevertheless presents his scientists critically; they are typically shown as being deficient in human emotions and social consideration as a result of their scientific perspective.

The two extended portraits of scientists in his novels are those of Henry Knight, an amateur geologist in *A Pair of Blue Eyes* (1873), and Swithin

St. Cleeve, an astronomer, the protagonist of *Two on a Tower* (1882). In the tradition of the Romantic stereotype, both are presented initially as observers, detached from any emotional involvement in life. But the reason for this noninvolvement is attributed to a different source from that of the Romantic scientist: it is their intense awareness of the vastness of geological time and astronomical space that causes them to view the individual affairs of men as trivial and manipulable.

We are introduced to Knight first in his rooms in London, where he has imprisoned nature in an aquarium in order to study it objectively and in isolation. He is thus the rational scientist-as-observer, distanced from the object he is studying even when this subject is humanity.[49] However, he is soon to be disabused of the notion that an individual may assume such an attitude of objectivity. Secure in his assumption of detachment and superiority over nature because he can explain natural phenomena in terms of physics, he performs an experiment to demonstrate his theory concerning updrafts at a cliff face; that is, he attempts to manipulate nature. However, he is rapidly forced to realize that nature is capricious (it sends his hat down the cliff, not up) and that he cannot remain merely a spectator. Obliged to climb down the slippery cliff face to retrieve his hat, he finds that he cannot climb up again and is thus jolted out of the secure role of observer into the precarious position of one involved in a personal struggle for existence. This is emphasized by his visual confrontation with the fossil of a trilobite embedded in the cliff, the victim of just such a struggle as his own. Geology ceases to be, for him, an abstract study and instead becomes alarmingly personal: "The eyes, dead and turned to stone, were even now regarding him. . . . Separated by millions of years in their lives, Knight and this underling seemed to have met in their place of death."[50] Just as both Knight and the fossilized trilobite have been deposited on the surface of the cliff by natural forces, so the whole of organic nature is but a recent accretion on the inanimate nature beneath, and human life is as precarious and temporary as that of its extinct ancestor. Knight, a geologist, thus enacts Hardy's own evolutionary pessimism. However, Hardy does not allow Knight to accept death with Darwinian resignation. Instead there is considerable irony directed against this scientist who affects to accept with complacency a theory of nature in which man is infinitesimal, transitory, and unimportant but who, when his own existence is threatened, becomes as fearful and determined to cheat nature of its victim as any of the uneducated country folk he looks down upon.

In his later novel *Two on a Tower* Hardy extends this idea into a major theme, expanding the time scale from the geological range to the astronomical. Swithin St. Cleeve is the first astronomer to feature as the protagonist of a full-length novel, and Hardy took some trouble to obtain accurate information for the astronomical details in the novel.[51] As the title suggests (the tower of the title has been fitted out as an astronomical observatory), the novel explores an astronomer's perception of the relationship between the individual and the universe, of which he is both part and victim.[52]

Swithin is introduced from the outset as exhibiting two apparently contrary moods contingent upon these two perspectives. On the one hand, he feels intense elation about his future career in astronomy, seeing no anomaly in his ambition to become another Copernicus. On the other hand, he is prey to a deep melancholy and preoccupation with death. Almost his first words in the novel reveal this juxtaposition: "I aim at nothing less than the dignity and office of Astronomer Royal, if I live. Perhaps I shall not live" (9).[53]

Through his protagonist, Hardy suggests from the outset that astronomy necessarily engenders pessimism, as Swithin appropriates the *ars longa vita brevis* theme for this purpose: "Time is short, and science infinite,—how infinite only those who study astronomy fully realize" (10). Swithin's melancholy arises from three corollaries of his observations: the immensity of space and time compared with the scope and brevity of human life; the consequent insignificance of human life; and a profound "mistrust of all things human." The first of these is referred to several times in some of the most powerful passages of the novel. Demonstrating his telescope to Lady Viviette Constantine, legal owner of the tower, he directs her vision to the sun, which appears as "a whirling mass, in the centre of which the blazing globe seemed to be laid bare to its core. It was a peep into a maelstrom of fire, taking place where nobody had ever been or ever would be" (8). Paradoxically, the telescope, which is intended to make remote objects appear closer, has the opposite effect: the overriding impression is that the magnitude of space annihilates the individual. Swithin tells Viviette, "The actual sky is a horror. . . . You would hardly think, at first, what horrid monsters lie up there. . . . Impersonal monsters, namely Immensities. . . . Those are deep wells for the human mind to let itself down into, leave alone the human body! and think of the side caverns and secondary abysses to right and left as you pass on!" (34–35).[54]

In 1863 a nova had been detected in Scorpio and Hardy would certainly have read accounts of its decaying brightness, even if he had not observed it personally. The immense wilderness of space was, for Swithin as for Hardy, made yet more terrible by this knowledge of its impermanence. "For all the wonder of these everlasting stars, eternal spheres, and what not, they are not everlasting, they are not eternal; they burn out like candles. . . . Imagine them all extinguished, and your mind feeling its way through a heavens of total darkness, occasionally striking against the black, invisible cinders of those stars. . . . If you are cheerful, and wish to remain so, leave the study of astronomy alone" (35–36).[55]

From a perception of the transience of even the apparently stable heavens (it is stressed that Swithin's interest lies not in the "fixed" stars but in the "variable" stars)[56] it is not far to the awareness of human mortality and insignificance. The parallel between the stars and human beings is drawn early in the comparison between the two figures on the tower and the constellation of the Twins (which, significantly, includes a number of "variable" stars).[57] This awareness induces, in turn, a sense of the centrality of chance and accident in human affairs, a common feature of all Hardy's novels but made more credible here in the context of the astronomical perspective. The triumph of chance and accident over man's reasoned intentions is here not merely a restatement of Darwinian theory but also Hardy's comment on the limitations of scientific method. Scientists endeavor to explain observed phenomena in terms of universal laws, which presuppose regularity of occurrence and an order underlying the apparent diversity. Swithin is attempting to do for the universe beyond the solar system what Kepler had done for the solar system, and Newton for physics, namely, to formulate the laws of its behavior. Yet, unlike his predecessors, Swithin finds not celestial order but chaos—unpredictable stellar outbursts, variable stars, and inexplicable immensities.

That chance and accident are the norm rather than the exception is made explicit when, near the end of the novel, Swithin spends some time observing in South Africa, only to find the same disorder in the farthest reaches of the universe: "There were gloomy deserts in those southern skies such as the north shows scarcely an example of; sites set apart for the position of suns which for some unfathomable reason were left uncreated, their places remaining ever since conspicuous by their emptiness" (317–18).

Thus astronomy, like Darwinism, reveals a world of chance and accident at the heart of even the most apparently stable phenomena. Indeed, if Hardy qualifies at all the ubiquity of chance, it is not to posit a benign Providence overruling accidents in order to spare humanity, but rather to suggest a wanton, malevolent power stacking the odds even more steeply against us. The maelstrom on the sun extends the terrestrial cataclysms revealed by geology to the celestial bodies, and the repeated references to the animal names for a number of the constellations keep before us the parallel between the heavenly and the earthly domain.[58] It is in this context that Hardy questions the alleged objectivity of scientists, the apparent complacency with which they discuss the immensities of time and space and the transitoriness of man's existence. Swithin is forced to acknowledge his suppressed fears and to espouse a more subjective position—to accept his humanity.

Swithin also displays the characteristic failings of the Romantic stereotype of the scientist—emotional deficiency and preoccupation with his work rather than with human relationships. Naive and egocentric, he never overtly becomes the cold-hearted villain deliberately exploiting others, but the effect is much the same. Symbolically, he lives (sleeps and eats) in his elevated observatory, looking down literally and metaphorically on those mortals who view the heavens unscientifically. Moreover, his preoccupation with astronomy leads to a certain degree of mental rigidity. Although minutely observant of the behavior of the planets, he is strangely insensitive to the emotional states of those around him and interprets everything literally, oblivious of the implicit messages that social convention prevents being stated. When Swithin leaves Viviette to observe at the Cape, Hardy remarks caustically: "Her unhappy caution to him not to write too soon was a comfortable licence in his present state of tension about sublime scientific things. . . . In truth he was . . . too literal, direct, and uncompromising in nature to understand such a woman" (310). Even in the final encounter between the lovers, Swithin fails to distinguish her real feelings from her conventional words. Thus Hardy uses the character of the astronomer to explore both the anguish engendered by a scientific understanding of the vastness of the universe and the pain an insensitive scientific viewpoint can produce in the realm of personal relations.

The Exploiters

The mechanistic principles and reductionist philosophy implicit in Darwinism led also to a resurgence in literature of the Romantic stereotype of the scientist as emotionally deficient and socially irresponsible, and for parallel reasons. By insisting that organisms could be explained in purely physical terms without reference to emotional or spiritual values, scientists unwittingly suggested that they themselves were lacking in these attributes.

In her short story "Cousin Phillis" (1865), Elizabeth Gaskell created an otherwise charming character who displays this insensitivity and amorality. Holdsworth, an engineer designing the new railway, is an intelligent, urbane, but basically superficial man who has traveled widely and takes new ideas and values for granted. He trifles with an unsophisticated country girl, Phillis, who assumes from his behavior that he intends marriage; when she hears that he has married in Canada, she falls seriously ill. Much of the moral emphasis and interest of the story hinges on the contrast between Phillis's sincerity and trust and Holdsworth's inability to comprehend such values or to treat her as anything more than a pretty child.[59] The scientifically trained man is associated throughout with the new, with the future, which to him is by simple definition superior to the past. His whole lifestyle embodies change—physical (travel, railways), mental (new ideas and theories), and moral (a different view of responsibilities)—and from this sophisticated perspective he sees no harm in paying Phillis elaborate and insincere compliments. In Gaskell's treatment, both the superficiality of Holdsworth and his insensitivity to Phillis's feelings are closely linked to his scientific background.

A similar but more reprehensible character is Dr. Edred Fitzpiers, of Hardy's novel *The Woodlanders* (1887), a scientific character as cold and amoral as any of Hawthorne's scientists. Although he has set up as a village doctor, Fitzpiers has little interest in general practice and spends more time in reading, study, and scientific experiments. One experiment involves the examination of the unusually large brain of Grammer Oliver. Fitzpiers tries to induce the old woman to sell him her brain for postmortem analysis, a project clearly intended to indicate his detached view of humanity insofar as it ignores the feelings of the old woman.[60] Like Faustus and Frankenstein, Fitzpiers has studied at a German university, in his case Heidelberg, but unlike his predecessors, he is presented as an intellectual dilettante, flitting between metaphysics, astrology, alchemy, and medical science. Hardy thus suggests that Fitzpiers has no commitment to the search for truth, but merely indulges his idle curiosity. Compared with the other characters of Little Hintock, he is a cold and calculating personality, and it would seem clear that, in this novel at least, Hardy intends to associate such traits with a scientific training.

Godwin Peak, the protagonist of George Gissing's novel *Born in Exile* (1892), is one of the first professional geologists to feature in fiction; he too displays the dehumanizing effect of scientific study on the nineteenth-century mind. Gissing stresses that Peak's emotional deficiency and unethical behavior are contingent on his assumption that there is no moral dimension or meaning in the universe.[61] What is new in this presentation is not merely Peak's assumption that no scientifically educated person can accept religious beliefs but his complete lack of regret that this should be so and his failure to see any reason for retaining a Christian ethic after renouncing its creed. Indeed, the implication in this novel, as in Hardy's *A Pair of Blue Eyes*, is that, seen from the perspective of geological history, man's life is indeed trivial. On observing the rock strata disclosed by a quarry, Peak reflects that "imagination wrought back through eras of geologic time, held [man] in a vision of the infinitely remote, shrivelled into insignificance all but the one fact of inconceivable duration" (67). Predictably, Peak instinctively applies the principles of Social Darwinism to justify his ruthlessness: "Life is a terrific struggle for all who begin it with no endowments save their brains" (177).

It is apparent from this survey that the realistic novelists of the late nineteenth century were forced to develop new images of scientists and new ways of presenting them. Most of these involved struggle—the struggle to survive financially, the struggle to reconcile scientific knowledge with faith, the struggle, in the face of all the apparent evidence to the contrary, to find meaning in the universe. Despite the early favorable depictions of scientists, such as Kingsley's Tom Thurnall and Eliot's mainly sympathetic portrait of Lydgate, the image of scientists in the Victorian novel became increasingly bleak. The discoveries of geology and astronomy and the popularizing of evolutionary theory, especially in its distorted form of Social Darwinism, produced a prevailing image of a cold, unemotional, and, by extension, amoral scientist. Such a character looks back to the Romantic stereotype of the scientist as emotionally deficient and prefigures one of the most common twentieth-century images of the scientist, as arrogant, unfeeling, and indifferent to the sufferings that may result from his research.

These attitudes can in fact be seen as arising from two curiously contradictory estimates of science itself. On the one hand, science is assumed to be more powerful than man, since it is invoked to explain him to himself in wholly scientific terms. On the other hand, it is assumed to be so far inferior to man that he believes he can control it. It is the same paradox that Mary Shelley described underlying the tragedy of Frankenstein, who also embraced a mechanistic and reductionist view of man (attempting to construct one from dead or inanimate pieces) and found to his horror that the power of the created Monster was beyond his control.

CHAPTER NINE

THE SCIENTIST
AS ADVENTURER

> As for my uncle, who never forgot his work, he was carefully examining the nature of the terrain, torch in hand, trying to discover where he was from observation of the strata. . . . a scientist is always a scientist as long as he retains his composure, and Professor Lidenbrock certainly possessed this quality to an extraordinary degree.
> —Jules Verne

The mathematician and philosopher Alfred North Whitehead once remarked that "the greatest invention of the nineteenth century was the invention of the method of invention."[1] Certainly the cult of invention and the technology to which it gave rise were to have such an impact on public attitudes towards science in the latter half of the century that scientists, who had collectively been held responsible for revealing the terrifying immensities of time and space and the fragility of man's existence, were hailed in late Victorian times as offering the major bulwark against human insignificance. This changed perception of the role of science as empowering rather than diminishing became possible as technological progress brought a new era of economic prosperity to the industrialized nations and ushered in commercial imperialism as a consequence (see fig. 17). In such a climate, where the close nexus between capitalism and cultural superiority was accepted without question, science was no longer seen as godless or disruptive. On the contrary, the exploitation of technology to produce more wealth was regarded favorably in moral terms as evidence of due adherence to the Protestant work ethic. There emerged an attitude of complacency, even benign patronage, towards science as an obedient servant that could be relied upon to be both useful and entertaining. Soirees at which the scheduled entertainment was a series of scientific demonstrations were as popular in upper-class society as public lectures on science at the mechanics' institutes were with the working class (see fig. 18). Technological devices and products became commonplace in middle-class homes, and it seemed safe to sport with the technological monster, to regard it as reliable, even pictur-

FIGURE 17.
The Wise Scientist.
By the late
nineteenth century
the scientist had
achieved
respectability as an
authority on almost
every subject. Mary
Evans Picture
Library.

esque. It was, after all, the age when locomotives were called *Pegasus* or *Black Bess* or named after Sir Walter Scott's novels—*Rob Roy, Ivanhoe, Waverley*.

The literary representative of this new attitude was the figure of the adventurer-scientist, a modern counterpart of the Romantic hero, but now allied with science rather than opposed to it. Heirs to the optimism of both the Utopian tradition and the wonderful-journey stories, these characters entered into the popular culture of their time as humanity's advance guard, extending the frontiers of experience, whether in space or time, confident of subduing to their will whatever they found there and transcending mankind's former limitations.[2] The test whereby these technological knights earned their spurs involved not only courage but resourcefulness; it was no less a feat than subduing the dragon represented by the natural world. The reward was not only the satisfaction of overcoming human limitations but the honor of having been the first to do so, a criterion that was becoming singularly important in the increasingly competitive environment of industrial science. Their right to dominate nature, the universe, or whatever alien

FIGURE 18. *A Scientific Conversazione,* by William McConnell, from Sala's *Twice Around the London Clock.* Mary Evans Picture Library.

societies they encountered was no more questioned than the imperialist regimes, of which they were the literary and scientific champions, doubted their self-proclaimed mission to bestow their civilization, by force if necessary, on the benighted areas of the globe. Technological might was regarded, almost by definition, as morally right. In such a climate, progress was the supreme good. As Lewis Mumford has observed of this period, "One could not have too much progress, it could not come too rapidly, it could not spread too widely and it could not destroy the 'unprogressive' elements in society too swiftly and ruthlessly: for progress was a good in itself independent of direction or end."[3]

The belief that scientific discovery was the greatest of all adventures and that European man would progressively master nature was encapsulated preeminently in the early novels of Jules Verne, collectively (and significantly) entitled *Les Voyages extraordinaires* and later subtitled *Les Mondes connus et inconnus.* Verne's debonair and irrepressibly optimistic heroes are bent on adventure, courageously risking their lives for the delight and the honor of the quest. But this intrepid exploration is presented as more than entertainment. These myths of conquest, wherein the marvels of science engage with and overcome the marvels of nature, are an expression of logical positivism with a strongly didactic subtext, intended to elicit in the youthful readers of Pierre-Jules Hetzel's *Magasin d'éducation et de récréation* a spirit of bravery, optimism, and reverence for scientific knowledge (see fig. 19).[4] Just as Newton's intellectual voyages were endowed with moral significance, so Verne's explorer-scientists function also as intellectual and moral guides to both the known and unknown worlds. Not only do they lay claim to darkest Africa, the interior of the earth, the depths of the oceans, or outer space for human domination but, as Roland Barthes has suggested, they also define the familiar, which they symbolically carry with them inside their vehicles, by means of its opposite.[5]

Himself an engineer, Verne was adept at including, with an air of assurance, scientific or quasi-scientific explanations of technological inventions, but his particular focus on various novel means of transport has deeper significance. Motion, especially speed, becomes a metaphor for the assertion of individual freedom and environmental domination, a notion encapsulated in the motto of Verne's most famous character, Captain Nemo, *Mobilis in mobili*. Thus the projectiles shot into space, the balloons, and submarines are repeatedly characterized by their velocity. However, these various means of transport, which seem to catapult the adventurers into the unknown, also provide a safe shelter, a cocoon in which the intrepid individual may investigate and explore new realms while effectively remaining enclosed within the comfort of his house. Barthes comments that Verne "has built a kind of self-sufficient cosmogony, which has its own categories, its own time, space, fulfillment and even existential principle."[6] Indeed, these vehicles display as much ingenuity in the service of material comfort as in technological efficiency. Their ornamental elegance, liberally furnished with the cultural icons of the Second Empire, both removes the threat of the inhuman machine and celebrates the superiority of French artistry.[7]

Overburdened with their representative function, Verne's individual scientists are rarely developed psychologically, lest they detract from his primary concern, the journey itself. Otto Lidenbrock, Pierre Aronnax, Paganel, and Palmyrin Rosette are distinguished in character from the explorers, Captains Hatteras, Servadac, and Grant, chiefly by their determination to explain some device or expound a catalog of marvels in scientific terms. Verne's first novel, *Five Weeks in a Balloon* (1863), set the pattern for the majority of its successors.[8] Drawing on a recent actual invention, Nadar's balloon *Géant*, which had crashed during its first flight, he imagined an improved version, the *Victoria*, which, powered by hydrogen gas and equipped with an effective navigational mechanism, was capable of traveling virtually anywhere.[9]

This early novel introduced the trio of characters who, with minor variations, were to feature in nearly all Verne's adventures—the resourceful but eccentric scientist, in this case Dr. Fergusson, obstinately pursuing a

FIGURE 19.
Illustration by
Hildebrand for
Jules Verne's *From
the Earth to the
Moon,* 1865.

dangerous obsession; his foil, the hotheaded but honorable Kennedy; and the phlegmatic servant, Joe, loyal unto death if necessary. Once the *Victoria* is aloft, the main purpose of the balloonists is to afford Verne the opportunity of describing the journey across Africa, geography being the one science in which, as a member of the Société de Géographie, he could claim expert standing. The intrepid Fergusson delivers detailed lectures on all matters of interest in between the adventures, the outcome of which, given his resourcefulness, is never in doubt. In *The English at the North Pole* (1864) again the details of the journey achieve prominence.[10] The descriptions of life aboard ship and the difficulties of crossing the Arctic Ocean are scrupulously derived from actual explorers' accounts, but Verne has little interest in assessing the value of such a trip or the motivation of the explorers.

Verne's most popular novel, *A Journey to the Centre of the Earth* (1864), was based largely on his imaginative synthesis of some then-current theories of terrestrial structure. Verne allows his protagonists, Professor Lidenbrock, a geologist, and his nephew Axel, an apprentice mineralogist, to discuss these at some length, Lidenbrock invoking Sir Humphry Davy's theory that the core of the Earth could not be liquid, because if it were, the attraction of the Moon would cause twice-daily internal tides, with consequent periodic earthquakes.[11] The journey involves a conflation of the idea of John Cleves Symmes that the Earth was essentially hollow, containing five concentric spheres all with openings at the poles, and Edmund Halley's theory of the aurora as emanating from the interior of the Earth. Eventually, however, all is resolved; divergent theories are not permitted to threaten the intellectual preeminence of Lidenbrock or the ability of science to resolve theoretical problems.

It is characteristic of Verne's stories that nature invariably yields up her secrets to the resourcefulness and determination of the scientists, who indefatigably name, classify, and codify everything they encounter, thereby intellectually colonizing the hitherto puzzling universe and making it safe for humanity. They are rewarded by achieving a clarity and a certainty that twentieth-century physics was soon to undermine. From his first encounter with the Icelandic volcano of Sneffels, Lidenbrock never doubts the successful outcome of his contest with nature: "The Professor never took his eyes off it, gesticulating as if he were challenging it and saying, 'So that is the giant I am going to defeat!'" (85). At a climactic moment of danger he cries, "Air, fire and water combine to block my way! . . . I won't give in . . . and we shall see whether man or Nature will get the upper hand!" (206). This "defeat" of nature is a dual one; at the physical level it involves the survival of the party against all the violent forces and hardships nature can muster, even that of a volcanic eruption, but it is also an intellectual defeat, signified by the imposing on nature of a scientific explanation for whatever she can produce. When the party encounters a subterranean forest of strange umbrellalike trees, Axel reports:

> I hurried my step, anxious to put a name to these strange objects. Were they outside the 200,000 species of vegetables already known, and had they to be accorded a special place among the lacustrian flora? No; when we arrived under their shade, my surprise turned to admiration. I found myself confronted with the products of the earth, but on a gigantic scale. My uncle promptly called them by their name.
> "It's just a forest of mushrooms," he said.
> And he was right.   (166)

In no time Axel has assigned them a Latin classification, *Lycopodon giganteum*, measured them, compared them with their domestic counterparts, and made the colonizing process all but explicit by declaring their similarity to "the rounded roofs of an African city."

This obsession with definitive explanations continues to the very end of the novel. Even after the safe return of the adventure party, to a chorus of universal scientific acclaim, Axel and his uncle are mortified by their failure to explain why they were apparently traveling in a direction opposite to that which they had predicted. "One aspect of the journey—the behaviour of the compass—remained a mystery, and for a scientist an unexplained phenomenon is a torture for the mind." When, six months later, Axel notices that the poles of their discarded compass are reversed, Lidenbrock is overjoyed. All is explained; the answer comes out as neatly as a syllogism:

> My uncle . . . gave a leap of joy which shook the whole house. Light dawned at the same time in his mind and mine.
> . . . "Then that explains our mistake. But what phenomenon could have caused this reversal of the poles? . . . Explain yourself, my boy."
> "During the storm on the Lidenbrock Sea, that fireball which magnetized all the iron on the raft simply reversed the poles of our compass!"
> "Ah!" cried the Professor, bursting out laughing. "So it was a practical joke that electricity played on us!"
> From that day onward, my uncle was the happiest of scientists.
>
> (253–54)

The unacknowledged aim of all the *Voyages extraordinaires* is to convert the bewildering diversity of the infinite universe to a state of tame finitude. Barthes finds the culmination of this aim in *Mysterious Island*, "in which the man-child re-invents the world, fills it, closes it, shuts himself up in it, and crowns this encyclopaedic effort with the bourgeois posture of appropriation: slippers, pipe and fireside, while outside the storm, that is, the infinite, rages in vain."[12]

The same obsession with order also underlies the fanaticism with which Lidenbrock and Axel categorize and name every geological stratum, every specimen of rock, every species of underground plant and animal they find. This classifying zeal is a multifunctional aspect of characterization, for, besides attesting to the mental rigor of these dedicated scientists, whose urge to know and explain supersedes all discomfort, it demonstrates their desire to hammer out a definitive theory that will finally eliminate uncertainty.[13] This determination to impose order and establish predictability represents the Baconian ideal of power through knowledge and consequent domination over nature. At the moments of gravest physical danger, Lidenbrock justifies his courage and optimism by expounding his theories and asserting the power of the mind over the obstructions of matter. Like their predecessors in New Atlantis, Verne's scientists are compulsive collectors of facts and natural objects. Museums and encyclopedias feature repeatedly as both the result and the symbols of this systematizing process. Captain Nemo's collections have converted the *Nautilus* into a vast submarine museum. Professor Lidenbrock's study represents a private museum. And numerous other scientists in the *Voyages extraordinaires*, if they do not actually have physical collections, are themselves living encyclopedias: the geographer Paganel has the complete classification of South American birds at his fingertips; Axel can recite the catalog of dinosaurs; and Conseil can reel off the taxonomic identification of fish.

Not only are Verne's scientists armed with all available measuring devices—clocks, compasses, sextants, and maps both ancient and modern—but they characteristically stamp the geographic features they encounter with their own names, thereby announcing their appropriation of the hitherto unknown in a form of intellectual as well as geographic imperialism. In the act of naming Hansbach, Axel Island, Port Gräuben, Cape Saknussemm, and the Lidenbrock Sea, Professor Lidenbrock and Axel not only lay claim to fame but establish their continuing possession of nature. In this process of colonizing and humanizing the alien, Verne's scientists not only explain objects and events to their own satisfaction but adopt, with almost missionary zeal, the responsibility of enlightening others. Like Lidenbrock, they indefatigably embark on a program of lecturing, explaining, and decoding the cryptogram of nature. Axel affirms Verne's article of faith: "However great the wonders of Nature may be, they can always be explained by physical laws" (208). Cyrus Smith, the engineer of *The Mysterious Island* (1875), who has transformed a desert island into a mechanized utopia, perpetually lectures and instructs his companions in the practical applications of science. "The colonists had no library at their disposition; but the engineer himself was like a book—always ready, always open to the exact page they needed, a book that solved all their problems for them and that they leafed through regularly."[14] Professor Palmyrin Rosette has prepared a lengthy treatise on comets, which is included in full for the edification of the reader. In line with Hetzel's didactic intention, these scientist-pedagogues are lionized by their societies. Otto Lidenbrock's book, whose title is also *Journey to the Centre of the Earth*, allegedly creates a sensation throughout the world, being printed and translated into every language, and its author becomes the "corresponding member of all the scientific, geographical, and mineralogical societies in the world" (254).[15]

Lidenbrock represents one of the earliest examples of the eccentric, irascible scientist who is also fundamentally moral—a happy conflation of the Restoration virtuoso and Bacon's noble scientist. In the twentieth century such eccentric characters were to be metamorphosed first into the absent-minded professor, out of touch with his surroundings, and later into the prototype of the mad and dangerous fanatic, prepared to sacrifice himself and everyone else for the cause of science. Verne, however, does not permit his characters to suggest any sinister overtones. "Otto Lidenbrock was not, I must admit, a bad man; but, unless he changes in the most unlikely way, he will end up as a terrible eccentric," confides Axel at the outset; and Lidenbrock's willingness to sacrifice his nephew and Icelandic guide along with himself in a maniacal bid for fame is defused both by comedy and by the reader's realization that at least Axel, the narrator, has lived to tell the tale. Lidenbrock has characteristics that are potentially more malevolent. He is endowed with a furious temper, which he directs against anyone who crosses him, and he is unconcerned about his pupils: "His teaching was . . . intended for himself and not for others. He was a selfish scholar" (8). Yet he is presented with a beguiling humor, which renders him nonthreatening. For this purpose Verne modifies several of the devices used by the seventeenth-century wits to satirize the virtuosi, recasting them as endearing eccentricities. Like Shadwell's Gimcrack, Lidenbrock has an attractive young ward in the house but cannot see the obvious fact that his nephew is in love with her. Like the virtuosi, he has a conglomeration of apparently ill-assorted facts at his disposal, but unlike theirs, his facts are relevant, even essential, to the occasion. His knowledge of obscure Icelandic history and geography and his acquaintance with the work of the alchemist Arne Saknussemm and the runic alphabet all turn out to be vital for the success of the adventure.

Lidenbrock's fundamentally ruthless determination to carry out his incredible plan is also offset by his frontier-style virtues—courage and optimism when all seems hopeless and the ability to continue his scientific investigations whatever the personal dangers involved. When the hapless Axel asks incredulously, "What! You still think there's a chance of escape?" he replies firmly, "Yes, I do. As long as this heart goes on beating, I can't admit that any creature endowed with will-power should ever despair" (235), and of course his faith is vindicated, whatever the cost to our credulity. Axel affirms: "A scientist is always a scientist as long as he retains his composure, and Professor Lidenbrock certainly possessed this quality to an extraordinary degree" (237). As the party is caught up in the eruption of Mt. Etna, which, unknown to them, is finally to fling them to safety on Stromboli, Lidenbrock remains calm and smiling in the midst of his calculations until finally Axel admits: "My uncle was absolutely right; and never had he struck me as bolder or more self-assured than at that moment when he was calmly working out the chances of being involved in an eruption" (240). This *sang-froid* was to become a trademark not only of Verne's scientists but also of the heroes of twentieth-century pulp science fiction.

Professor Pierre Aronnax, of the Paris Museum of Natural History, the narrator of *Twenty Thousand Leagues under the Sea* (1870), shares several of Lidenbrock's ambiguous traits, including a readiness to sacrifice his assistant Conseil (the counterpart of Hans in *Journey to the Centre of the Earth*) to his own scientific curiosity and a phlegmatic acceptance of imprisonment aboard the submarine *Nautilus* in return for the opportunity to observe and photograph underwater phenomena for the first time.

Unlike most of Verne's scientist characters, the anti-social and enigmatic Captain Nemo, an engineer and the first serious oceanographer in literature, is a complex character whose genesis appears to owe something to a number of Verne's contemporaries, including Colonel Charras and Albert I of Monaco, a great nineteenth-century oceanographer.[16] Yet even Nemo's interest for us depends less on his being a scientist than on his mysterious past, which is presumed to account for his insatiable desire for vengeance, his passion for music, and his hatred of imperialism, specifically British imperialism. This latter characteristic, while it doubtless chimed in with Verne's own Anglophobia, owes something also to the American Robert Fulton, who in 1801 had built an actual submarine called the *Nautilus,* which he offered to Napoleon for placing powder mines beneath the hulls of unsuspecting British warships. Fulton's motto, *Libertas maris, terrarum felicitas,* is not dissimilar to Nemo's *Mobilis in mobili,* painted around an *N.*[17] Nemo combines compassion for the oppressed with the ruthlessness of the avenger, who, motivated by a "fierce, implacable defiance towards human society," can cold-bloodedly drive through a British ship and watch its passengers sink. Nemo is, in effect, an underwater Robin Hood, distributing to suffering races the wealth he extracts from submerged wrecks. Aronnax comments, "His heart still beat for the sufferings of humanity, and his immense charity was for oppressed races as well as individuals" (191).

An interesting extension of this pity for the oppressed is Nemo's advanced environmental attitude towards the oceans and endangered species. He is far ahead of contemporary thinking in his refusal to countenance the killing of the southern whales as urged by the harpoonist Ned Land. "To what purpose?" replied Captain Nemo; "only to destroy! We have nothing to do with whale-oil on board. . . . Here it would be killing for killing's sake. I know that is a privilege reserved for men, but I do not approve of such murderous pastime. In destroying the southern whale . . . your traders do a culpable action, Master Land. They have already depopulated the whole of Baffin's Bay, and are annihilating a class of useful animals. Leave the unfortunate cetacea alone. They have plenty of natural enemies" (214).

The enormous popularity of Verne's writings during his lifetime indicates how competently he reflected current attitudes towards a science dealing primarily with technology and engineering marvels. The scientists of Verne's early novels are neither stupid nor sinister, but heroic in their acceptance of danger and always resourceful in extricating themselves. Borne along on a lava flow or imprisoned in a submarine, they contrive to remain absorbed in their scientific discoveries, never doubting their eventual safe deliverance or their intellectual conquest of the phenomena. Embarked on a physical journey, a metaphor for the intellectual journey they believe will reveal a marvelous future, they embody the limitless nineteenth-century confidence in science to overcome mere physical dangers and to plumb the depths of the hitherto unknowable. In these stories complex technological marvels also guarantee the superiority of the scientist-protagonists who manipulate them so calmly, controlling the power currency of the future.[18]

In many of these stories the ethical qualities appropriate to some swashbuckling adventure—bravery, coolness in the face of danger, and self-sacrifice—are manipulated to vindicate more sinister and more socially destructive traits. The success of Verne and others in making scientific ruthlessness acceptable in a dangerous situation, defusing it by an element of noninjurious comedy, was to have dangerous cultural repercussions. It impeded a critical appraisal of the potential danger to society of arrogance and rampant individualism, both of which were made to seem essential to success in the technological and capitalist enterprise.

From the journey through unknown space it seems, in hindsight, only a small step to postulating a journey in time, yet it was not until 1888, when Wells published his early story "The Chronic Argonauts," the forerunner of "The Time Machine," that such an idea was seriously considered. Although, as we shall see in chapter 10, the majority of Wells's scientific romances are more ambiguous in their presentation of scientists, in "The Chronic Argonauts" the adventurer motif is still prominent in Dr. Nebogipfel, the exotic traveler who transcends all the boundaries accepted by others, even the last frontier of time. Wells draws on many time-honored features of the alchemist stereotype as a guarantee of his argonaut's strangeness and intellectual obsession: "thin lips, high cheek-ridges, and . . . large, eager-looking grey eyes that gazed forth from under his phenomenally wide and high forehead . . . [and] glowed like lights in some cave at a cliff's foot."[19] Physical comfort and social intercourse are alike irrelevant to Nebogipfel, who appears arrogant and secretive to the point of being reclusive. Nevertheless, we are induced to feel considerable sympathy for him, because his motivation is not desire for financial gain but the hope of finding in a future, wiser generation the personal fulfillment that the present world cannot sustain.

The Professor Challenger stories of Arthur Conan Doyle, although written a decade and more into the next century, are effectively variations on this late-nineteenth-century stereotype of the scientist as adventurer. The first novel of the series, *The Lost World* (1912), introduces the eminent biologist Challenger as an engaging caricature, his towering stature and impressive scientific reputation exceeded only by his colossal arrogance. Many of his physical attributes were doubtless based on those of William Rutherford, Doyle's professor of physiology at Edinburgh University—the booming voice, the powerful figure, the Assyrian black spade beard, and the macabre sense of humor[20]—but Doyle went on to develop Challenger's eccentricities for his own purposes, both stylistic and didactic.

In *The Lost World* Challenger is still mainly comic and presented in a situation reminiscent of Verne's stories. Like Lidenbrock and Nemo, he is equipped with the obligatory character foils—the Axel-like Malone, the admiring narrator who professes objectivity if not skepticism; the intrepid young Lord Roxton; and the irascible Professor Sommerlee, Challenger's sworn opponent on every subject except the reputation of a third scientist, Dr. Illingworth, whom they both despise. Like Lidenbrock, Challenger is determined to risk his own life and that of everyone else concerned, in this case on an expedition to the upper Amazon to verify the report of an inaccessible plateau, allegedly unchanged since the Jurassic Age. True to his name, Challenger contrives to place the burden of proof on his opponents, on whom he vents his aggressive sarcasm, even physical violence, without paying even lip service to scientific objectivity in debate. Doyle thus uses the rivalry between Challenger and Summerlee to satirize the arrogance of scientists and their pretense to objectivity. Each believes himself to be the greatest zoologist in the world and regards the other as a retarded trickster. "Microcephalous idiot!" and "Simian survival!" are among Challenger's more colorful insults. Doyle characterizes the whole scientific fraternity in much the same way. His set piece is the meeting of the Zoological Institution, where the whole assembly degenerates into pandemonium, leaving "the students rigid with delight at seeing the high gods on Olympus quarreling among themselves" (89).

Once on the expedition, however, Challenger becomes more than a bombastic eccentric as his courage and integrity and his use of scientific method predominate over the less endearing aspects of his character. "All day amid that incessant and mysterious menace, our two professors watched every bird upon the wing, and every shrub upon the bank, with many a sharp contention . . . but with no more sense of danger . . . than if they were seated together in the smoking-room of the Royal Society's Club in St. James's Street" (92–93).

In the next Challenger story, *The Poison Belt* (1913), Doyle capitalized on the rampant speculation and fears that had attended the 1910 appearance of Halley's comet. Intellectual arrogance is still much in evidence as Challenger predicts that the earth is about to pass through a belt of poisonous ether, the effect of which will be to render people at first jolly, then frenetic, and finally dead. He alone knows how to take avoiding action—by laying in a store of oxygen sufficient to outlast the results of the poison. "'Alone of all mankind I saw and foretold this catastrophe' said he with a ring of exultation and scientific triumph in his voice," graciously permitting his wife, Malone, Roxton, and Summerlee to share his shelter.[21] What distinguishes this novel is the equanimity with which Challenger views the seemingly imminent extinction of humanity and the lack of authorial criticism of this position. It appears that Doyle regards Challenger's fatalistic stance as a proper or at least inevitable outcome of scientific objectivity. "'The world was empty before,' Challenger answered gravely. . . . Why may the same process not happen again? . . . the true scientific mind is not to be tied down by its own conditions of time and space. . . . As to death, the scientific mind dies at its post working in normal and methodic fashion to the end. It disregards so petty a thing as its own physical dissolution as completely as it does all other limitations upon the plane of matter" (101–3).

Challenger's popularity soon suggested to Doyle a way to enlist his character in the defense of spiritualism, a movement of which he was himself a devotee. In *The Land of Mist* (1926), Challenger, whose arrogance has hitherto been merely comic, is resurrected as the prototype of the worst form of scientific pride, refusing to consider any explanation other than a strictly materialist one. Doyle sets out to demonstrate that such an attitude is itself profoundly unscientific, as Challenger condemns spiritualism without having examined the evidence and effectively loses the debate with a spiritualist because he has been too arrogant to study the case of his opponent. In the earlier novels, Challenger's scientific integrity was subordinated in the interests of comedy; here it is sacrificed on the altar of spiritualism. He is publicly castigated by his adversary in the terms that hurt most: "The fact that a man was a great physiologist and physicist did not itself make him an authority upon psychic research" (473–74). Challenger is readmitted to a state of scientific grace by confessing his error, and upon witnessing an example of psychic empathy between his daughter and his dead wife, he capitulates completely, enrolling in the ranks of the spiritualists to undertake an adventure into the unknown psychic realm. Thereupon he proves as intrepid in espousing an unpopular theory as in undertaking the physical dangers of the Amazon, both, we are asked to believe, in the name of true science.[22]

Doyle was careful to preserve the more attractive of Challenger's characteristics from his preconversion days. With as much gusto as in the earlier debates, he insults his new adversaries, calling them "mental fossils dug from some early Pliocene horizon" (516). It is noticeable, however, that for most of the novel, Doyle ascribes the true scientific attitude to the proponents of psychic research, insistently claiming that it is a subject to be investigated with the tools of science and arraigning as nonscientific the skepticism of the scientist fraternity. Challenger seems not to notice that in his convert's zeal to dignify psychic research as a science and to affirm the possibility of investigating spiritual phenomena in material terms he is effectively denying its basic postulate, namely, that there is a metaphysical dimension beyond the reach of reductionist science. Challenger's confusion on this issue, which is of course Doyle's, is reminiscent of Bulwer Lytton's attempts to prove the existence of a metaphysical principle by recourse to the evidence of physics.[23]

One of the most unusual features of Challenger qua scientist is his highly emotional nature. Unlike the Romantic stereotype of the scientist, and equally unlike Doyle's even more famous creation, Sherlock Holmes,[24] Challenger is never the cold, rational figure, but a cauldron of emotions, for the most part violently expressed. His notorious fits of rage directed at his scientific opponents, the press, and the population in general are counter-balanced by his equally deep affection for his daughter and his dead wife, through whom he is finally led to embrace spiritualism.

Having converted Challenger to the spiritualist cause, Doyle could not very well use him again for the same purpose. In *The Maracot Deep* (1928), therefore, he created another scientist of similar attributes—the same eccentricity, crusty demeanor, even ferocity, and complete devotion to his research, no matter what the dangers involved for himself or others. Like Verne's heroes and Challenger, the marine biologist Dr. Maracot has a supporting cast in his journey to the floor of the Atlantic at a point called the Maracot Deep. When the steel cable holding their diving ball breaks and the ball settles thirty thousand feet down, Maracot, like Challenger, seemingly faces certain death with serenity and scientific integrity, busily taking notes to what seems the last. Much of the novel is concerned with the submarine civilization of Atlantis, which they encounter, but more important for Doyle's evangelistic purposes is Maracot's conversion from a position of scientific rationalism to a spiritualist perception. Empowered by Warda, a spirit of good, he faces, on behalf of the citizens of Atlantis, the evil Lord of the Dark Face, Baal-Seepa, and destroys him. Subsequently Maracot, like Challenger, becomes an evangelist for spiritualism.[25]

It is clear, even from this brief account, that Doyle used his scientist characters for a didactic purpose, paradoxically to further an apparently unscientific cause. Challenger the great biologist is employed to challenge the scientists' rejection of spiritualism. This in itself was a daring technique, carrying the battle into the enemy's camp. Doyle succeeded as well as he did only because of the implicit parallel between physical, mental, and spiritual adventures: Challenger is ready for any and all of them. Hence, while in one sense he undertakes the least scientific cause, in another he can be promoted as undertaking the most extreme adventure, voyaging forth into all realms of experience, even that of psychic phenomena.[26]

In all these cases it is essential to the conception of the adventurer-scientist that he should be an isolated individual or at most supported by a small band of assistants, necessarily inferior to him in mental ability. (At this stage there is, of course, no question of a female adventurer-scientist.) Certainly there is no technostructure to detract from the triumph of the protagonist. These adventurer-scientists are the precursors of the heroes of popular twentieth-century pulp science fiction—the space travelers who journey forth intrepidly where no man has gone before, unfailingly noble in their efforts to repel evil Martians and other galactic conspirators bent on invading the good planet earth or mistreating her well-intentioned ambassadors.

Even before the turn of the century, however, it became apparent that the progress of science in the real world was increasingly dependent on the backing of scientific institutions and hence on the cooperation and support of society. It was no longer merely a matter of individuals risking their own lives; scientists were beginning to wield considerable power over the whole society. While some writers believed that this would usher in a Baconian utopia, many of the most perceptive writers in Britain and Europe, including Verne himself in later life, expressed increasing mistrust of technological progress. The atmosphere of Verne's later novels became much darker, more pessimistic, and the character of his scientists changed accordingly.

CHAPTER TEN

EFFICIENCY AND POWER:
THE SCIENTIST
UNDER SCRUTINY

The Machine develops—but not on our lines. The Machine
proceeds—but not to our goal. We only exist as the blood
corpuscles that course through its arteries, and if it could work
without us, it would let us die.
—E. M. Forster

To this day I have never troubled about the ethics of the matter.
The study of Nature makes a man at last as remorseless as
Nature.
—H. G. Wells

By material standards, the wealth and power that had continued to accrue
to Britain as a result of her industrial revolution confirmed the victory of the
machine and its methods. Insidiously the criteria of the machine-oriented
system—uniformity, competency, material productivity, unfailing "obedience" to orders—became the tacitly accepted values of a grateful society,
and hitherto neutral terms such as *order, stability,* and *efficiency* acquired
moral overtones. But at the same time as the writers considered in chapter 9
were extolling the high-technology adventure, there were also sporadic
protests against the cult of efficiency and uniformity engendered by the
machine age. These protests were diverse and often radical, including, for
example, the proposals by William Morris in his utopian *News from Nowhere*
(1890) for a return to an essentially medieval, socialist system, with an
economy based on handicrafts and organized through guilds. In this chapter we shall examine the image of the scientist in turn-of-the-century literature that attacked the philosophy of scientific materialism and its technological superstructure.

The principles most commonly disputed included those already attacked by the Romantic writers: the preeminence of matter as defined by the
so-called primary characteristics; the supremacy of objective observation as

a method of attaining knowledge about the world; and in particular the mechanistic view of man, which had been gaining ground since the publication of La Mettrie's *L'Homme machine* in 1747.[1] This latter thesis had increased in complexity from the mathematical and mechanical explanations of the seventeenth and eighteenth centuries to include the biological evidence marshaled by Charles Darwin in *The Descent of Man* (1871) and the work of Francis Galton, which effectively professed to do away with any essential physical distinction between man and the other animals. The emerging science of psychology further eroded the claim of man's uniqueness, and even the long-held distinction between living and nonliving, the doctrine of vitalism, was discarded by chemists after 1828, when Friedrich Wöhler succeeded in producing the organic substance urea from the inorganic compound ammonium cyanate. The eminent biologist John Tyndall publicly asserted that although there was as yet no clear proof of a link between consciousness and molecular activity, all mysteries of nature would ultimately be explained in mechanistic terms.[2] As far as science was concerned, the centuries-old controversy over mind versus matter was finally resolved by discarding the distinction altogether.

Despite the apparent unanimity in the ranks of science, however, there was considerable criticism of this view from other quarters. The cult of the machine was blamed for weakening one or all of man's body, individuality, mind, character, and morale and fell under attack on philosophical as well as social and aesthetic grounds. This end-of-the-century offensive was not a repetition of the Romantic onslaught a century before. The Romantic writers had opposed scientific materialism on the grounds that its procedures and indeed its very terms of reference were inimical to the creative process and to the natural expression of the emotions; and their literary challenge had been expressed dramatically, in terms either of inner struggle or of a struggle between the individual and society. But rather than opposing the intellect with the emotions, the writers to be discussed in this chapter fought their battle on intellectual grounds, that is, within the parameters acceptable to science. Their criticism varies from the more obvious attacks on the machine and mechanism per se, such as Samuel Butler's *Erewhon*, to the more philosophical explorations of scientism in H. G. Wells's *Island of Doctor Moreau.*

Most frequently it was the wider social implications of mechanization that came under scrutiny. One of the first sustained literary assaults on the machine was delivered by Samuel Butler, who on first reading *The Origin of Species* had been so impressed that he wrote to Darwin expressing his admiration. Later, however, he attacked the philosophical validity of postulating a mechanical system of evolution, while relying upon chance to set it in train. Like Darwin himself, he soon realized that the weakest part of Darwinian theory was its inability to account for the chance variations upon which it hinged,[3] and he thereupon defected to the Lamarckist camp. As a vitalist, Butler came to abhor the mechanistic implications of evolutionary theory, which he now saw as a further extension of the society's obsession with technology.

In chapters 21 to 23 of his utopian fantasy *Erewhon* (1872), collectively entitled "The Book of the Machine," Butler describes a land, Erewhon (an anagram of *nowhere*), where machines have been outlawed because they threatened to enslave the populace. The Erewhonian philosophers had argued that, far from making the machines work for them, people would soon become merely tenders of machines, grooming, feeding, and nurturing their machines, subservient to their every "wish." Butler's satire rests upon the implicit premise that if animals are *only* machines, it is valid to apply the theories associated with animals to machines. He therefore transfers the concept of evolution to machines. Just as simple forms of life had developed into more complex organisms characterized by increasing levels of consciousness and purpose, which allowed them to supplant earlier forms, so machines would evolve in complexity and consciousness and overthrow their human masters. As an example of such a process, Butler pointed to the calculator and thereby gave uncanny force to his argument for the twentieth-century reader, who sees many of the Erewhonian postulates already come to pass in a computer-dependent society. He argued that since machines could now "do all manner of sums more quickly and correctly than we can," and do them without tiring or slacking, they were already more successful, that is, more evolved, than humans.

The narrator points out that machines have already taken on many of the characteristics formerly considered unique to living organisms, even reproduction, for machines can produce new machines. Man, by contrast, becomes increasingly helpless, relying upon the machine to do all his hard labor, until his organs atrophy through disuse. Thus, as the machine develops, man regresses to an enfeebled lump of flesh, wholly supplanted by the machines, the real rulers of the world. After hearing these arguments from their learned philosophers, the Erewhonians decided to destroy all machines invented during the preceding 271 years, that is, since the industrial revolution. Even watches were forbidden, since their complex mechanism posed a similar threat.

Although it is presented as a fable rather than realistically, "The Book of the Machine" constitutes one of the most sophisticated arguments of the time against the galloping mechanistic assumptions of Victorian society. It points to what H. L. Sussmann has called "the central paradox of Western philosophy, the conflict between the deterministic implications of science and the inward apprehension of volitional freedom."[4]

The idea that men would become the willing and eventually helpless slaves of the machines that answered their every need was expanded by

Rudyard Kipling in "With the Night Mail" (1905) and its sequel, "Easy as A.B.C." (1912), and by E. M. Forster in "The Machine Stops" (1909). This latter story describes a world in which virtually the whole population of the earth lives underground, totally dependent on the Machine for their survival. "The Machine develops—but not on our lines. The Machine proceeds—but not to our goal. We only exist as the blood corpuscles that course through its arteries, and if it could work without us, it would let us die."[5] This idea was to become the basis for a number of twentieth-century dystopias, including Yevgeny Zamyatin's *We* and Aldous Huxley's *Brave New World*, which will be discussed in chapter 13.

Apart from the fear that man would become dependent on, and eventually subservient to, his machines, the other major anxiety arising from the mechanistic assumptions of nineteenth-century biology and psychology was that human beings might indeed be no more than a mechanism, either an engineering model of pumps and levers or a complex chemical laboratory in which the relative quantities of various organic reagents determined even the subtleties of personality and behavior. While such a concept might be a triumph for science, it engendered a deep fear in those who wished to believe in a system that could accommodate emotions and abstract values and who wished to believe that the mind was more than the brain that housed it. This apprehension was expressed in several ways in literature, one of which involved the use of scientists as creators of machines indistinguishable from people. These scientists are invariably portrayed as culpable, being either arrogant or negligent concerning the consequences of their work.

An obvious test case was the proposed construction of a chess-playing automaton, since chess was considered the severest test of the intellect, requiring a high degree of both logic and creative thinking. One of the first such stories was Edgar Allan Poe's "Maelzel's Chess-Player" (1836). Given Poe's predilection for mystery and the powers of the imagination, it is perhaps not surprising that the automaton chess player in this story is a fake, being activated remotely by a concealed master chess player.[6] But in Ambrose Bierce's more interesting story, "Moxon's Master" (1894), the automaton chess player is genuine and, piqued at being beaten at the game, it kills its creator. Bierce's story is important because it explicitly discusses, through a dialogue between the scientist-inventor Moxon and the narrator, what, if anything, distinguishes living systems from machines. Instead of propounding the mechanist view that all organisms are merely complex machines, Moxon proposes the contrary view, namely, that "all matter is sentient, that every atom is a living, feeling, conscious being."[7] He claims that his experiments have shown that plants think, and he now believes that even the constituent atoms of minerals think, since they arrange themselves into mathematically perfect patterns. The satire implicit in the ambiguity of the title is a continuation of that inherent in *Erewhon*. Moxon has made a machine to entertain himself and to demonstrate his theories; but he has not worked through the consequences, for it is the validity of those very theories that is his undoing. In accordance with his own postulates, the machine is alive and therefore not content to accept a subservient, machinelike role; experiencing frustration at losing the game and anger against its opponent, it reaches forward and strangles Moxon with its iron hands. This ending looks back to *Frankenstein* and forward to the twentieth-century stereotype of the scientist unable to control his inventions, especially computers, the twentieth-century successors to chess-playing automatons, but "Moxon's Master" differs from both its predecessors and its successors in its highly intellectual examination of the consequences of mechanistic theory.

The intellect was not the only human characteristic to be explored in relation to mechanistic control. The emotions, the soul, and psychological states were also fair stakes in the fashionable game of hunt-the-mechanism. Edward Bellamy's short story "Dr. Heidenhoff's Process" (1880) describes a procedure, analogous to hypnotism, for eradicating unpleasant memories and hence all sense of guilt, his assumption being that such psychological states are merely the result of physicochemical processes that can be manipulated as though they were part of a machine circuit. Later versions of the idea—H. G. Wells's "Story of the Days to Come" and *The Sleeper Awakes*, Aldous Huxley's *Brave New World*, and George Orwell's *1984*—show the use of hypnotism and hypnopaedia to instill the values compatible with those of a mechanistic society and to remove those contrary to it.[8]

Robert Louis Stevenson's classic story, *The Strange Case of Dr. Jekyll and Mr. Hyde* (1886), is based on a similar thesis, that even moral character might be changed by material (in this case chemical) means. Like so many erstwhile fables of science, this once exotic tale was soon to acquire an uncomfortable degree of verisimilitude in a world where drugs and lobotomies were used to modify the personality of social misfits. Stevenson, who had studied engineering and law before turning to literature, was greatly interested in contemporary psychology, and his story is usually considered as a parable about human nature, which of course it is. A story about the splitting of a personality into distinguishable but, as it turns out, inseparable components has universal relevance, and the fact that Jekyll is a scientist may seem incidental. Jekyll's research in chemistry is, however, more than a plot device, and the parallels with Frankenstein are clear and intentional. Stevenson's particular censure falls on the confidence of contemporary science in its ability to subdue and improve upon nature, including human nature, to usher in a utopian future that will supersede the rosiest visions of Eden. Like Faust and Frankenstein before him, the basically good but overcomplacent Dr. Jekyll believes that he can refine nature, in this case morally rather than intellectually or in terms of longevity. Like Bellamy, Jekyll assumes that technological progress is at least parallel to, and perhaps even synonymous with, ethical progress and that sin and guilt are mere atavistic remnants from an earlier stage of evolution. He expects, by means of his chemical knowledge, to expunge the less desirable moral elements from his nature and thereby evolve a superior individual (see fig. 20). Stevenson is careful to stress that the procedure Jekyll uses is not at fault. Scientifically considered, Jekyll's experiment, like Frankenstein's, is a brilliant success. The flaw lies in the scientist himself, in his assumption of his own perfectibility and his consequent inability to see that the two parts of his nature are inseparable. G. K. Chesterton, who was later to be a trenchant critic of Wells's utopias also, pointed out the major irony of the story: "The real stab of the story is not in the discovery that one man is two men, but in the discovery that two men are one man. . . . The point . . . is not that a man *can* cut himself off from his conscience, but that he cannot."[9]

Thus Jekyll, so intelligent qua scientist, lacks self-knowledge. Bent on excising the evil from his nature and consigning it to oblivion, he fails to realize that it cannot be so discarded, but will, on the contrary, grow stronger without the constraints of its counterpart. The inescapable duality of Jekyll and Hyde was a metaphor not only for the nature of man but in particular for the nature of the scientist in his role as creator of another being. Jekyll, in attempting to reconstruct a new self, encounters, as Frankenstein did, his other self, a Doppelgänger, through the fission of his own personality.[10]

Automatons were also pressed into the service of the erotic as part of the comment on the mechanized view of Man. *L'Eve future* (1887), by the French writer Villiers de l'Isle-Adam, tells the story of an inventor, Edison, who constructs a lifelike robot woman, Hadaly, to console his friend Lord Ewald. Ewald loves the beautiful girl Alicia but is repelled by her bourgeois soul, so unworthy of her physical charms. The synthetic Hadaly is physically indistinguishable from Alicia but has the spirit of a virtuous young woman, Amy, over whom Edison has hypnotic control. Ewald's initial diffidence towards this model of artificial perfection elicits an admonitory lecture from her creator, who points out that what Ewald loved in Alicia was merely a projection of his ideals and that he can as easily breathe life into Hadaly. One of the many literary daughters of this Eve was the protagonist of *La Femme Endormie* (1899), by Madame B, in which an artificial doll accommodates without resistance the bizarre sexual requirements of her clients.

From the 1880s on, even Jules Verne, originally so sanguine about technological inventions, became increasingly pessimistic and cynical about science and scientists.[11] Unlike the early *Voyages extraordinaires*, his later writings show a technology shaped no longer for the benefit of society but for the purposes of self-aggrandizement and scientists who have become irresponsible, power-crazed maniacs, obsessed with the despoliation of na-

FIGURE 20.
Illustration for *Dr.
Jekyll and Mr. Hyde*
expressing the
double image of the
scientist.

ture and the destruction of humanity. In *The Begum's Fortune* (1879) a French
scientist, Dr. Sarrasin, and a German scientist, Herr Schultze, each construct
an ideal city, France-Ville and Stahlstadt, respectively. Schultze, who embodies French views about Germany in the aftermath of the Franco-Prussian
War, uses his science to revolutionize warfare and initiates a lucrative arms
trade to ensure long-term racial superiority for the German people. He is
prevented from destroying France-Ville only by an act of Providence such as
Verne, disenchanted with human morality, was to invoke with increasing
frequency against the evil machinations of successive scientists. Even initially moral scientists such as Robur are not proof against corruption. In
Robur the Conqueror (1886) the hero acts responsibly, if paternally, refusing to
impart his technological knowledge until the society, in this case the United
States, is morally prepared for it. "My opinion is, as of now, that nothing
should be rushed, not even Progress. Science must not get ahead of social
customs. . . . It [my secret] will belong to you the day that you become wise
enough to use it constructively and never abuse it." However, in the sequel,

Master of the World (1904), Robur has deteriorated into a maniacal terrorist (his new vehicle is called *Epouvante* [Terror]) wreaking gratuitous violence to prove his superiority until he too is disposed of by the power of Providence in the form of lightning. In *The Floating Island* (1895) Verne produced a bitter satire of technological materialism, suggesting that he had come to realize the sinister potential of the obsession with technological power.[12] In *Topsy Turvy* (1889) Barbicane, Maston, and the rest of the group in *From the Earth to the Moon* set out to alter the earth's rotational axis in order to melt the polar icecaps and access the mineral deposits beneath. Immune to the pleas and protests from all nations threatened with disaster by this exercise, they are prevented from carrying out their plan only by a mathematical error in Maston's equations. Other irresponsible scientists in Verne's late stories—Thomas Roch in *For the Flag*, Wilhelm Storitz in *The Secret of Wilhelm Storitz*, and Orfanik in *The Carpathian Castle*—are insane and hence even more dangerous because of the unpredictability of their actions.[13] Marcel Camaret, of *The Amazing Adventure of the Barsac Mission* (1920), is not actively evil; on the contrary, he is a dreamy, absent-minded scientist absorbed in abstract problems and unaware that he is being manipulated by Harry Killer, the evil tyrant of Blackland. Yet Verne makes it clear that the social responsibility of scientists cannot be waived, or their failures pardoned, merely because they had not intended any harm. Camaret is pardoned only after he has destroyed the evil city of Blackland.

Ironically, the most influential and systematic critic of scientists in the last decades of the century was Herbert George Wells, the first English novelist to have received a formal training in science before he began to write fiction and the writer who was later to be most completely identified with the new scientific utopia, the twentieth-century counterpart of Bacon's New Atlantis. Through his experience at the South Kensington Normal School of Science, where he studied biology under T. H. Huxley, Wells was able to expand considerably on the range of scientists depicted in literature, introducing entomologists, geneticists, ecologists, and microscopists. The development of the scientist character in Wells's novels functions both as an index of the increasing influence of actual scientists in his society and as a dramatization of his own changing perception of their potential power, which he viewed with varying degrees of apprehension and optimism. Through his successive scientist figures, Wells explored the social implications of the impending technological boom and the inadequacy of the earlier code of scientific ethics—strict objectivity and refusal to acknowledge responsibility for the results of one's research, the criteria of so-called value-free research.

Wells's most famous story, "The Time Machine" (1895), a reworking of "The Chronic Argonauts,"[14] introduces a range of complex attitudes towards science. It embodies both the optimism Wells associated with contemporary scientists and the reservations he himself felt at this stage about science and its values. Our first impression of the nameless Time Traveller is of someone open to new ideas. This immediately distinguishes him from his guests, who embody the normal skepticism of society towards anything mentally or physically disturbing. This framing scene inclines us to sympathize with the Time Traveller and, provisionally at least, to accept his reasoning. However, the Time Traveller is no simple hero, and the story he relates carries its own trenchant criticism of the intellect, which, it seems, is a shaky reed to rely on in the year 802,701. The main section of the story concerns the Time Traveller's stopover in the year 802,701, when England, at least, has become literally the "Two Nations" that Benjamin Disraeli had symbolically called it. One the one hand there are the effete, childlike Eloi, the beautiful people who inhabit an apparently idyllic paradise in the sun, laughing their way through a vegetarian, close-to-nature existence; on the other, the carnivorous, subterranean Morlocks, the real rulers of the upper world, who prey upon the Eloi as though upon cattle. In Wells's mythology, the Eloi represent one obvious facet of the Romantic ideal; the Morlocks, by contrast, are presented as hideous, ruthless, inhuman, and it is they who are identified with technology. This association, simplistic as it necessarily is in Wells's treatment, generated a powerful image that was visually reinforced in the illustrations and in the subsequent films of "The Time Machine." Technology is linked with a dark, subterranean existence (an image derived from both the alchemist's cave and the working-class underworld of late Victorian London), with ruthless, stunted creatures who prey upon the beautiful children of nature. It was not too exaggerated an image of the way the Victorian industrial system preyed upon the lives of its workers, who for virtually their whole working life, were little more than machine-fodder, pent up in dark factories, sealed off from the natural rhythms of light and dark.[15]

Overpowering as this visual imagery is, along with the unforgettable picture of the dying planet (the Time Traveller is no more fortunate in his choice of temporal venues than Nebogipfel was in "The Chronic Argonauts"), the Time Traveller himself is also an important contribution to the character of the scientist as perceived by Wells at this point in his career. Whenever he feels secure in his ability to solve a problem by reasoning, the Time Traveller is invariably on the brink of disaster. Thus he complacently leaves his time machine, having cleverly, as he thinks, removed the crystal bars essential for driving it through time, only to find on his return that it has disappeared, the Morlocks having simply lifted it up by brute force and transported it through space. He departs from the Palace of Green Porcelain confident in his ability to defeat the heliophobic Morlocks with his matches, camphor, an iron bar, and a bundle of firewood, the equipment suggested by his reason; instead, he starts a forest fire, thereby killing Weena, and loses his sense of direction. Witnessing the "decaying vestiges of books" in the deserted museum, he is forced to acknowledge the futility of his "own seventeen papers upon physical optics."[16]

Yet, despite his repeated failures, the Time Traveller continues to trust in rationalism. Just as he had argued logically with his unreceptive guests, so he tries, equally unsuccessfully, to teach the Eloi of the future world the elements of a grammatical language. By a series of successive approximations, he evolves a rational theory to fit the observed facts of the year 802,701; and at the end of the story, still a firm believer in scientific method, he again sets off on his time machine, equipped, as he thinks, with the requisite apparatus to furnish incontrovertible "proof" to his skeptical friends.

This ambivalence about the value of the intellect is reflected and amplified in the Time Traveller's ambiguous attitude towards the future. The narrator, who is given the last word in the epilogue, maintains that the Time Traveller is fundamentally a pessimist about the future: "He, I know—for the question had been discussed among us long before the time machine was made—thought but cheerlessly of the Advancement of Mankind, and saw in the growing pile of civilization only a foolish heaping that must inevitably fall back upon and destroy its makers in the end" (83). Certainly, his experiences in 802,701 and, even more dramatically, at the demise of life on earth should have confirmed this view. Yet he sets off with as much enthusiasm as ever on his next voyage into time. In other words, he, like the narrator, acts "as though it were not so."

The Time Traveller in fact is the victim of considerable irony on Wells's part. Having spent years constructing a time machine and having endured the extreme discomfort of time traveling in order "to get into the future age," he arrives in 802,701 only to find himself terrified at the "possibility of losing my own age" (34). Yet, even knowing the risks involved, he remains determined to endure the whole procedure again. It seems likely that the ambiguity that underlies the story reflects Wells's own uncertainty about the future at this stage of his career. Certainly he himself inclined to a pessimism that may have been partly a reflection of the prevailing *fin-de-siècle* atmosphere[17] but probably had its more immediate source in the comparable ambivalence of T. H. Huxley, Wells's most revered mentor, who had painted a relatively pessimistic view of humanity's future in his recent Romanes lecture, "Evolution and Ethics."[18]

Despite these ambiguities, Wells has ensured that the Time Traveller emerges as a sympathetic character, especially in the superb scene of the dying world, where the Time Traveller experiences total isolation and the most extreme helplessness. He may struggle, albeit with difficulty, against the Morlocks; he cannot struggle against the second law of thermodynamics. In this scene, indeed, he emerges as the protagonist of a cosmic tragedy, with all the associated nobility of defeat by an insuperable fate.

Thus the Time Traveller, Wells's first developed fictional scientist, is a sympathetic but fundamentally powerless figure whose intellect does not enable him to overcome either the forces of evil in society or the inevitability of nature's laws; on the contrary, he errs and suffers, always within the limitations of the human condition.

In the volume of short stories *The Stolen Bacillus and Other Incidents* (1895) Wells explored several examples of scientists, often themselves ruthless, caught up in circumstances beyond their control. "The Diamond Maker" and the title story are simple studies of the scientist exploiting his discovery to gain power, a figure Wells was later to develop in Griffin, the Invisible Man. But two stories, "The Moth" and "The Lord of the Dynamos," are more complex and deserve particular attention. Underlying the overt comedy of "The Moth" is an analysis of the motivation of an entomologist who appears to his colleagues a dedicated researcher. We discover, however, that Hapley's twenty-year devotion to science has sprung exclusively from his hatred of his rival Pawkins, whose death leaves a void that nothing can fill, nothing, that is, except Pawkins himself.[19] Pawkins indeed returns to Hapley's deranged mind in the form of a rare moth, whose capture becomes the sole purpose of his life, even though he knows with half his mind that it is an illusion. Thus in the superficially comic character of Hapley Wells explores more profound implications of dedication to research, notably the hypocrisy whereby, on the pretext of scientific argument, a scientist may vent his own subjective passions and obsessions. The suggestion remains that Hapley is not a unique case.

"The Lord of the Dynamos," in the same volume, is one of Wells's most effective short stories, depicting, among other things,[20] several prevailing attitudes towards science and technology. The shed housing the dynamos that supply the electric railway at Camberwell is a model of the technological society, its chief and most obvious attribute being power—in both senses of the word. To it come representatives of three diverse attitudes towards technology. Holroyd, the "practical electrician," reveres the dynamo, partly because it is more powerful than him and partly because it lends support to his innate desire to exercise physical power over his subordinates. "To James Holroyd, bullying was a labour of love."[21] He embodies the attitudes of those Social Darwinists who professed to find in evolutionary struggle and the technological superiority of one race or group a justification for laissez-faire policies and the exploitation of workers. Finding traditional religion inadequate to this purpose, he has transferred his worship to the most immediate representative of technological power. When Holroyd is propelled to his death on the live terminals of the dynamo, he becomes literally what he has already been symbolically, a part of the machine.

Holroyd is not the only one to revere the machine's power; the primitive Azuma-zi worships the dynamo for the powerful sense impressions it evokes. His voluntary death—he is fused to the dynamo as part of its circuit—enacts literally his sense of mystic communion with his Lord. The parallel nature of the two deaths underlines the fundamental similarity between these apparently disparate worshipers of technological power, Azuma-zi parodying Holroyd and, by extension, the worship of technological power that Wells observed in so many of his contemporaries. If these two characters were the only representatives of science in this fable, we might well assume that Wells's attitude towards the technological society was entirely pessimistic. However, there is a third character, the scientific manager, who unites efficiency and feeling within a framework of social responsibility, providing the first example of a combination Wells was later to develop in the "noble" scientists of his utopias.

Wells's next major novel, *The Island of Doctor Moreau* (1896), was his most complex and provocative critique of scientism and arguably one of the great modern myths in the tradition of Faust and Frankenstein. Moreau himself is at first presented realistically enough as a somewhat fanatical biologist. Exiled from Britain after a scandal involving vivisection, still a topical issue in the 1890s,[22] he now lives, with his assistant Montgomery, in almost complete isolation. It is important that Moreau, like Frankenstein, seems at first a charismatic figure. Not only is his physical appearance impressive (Prendick remarks upon "his serenity, the touch almost of beauty that resulted from his set tranquillity, and from his magnificent build")[23] but there is evidence of that noble dedication to research and contempt for ease and social rewards that were to distinguish the noble scientist stereotype of the early twentieth century.

Yet Moreau also functions on a number of symbolic levels, which, while they do not undermine his credibility, are crucial in universalizing the meaning of the novel. It is clear, for instance, that he embodies several characteristics of the alchemist. Living in rigorous seclusion on his island (itself a further symbol of isolation), he conducts his research in the strictest secrecy. Like Frankenstein, he directs his research towards the creation of human beings, and, ironically, like him he is so obsessed with his work that he finds any human interruption almost intolerable.[24] Just as Frankenstein paradoxically creates his living Monster from the materials of death, collecting bones and sinews from graves and charnel houses, Moreau's experiments, too, are associated with death—with the literal death of the "failures" (Moreau himself facetiously but not inaccurately calls his laboratory "a kind of Bluebeard's Chamber") and with the baptismal bath of pain through which the animals must pass, as though through death, to be made "human." This is further reflected in Moreau's own name, a condensation of "water of death,"[25] and in that more visible water of death, blood, the most powerful and pervasive symbol of the novel and the one largely responsible for the widespread distaste and rejection it immediately evoked on its publication.

Moreau resembles Frankenstein also in cutting himself off from the structures and values of society, dismissing as stupid those who do not approve of his methods. He thus reflects in his own character the profound hubris implicit in his experiment and in the science he typifies. As Frankenstein embodied the idea of the Creator current in Mary Shelley's generation, Moreau represents just such a Creator as might be derived from the evolutionary process. The almost arbitrary succession of beasts passing through Moreau's hands is a reenactment of the idea of the evolutionary process as nature's giant experiment, wherein much "material" must necessarily be lost for the sake of a few "successes." Moreau, who uses these same terms when describing his work, is as ruthlessly amoral as the Darwinian picture of nature; indeed he dramatizes almost exactly the dangers Huxley had warned would result from an "imitation" of the cosmic process and from the attempt to derive a social ethic from it. "Cosmic evolution may teach us how the good and evil tendencies of man came about; but in itself it is incompetent to furnish any better reason why what we call good is preferable to what we call evil than what we had before. . . . Let us understand once and for all, that the ethical progress of society depends not on imitating the cosmic process, still less on running away from it, but in combatting it."[26]

Moreau justifies his relentless pursuit of his work by just such an appeal to a "natural" philosophy: "I am a religious man, Prendick, as every sane man must be. It may be, I fancy, I have seen more of the ways of this world's Maker than you—for I have sought His laws, in *my* way, all my life" (107). The "laws" of nature that Moreau exemplifies are those that underlie the evolutionary process—chance, waste, and pain—and it is repeatedly insisted that Moreau's creations involve all these aspects.[27] Thus Moreau functions both as a parody of the Old Testament Creator and as an allegory of evolution itself. In this latter role, he implicitly exposes some of the problems raised by Darwinism not only for religious orthodoxy but, equally, for the humanist belief in the essential nobility and goodness of man.[28]

Moreau also embodies both the arrogance of science in the social context and the ruthlessness of the Social Darwinists, who translated the same ideology from biology to a system of economic rationalism. Contemptuous of social condemnation, Moreau insists on his right to inflict pain on his experimental creatures,[29] justifying his procedures by appeal to the frames of reference of geology and biology, cosmic dimensions of space and eons of time: "A mind truly opened to what science has to teach must see that it [pain] is a little thing" (106). "After all, what is ten years? Man has been a hundred thousand in the making" (113).

Typically, Moreau's criteria are impersonal; here again Wells expresses his reservations about the scientific perspective, namely, that by its standards of objectivity, human feelings, sympathy, and ethical values are indeed irrelevant. Moreau denies the relevance of a moral code in scientific research: "To this day I have never troubled about the ethics of the matter. The study of Nature makes a man at last as remorseless as Nature" (108). His only guidelines are those of the experiment itself: "I went on with this research just the way it led me. That is the only way I ever heard of research going. The thing before you is no longer an animal, a fellow-creature, but a problem" (107).

However, Moreau does not escape Wells's irony. Just as the Time Traveller, having invented a time machine in order to overcome the restrictions of time, finds himself still the prisoner of time, in each case a time more inimical than his own age, so Moreau, obsessed with achieving mastery over evolution and the limitations of human creativity, falls victim to the very plasticity he insisted upon, being killed by one of his hybrid creatures. Similarly, just as the Time Traveller finds that the further he travels into the future, the more that future resembles the remote past, so Moreau finds that the more he attempts to overcome the limitations of the human condition, the more he uncovers those limitations; the more he determines to humanize his beasts, the more he demonstrates the bestiality of man.

Despite the dominance of Moreau, both on the island and in the novel, the two other scientists of the novel, Moreau's assistant Montgomery and the narrator Prendick, both contribute to the composite picture of the scientist that emerges. Montgomery is judged by Prendick to be a fundamentally weak character, assisting Moreau even while disapproving of his methods. He represents the scientist caught up in a morally repugnant situation from which he feels unable to escape; indeed, Montgomery is literally trapped on the island (although Wells implies that this was his own choice). It is he who, taking over where Moreau's surgery left off, has attempted to teach the Beast Folk to speak. Unlike the research biologists Moreau and Prendick, Montgomery has been a doctor and retains something of the compassion associated with his calling. It is one of the many somber considerations of the novel that in attempting to retain humane feelings while working on Moreau's project Montgomery has virtually become insane. Like his increasingly frequent bouts of drunkenness, his last words express his inability to cope with the paradox: "The last of this silly universe. What a mess" (162); but they comment equally well on the Darwinian picture of a universe governed by chance and accident. In this Montgomery reflects the response of many late-nineteenth-century minds to Darwinism. In their diverse ways, Tennyson, Arnold, Clough, and Hardy all made a similar comment at some stage.

The third scientist, Prendick, is an important and subtly drawn figure whose attitudes change considerably during his stay on the ironically-named Noble's Island. After overcoming his initial fear and disgust of the Beast Folk, he pities them,[30] but after Moreau's death Prendick takes over his godlike role on the island, employing both physical and moral force (symbolized by the whip) to keep the Beast Folk under control. On his return to England, he buries himself as effectively in his private research as Moreau had done on his island. Significantly, he takes up chemistry and astronomy: "There it must be, I think, in the vast eternal laws of matter, and not in the daily cares and sins and troubles of men, that whatever is more than animal within us must find its solace and its hope" (191–92). The parallel with Gulliver's pretensions to be above the rest of humanity, to live on a higher plane of pure reason, remote from physical considerations, is strikingly clear.[31]

Thus *The Island of Doctor Moreau* contains, not just the element of "theological grotesque," which Wells himself pointed out, but equally a trenchant satire on scientism, on the isolationism of scientists and their contempt for the layman and, ultimately, for mankind. By portraying the same fundamental attitude, not only in the mythological figure of the fanatical Moreau but also in the apparently ordinary, decent Prendick, Wells extends his implied criticism to include even the respectable gentleman scientist who has "taken to natural history as a relief from the dullness of [his] comfortable independence" (15) and who has "done some research in biology under Huxley" (41–42).

In Wells's next novel, *The Invisible Man* (1897), the main interest lies in the transformation of character when great power is suddenly placed within its grasp; it is interesting therefore that Wells uses a scientist, Griffin, for such a role. Like Faust and Frankenstein, Griffin began his researches in optical density idealistically, for the sake of knowledge, but he became corrupted with two countervalues, secrecy and the desire for fame. His account of his subsequent accidental discovery whereby he could make an animal transparent and hence invisible shows that he was immediately aware of the power factor. "I beheld, unclouded by doubt, a magnificent vision of all that invisibility might mean to a man,—the mystery, the power, the freedom."[32]

One of Griffin's most chilling aspects is his emotional deficiency, which is explicitly presented as an extension of his scientific objectivity. Believing that society is stupid and obstructionist, Griffin does not scruple to dispose of those in his way and propose a Reign of Terror.[33] The novel is thus essentially a tale of scientific hubris in which Griffin inevitably effects his own nemesis. Much of the irony of the story results from the paradoxical inversions whereby the protagonist, like the Time Traveller and Moreau, achieves his specific desires only to find that they restrict rather than increase his power.[34] Once he has achieved invisibility, Griffin's voluntary isolation becomes involuntary, and this arrogant scientist who believes he has progressed beyond "common humanity" is reduced to the most basic survival activities. His brilliant intellect, directed in the pursuit of one goal, becomes monomania; and what is most ironic, this man, who has striven to become invisible, now desires most of all to become visible again. This too he finally achieves, but only when he is battered to death like an animal by the villagers, rendered brutal by the fear he has evoked.

Like Moreau, Griffin also functions as a symbol of science itself. In an invisible man, who apparently represents a disembodied intellect, Wells has in fact parodied the primary claim of scientific method, namely, that reason is the ultimate authority.[35] Hence Griffin's career, from the pursuit of pure knowledge to the instigation of his Reign of Terror, becomes an implicit comment on the progress of science and a prediction of the likely consequences of adopting its values without question. Griffin represents the culmination of Wells's sequence of unmistakably reprehensible scientists, but before turning to the portrayal of those noble scientists who were to usher in his utopian society,[36] Wells depicted two, Cavor and George Ponderevo, whose interpretation remains ambiguous.[37]

Cavor, the chemist of *The First Men in the Moon* (1901), who produces a gravity-screening substance called cavorite, remaining blissfully intent on his discovery while disasters threaten, at first appears to be merely another absent-minded scientist.[38] But despite the considerable element of comedy, there is a serious moral statement implicit in the complex portrayal of Cavor, which was entirely lacking in Verne's story of a voyage to the moon. Compared with Moreau and Griffin, Cavor is untainted by the desire for power or personal gain, or even for scientific fame. Bedford, spokesman for the ordinary man, remarks incredulously, "It wasn't that he intended to make any use of these things, he simply wanted to know them."[39] Cavor's physical journey to the moon is also a symbol of his readiness to undertake adventurous mental journeys, and like Verne's scientists, he displays in the cause of science an impressive unconcern about his own safety. When imprisoned by the Selenites, the lunar inhabitants, he meditates calmly on the abstract correlations between the earth and the moon, and just before his recapture by them, and the certain death this implies, he is busily scribbling notes about the various forms of Selenite.

But despite these engaging qualities, Cavor is not presented as a hero; he is, on the contrary, morally suspect. If Bedford, the imperialist and entrepreneur, is prepared to colonize the Selenites in the cause of wealth (and the parallel with the genocide of the Tasmanian aborigines under British colonial rule is made explicit at the beginning of the novel), Cavor is prepared to sacrifice not only himself but anyone else in the cause of his research. When his first attempt to make cavorite results in the death of his three assistants, he merely comments, "That is a detail. If they have [perished], it is no great loss; they were more zealous than able" (32–33). He then proposes to pass the disaster off as a cyclone and collect a considerable share of any compensation.[40] Fundamentally, then, the disarmingly cheerful Cavor is as amoral as Moreau, a fact Bedford perceives early in the novel, commenting: "He had troubled no more about the application of the stuff he was going to turn out than if he had been a machine that makes guns" (22). This comparison of Cavor to a machine becomes most significant when, at the end of the novel, he radios back to earth his impressions of lunar society. Cavor finds fascinating the highly rationalized system whereby the Selenites condition their offspring for the jobs they are predestined to do, so that "each is a perfect unit in a world machine" (258). He regards his own initial instinctive distaste at the means of adaptation as an unfortunate lapse from scientific objectivity and expresses a hope that he will soon learn to appreciate these methods fully, that is, without any emotional involvement. In effect, the entire Selenite society functions as an extended parody of Cavor's belief that the scientific end justifies the psychological means and that efficiency is the preeminent criterion of morality.

Like the Time Traveller, Moreau, and Griffin, Cavor receives the nemesis he deserves. Having unreservedly admired the strict rationalism of the Selenites, he becomes its victim.[41] It would seem that at this stage Wells himself was ambivalent about the rationalist criteria of Cavor and the Selenites, for the satire underlying the comparison between earthly and lunar societies is clearly double-edged. Each falls short of a utopian ideal that would permit both freedom for the individual and an efficient organization of society. Indeed, Wells was the first to enunciate this fundamental logistical problem, which has been echoed in a plethora of twentieth-century dystopias, beginning with his own *When the Sleeper Awakes:* how to combine freedom and efficiency. Our own highly computerized society still has not resolved this question, as recurrent expressions of concerns about privacy of computerized data indicate.

Wells's other ambivalent treatment of the scientist in this decade was George Ponderevo, the narrator of *Tono-Bungay* (1909). Although this novel has been valued chiefly for its commentary on the social and intellectual climate of late Victorian and Edwardian England, George Ponderevo also marks Wells's first treatment of a scientist as the protagonist of a realistic social novel. George does not begin as a scientist; indeed much of the novel is concerned with his progressive rejection of the institutions and attitudes that obstruct science. These incidents themselves serve to define the role of the scientist by contrast. George writes, "I have called it *Tono-Bungay* but I had far better have called it *Waste*,"[42] and his story presents one instance after another of waste and confusion at both the social and the individual level. Science emerges as the only viable alternative to this all-pervading chaos against which George struggles, as Wells himself was to struggle all his life.[43] "Through the confusion something drives, something that is at once human achievement and the most inhuman of all existing things. . . .

Sometimes I call this reality Science, sometimes I call it Truth" (335).

The qualities George possesses or acquires that enable him to overcome this almost universal disorder, at least in his own life, include a certain detachment, both emotional and mental, in his perception of others, a desire for an overview, and a pervading skepticism, which prevents him from giving his allegiance to any cause or relationship. Temporarily, he participates in the grandiose marketing plans of his uncle or romantic relationships (with Marion and Beatrice), but his enthusiasm is always qualified, distanced. Love is seen as a distraction, a rival for the scientist's attention, and George is presented as victoriously detached after an unsatisfactory marriage and an enervating affair. Science is explicitly personified in terms of a mistress who, unlike her rivals, fulfills her promises: "Scientific truth is the remotest of mistresses, she hides in strange places, she is attained by tortuous and laborious roads, but she is *always there!* Win her and she will not fail you; she is yours and mankind's for ever. She is reality, the one reality I have found in this strange disorder of existence" (239). The pursuit of science, which demands discipline, both physical and mental, is also contrasted with self-indulgence, as embodied in both George's uncle and British society generally.

George is also skeptical of conventional religion. All the representatives of religion in the novel are treated with irony as reactionaries identified with traditions that are seen as irrelevant and powerless to adjust to the necessary changes in society. Instead, George substitutes science as a secular religion akin to a revelation and pursues it with the zeal of a convert, regarding all other systems of allegiance as mere illusion. For him the only surviving reality is science. In later writings, Wells was to make the connection more explicitly, even mystically. In the essay "Religion and Science" he specifically identifies the scientist with a religious pursuit: "The scientific worker, whatever his upbringing may have been and whatever sectarian label he may still be wearing, does in fact believe in Truth—which is his God—in a God who is first and foremost Truth and mental courage. His life business is unfolding the divinity in things, and the real conflict is between the Truth as he unfolds it and the priests and exploiters of the false Gods who still dominate most men's lives."[44]

Scientists were the only members of society whom Wells associated with an acceptance of change, but change almost inevitably involves destruction, and this, it seems, the scientist must accept without regret. George registers no nostalgia for the passing of tradition, just as he has no hesitation about shooting the native who threatens to interfere with the "quap" project. It is symptomatic that the final symbol of science in the novel is George's destroyer, X2, which he has sold without any patriotic scruples to a non-European power. Hence George, in the final analysis, shows a ruthlessness emanating from his belief in the principles of science that differs only in degree from that of Moreau and Griffin. The ambiguity with which he is presented, compared with the clear authorial condemnation leveled against the earlier characters, is indicative of Wells's own changing attitudes towards science.

It is noteworthy that concern to protect the environment from the abuses resulting from science and technology, a theme that has become so prevalent in the literature of the latter half of the twentieth century, is almost completely absent from nineteenth-century writing. An interesting exception is the prophetic *Pfisters Mühle* (1884), by Wilhelm Raabe, which examines the question of pollution emanating from a chemical firm and the problems confronting a young chemist, Dr. Adam Asche, who tries to deal with it. The mill of the title has been converted to a restaurant, but this is threatened by the dead fish and stench resulting from the industrial waste dumped upstream by a sugar factory. Asche, while committed to the development of chemical science, also takes a humanistic approach to social problems; in this particular case he achieves a partial solution, so that as an individual scientist he is presented in a favorable light, combining science and justice. However, the overall effect of the novel is pessimistic, suggesting that the problems of industrial pollution and their devastating consequences on individual lives and on the environment will proliferate in the technological society of the future.

Trenchant as they are, these skeptical treatments of scientists and technologists in the late nineteenth century are little more than a thoughtful pause between the unthinking exuberance of Verne's adventurer-scientists and the frank glorification, in the first decades of the twentieth century, of the new scientist hero, either as inventor or as leader-elect of a future utopia. These characters are the subject of the next chapter.

CHAPTER ELEVEN

THE SCIENTIST
AS HERO

The engineer himself neither sighed for earth nor prayed to
heaven—he was a man of science. His long white fingers did not
quiver, his pulse did not increase a beat. . . . "We are now outside
the Earth's atmosphere," Barnett said quietly.
—Robert Cromie

Scientia redemptor mundi.
—Martin Atlas

Despite the misgivings voiced by the writers discussed in chapter 10, the
beginning of the twentieth century was characterized by a wave of optimism. *Fin-de-siècle* weariness and pessimism were swept aside by a renewed
belief in progress that mushroomed along with the material triumphs of
technology. Carl Becker has commented that to nineteenth-century Europeans, "progress was not so much a theory to be defended as a fact to be
observed";[1] this was even more true in the first decades of the twentieth
century. The almost unquestioning faith in progress was inextricably bound
up with the belief that the products of science would inevitably lead to a
better world with better human beings, better government, and only as
much departure from perfect happiness as might be necessary to avoid
universal boredom.[2] This confidence in progress as unfailingly good was
most pronounced in America. Whereas Europeans, with their longstanding
literary tradition of a golden age located irretrievably in the past, were
aware of those cultural values that had been marginalized by science,[3] for
Americans scientific progress was basically a "conservative" idea insofar as
it was an extension of the beliefs and values that had characterized American culture from the beginning of European settlement.[4]

It was almost inevitable that the cult of progress should be associated
with science, for they shared the same values, namely, an orientation towards the future, the ideal of objectivity, rejection of external authority,
belief in the increase of knowledge, and the corollary that such knowledge
would inevitably be for the betterment of mankind—all Sir Francis Bacon's

ideals, decked out even more attractively and forcefully in modern dress. As early as 1872 Francis Galton, a prominent Darwinian, had envisaged "a sort of scientific priesthood" that would provide the teachers, researchers, and government administrators of the not too distant future.[5] During the 1870s and 1880s T. H. Huxley, the eminent biologist, and Norman Lockyer, the editor of *Nature,* used their public positions to insist that good government and sound politics should be based on the principles of science and carried on by those with scientific training.[6] Scientists were encouraged to pontificate on all matters social, political, religious, educational, and moral, as well as technological.

Within such a social framework, scientists became the new cult heroes. This uncritical acceptance was reflected in a large number of literary works spanning the period from the 1880s to the Second World War and later.[7] Most of the authors had little knowledge of science beyond what was reported in the popular press, but since this was equally true of their readers, there was little if any pressure on them to research the areas they wrote about so glibly. The scientists in these stories reflect the popular credulity about the limitless power of science for good, and few questions, if any, are asked.

The portrayals of heroism vary considerably, from the simple stories of inventors adventuring into space to psychological dramas in which the scientist hero agonizes over what is the best government for utopia. These heroic scientists will be discussed under four headings—the scientist as inventor, the scientist as world savior, the scientist as detective, and the scientist as utopian. Because of the enormous proliferation of science fiction and films in the twentieth century, it will be necessary in this chapter to refer to representative examples rather than to attempt a comprehensive survey. Again, because of the ephemeral nature of much science fiction and the films based on it, in some cases a brief plot summary is necessary to comment on the scientist character.

The Scientist as Inventor

The inventor whose discoveries prove to be of surpassing benefit to mankind is a feature of American rather than of European literature, where the theoretical or philosophical scientist continued to embody the scientific tradition. In the New World, which still espoused many frontier values, the inventor who typified the ideals of progress and adventure was perceived as a vital member of society, while the pure scientist was regarded as not only less useful but potentially less trustworthy, even sinister, tainted with the moral flaws of the alchemist.

Moreover, America's relatively brief history of science already boasted such prominent inventors as Benjamin Franklin, Alexander Graham Bell, George Eastman, Nikola Tesla, and Thomas Edison, all of whom had attained the status of national heroes. Edison, in particular, became a cult figure, his life story from poverty to fame carrying considerable appeal in the egalitarian climate of the time. Although he received only three months' formal schooling in his life, Edison eventually established the first industrial research laboratory, his "invention factory," as he called it, and by the end of his life had more than one thousand patents to his name, including patents for the electric light bulb, the phonograph, the carbon-resistance telephone transmitter, which greatly improved on the audibility of Bell's model, and the kinetograph, the first fully effective motion picture camera.[8] All these adjuncts to daily living kept Edison's name and fame before the general public to a degree that was unique at the time and has scarcely occurred since.[9]

Given this adulation of the inventor, it is not surprising that American writers at the turn of the century incorporated actual inventors as characters in their fiction. In *Edison's Conquest of Mars* (1898), for example, Garrett P. Serviss enlists not only Edison but Lord Kelvin and Wilhelm Röntgen as well. Authority was assumed to emanate from the mere mention of such characters.

Although their living namesakes were firmly grounded on the Earth, the fictional inventors quickly became associated with another turn-of-the-century preoccupation, Martians. The hypothesis that there was a Martian civilization had achieved popularity as a result of Giovanni Schiaparelli's discovery of the grooves, or *canali*, of Mars from the Milan observatory in 1877. Schiaparelli's word *canali* was mistranslated as "canals," thereby implying in popular understanding a construction by intelligent beings. From this suggestion, the astronomers Camille Flammarion in France and Percival Lowell in America developed elaborate theories about the purpose of such canals, suggesting that they might be an irrigation system for conveying water from the polar icecaps of Mars to the arid regions around the Martian equator. They also speculated about the nature of the putative Martians and the state of Mars itself. Lowell devised a scenario that was widely believed to be factual. According to this scenario, Mars was an aging planet whose inhabitants would soon have to flee and colonize another planet. What place more likely than their neighbor, Earth, with its favorable conditions? Within a short time two streams of Martian literature had sprung up, one an extension of the journey to the moon and one in which the Martians invaded Earth, the latter swelling the general torrent of "invasion" romances precipitated by Robert Cromie's *Battle of Dorking* (1871), an apparently realistic account of the invasion of England by aliens.

The journey-to-Mars romances assumed that man was still supreme in the universe; therefore in these stories the inventors of the new, improved spaceships could afford to be gracious galactic ambassadors. One of the most popular was Robert Cromie's adventure story *A Plunge into Space*

(1890), describing a journey to Mars in a spaceship driven by differential use of an anti-gravity shield. Cromie's hero, Henry Barnett, is a scientist who after twenty years of seclusion has discovered "the origin of force." Although he is introduced as somewhat comic and out of touch with reality, in the manner of the absent-minded professor, he soon emerges as an intrepid physical hero as well as an intellectual giant, directing the space flight with consummate bravery and dignity. "Some of those below drew a deep breath and half muttered a prayer. The engineer himself neither sighed for earth nor prayed to heaven—he was a man of science. His long white fingers did not quiver, his pulse did not increase a beat."[10] Barnett is also a moral hero. Of the whole party, he alone is able to understand and appreciate the ethically advanced social system of the Martians, who have outgrown aggression and competitiveness. He not only brings his party safely back to Earth but heroically grapples with one member who has become insane, saving the others at the cost of his own life. In this story Barnett is portrayed as a neo-Renaissance, courtly gentleman of science, the only Earthman worthy to be compared to the socially advanced and highly ethical Martians.[11]

In Garrett P. Serviss's romance *Edison's Conquest of Mars* (1898), Edison too is unfailingly chivalrous to his Martian enemies. When the commander of the expedition tells the emperor of the Martians, "We have laid waste your planet, but it is simply a just retribution for what you did to ours. We are prepared to complete the destruction, leaving not a living being in this world of yours, or to grant you peace, at your choice," Edison counsels mercy, and the "venerable Lord Kelvin, who, notwithstanding his age, and his pacific disposition, proper to a man of science, had behaved with the courage and coolness of a veteran in every crisis," is also a model of rectitude.[12]

Edison's putative nephew Frank features as the young hero of Weldon Cobb's *Trip to Mars* (1901), serving his apprenticeship as assistant to Nikola Tesla, another real-life popular inventor hero.[13] Tesla, a Croatian immigrant to the United States, not only achieved fame for his flamboyant and spectacular public demonstrations of high voltage effects but claimed to have received intelligible signals from Mars. Cobb was only reflecting popular opinion in his characterization of Frank, whose attitude is that of an adoring acolyte towards the "great wizard of science . . . the master genius of the twentieth century. . . . Frank felt he had reached the acme of his ambitions. He could not speak for awe, and he could not move . . . for sheer emotion."[14]

By contrast, in the stories of Martians arriving on Earth, the confrontation was naturally seen as intrinsically more threatening than when men from Earth landed on Mars, and writers of invasion stories were therefore less likely to extol the virtues of our planetary neighbors. In H. G. Wells's *War of the Worlds* (1898) the Martians, fleeing from their dying planet, are as ruthlessly bent on colonizing Earth as the Europeans had been in establishing an empire (see fig. 21). They easily overpower the disorganized English, who are intent only on saving themselves, and finally are defeated unheroically by a simple bacterial infection to which humans have built up an immunity but which kills the bloodsucking Martians. Wells's training as a scientist is apparent in any comparison of *The War of the Worlds* with other contemporary Martian stories. Given the premise of the existence of intelligent life on Mars, the sequence of events in his story is logically impeccable.

The prolific fiction of Martian confrontation after Wells was to have a long-lasting effect on the character of the scientist in space. The fear of invasion spelt the demise of the gentleman ambassador desirous of learning from his hosts. The adolescent readers of pulp fiction wanted human heroes with whom they could identify, and the lurid cover illustrations of Hugo Gernsback's magazine *Amazing Stories* and its later rival *Astounding Stories*, with their depictions of space cowboys slaying galactic Indians, both fulfilled and promoted this expectation.[15] The protagonists were required to battle against evil and aggressive Martians and, of course, to win. Most of these pulp stories employ as their heroes scientists who invent either novel means of traveling to Mars in order to subdue it or new and exotic weapons with which to combat the Martians.

Although scientists and engineers had been employed as weapons designers for centuries, this was the first time they had featured in this way in literature. It paved the way to the expectation prevalent during the First World War (and even more pronounced in the Second) that the scientist's patriotic duty was to direct his expertise towards the construction of ever more subtle forms of killing. Certainly in these early works no shadow of blame attaches to the scientists, whether they initiate an attack or recommend mercy.

In retrospect, these scientist-inventors seem crudely conceived, preoccupied with a relatively narrow range of devices—interplanetary travel and communication, new sources of energy (an emphasis triggered first by the Curies' discovery of radium in 1902 and later by accounts of the potential of Rutherford's splitting of the atom), and new kinds of weapons to defeat evil invaders, whether earthly or alien. But they answered the deeply ingrained desire of their (mostly young) readers to transcend the limitations of the material world, for in a time of relative peace scientists replaced military heroes as the conquerors of a new kingdom.

This was also an important factor in the phenomenal success during the 1930s of the science fiction pulp magazines, whose appeal resulted largely from their ability to convey a heady sense of the scope and grandeur of the world that science might unlock.[16] Science fiction told its readers that they too could transcend seeming physical limitations of space and time through

FIGURE 21.
Cover illustration
by Frank R. Paul for
Amazing Stories,
August 1927,
depicting a scene
from H. G. Wells's
War of the Worlds.

the power of science. In the tradition of H. G. Wells, their literary father, the
science fiction writers of the thirties invented the world that half a century
later would become reality, though at the time only their fans believed them.
They wrote about the increasing dependence of society on technology, nuclear and chemical weapons, mind-controlling drugs, the population explosion and its probable social consequences, and instantaneous global communications. It was not coincidental that the avid teenage fans of those
magazines (almost exclusively male) became the astronauts and rocket designers of the sixties and seventies, striving to emulate their former heroes.

The scientist characters of this pulp magazine format were presented
chiefly as adventurers, the successors of Verne's heroes, voyaging through
the galaxy, where no man had gone before, doing battle with the forces of
evil therein and breaking through spatial, temporal, and psychological barriers. Their latter-day equivalents are Dr. Who and his assistants and the

crew of the Starship Enterprise, among whom the scientists are indistinguishable from the rest. In these stories, science is both the metaphor and the rationale for adventure.

Almost without exception these scientists are aggressively male and represent a society of male elitism. If women feature at all in these pulp stories, they have a minor part, irrelevant to the real action and invariably subservient to male authority. Their role is usually that of the scantily clad victim menaced by some interplanetary monster, while the human hero is reduced to the role of voyeur, watching as the monster prepares to enact his suppressed desires (see fig. 22). Ironically this apparently radical and forward-looking genre was in fact highly conservative, even reactionary, in its presuppositions about society and human relations—never a strong emphasis in science fiction. Its strongly sexist and hierarchical structure was to become the prototype for American science fiction for half a century, leading Ursula LeGuin, now established as one of the greatest writers of science fiction, to describe it as "a perfect baboon patriarchy." "From a social point of view most SF has been incredibly regressive and unimaginative. All those Galactic Empires, taken straight from the British Empire of 1880. All those planets—with 80 trillion miles between them!—conceived of as warring nation-states, or as colonies to be exploited, or to be nudged by the benevolent Imperium of Earth towards self-development—the White Man's Burden all over again. The Rotary Club of Alpha Centauri, that's the size of it."[17]

The Scientist as World Savior

More frequently the scientist hero of early twentieth-century popular fiction not only produced some beneficial invention but used it to save the world—or at least the part of it that was regarded as important by the author and his reading public—from an evil enemy. The nature of the enemy varied greatly in different periods, from an interplanetary alien or hostile European, Middle Eastern, or Asian power to atomic war or a natural disaster, which might range from a sudden catastrophe to entropic death of the planet. It was this wide spectrum of hypothetical hostile situations that permitted the scientist-savior (always male) to endure as long as he did. The rapid social and technological changes and large-scale historical events of the first half of the century, including two world wars, offered continual scope for new kinds of heroism to emerge. The individual scientist locked in combat with the champion of evil, whether terrestrial or intergalactic, gave way to the scientist saving society from dirt, inefficiency, and war, and this figure, in turn, was overtaken by the purveyor of nuclear power guaranteed to drive the world's industry and usher in "atomic" peace and prosperity for all. Yet all these characters are essentially similar in their conception, representing both an acceptance of Bacon's dictum that knowledge is power and

FIGURE 22.
Illustration in the
first issue of *Marvel*
(October 1938),
showing an
extraterrestrial
attacking a woman
while the hero
watches,
temporarily
powerless to help.

his optimistic assumption that knowledge is also, by definition, good.

One of the first novels to feature a scientist-savior in this sense was Bram Stoker's Gothic horror story *Dracula* (1897), in which Professor Van Helsing, although somewhat upstaged by his more famous opponent, saves England and, by extension, Europe from the invasion by Count Dracula (significantly of Eastern European origin) and his vampire hordes. Van Helsing is a particularly interesting and unusual character in that while he is presented as both a medical doctor and a famous scientist, he is also critical of the narrow rationalist methodology of Victorian science. Another character introduces him as "a philosopher and a metaphysician, and one of the most advanced scientists of his day; and he has . . . an absolutely open mind."[18]

His "absolutely open mind" does not lead Van Helsing to reject the talismans of religion (as Victorian readers might have expected); rather, he employs them along with the remedies suggested by science, both old and new. He invokes the powers of hypnotism, blood transfusion, and the herbal remedy of garlic, along with the crucifix and the consecrated Host. Far from surrounding himself in the clinical atmosphere of the laboratory, he identifies his mission with a chivalric and essentially religious past: "We go out as the old knights of the Cross . . . to set the world free" (156). Such an alliance of science and religious tradition is atypical, especially in conjunction with a scientific savior such as Van Helsing ultimately proves to be, and reflects Stoker's own predilections rather than any contemporary scientific attitudes.

The scientist hero who demolished Martians and vampires soon evolved into the scientist as military hero, ensuring that the right side won in a world war and bringing peace through scientifically generated power. These heroes are almost exclusively American in origin, and the stories in which they appear frequently end with the setting up of an American world empire—a strange ideal for a republic so proud of its revolutionary past, but suppressed imperialist dreams could be justified on the pretext of ushering in a new world order and Pax Americana. An early example of such stories was Stanley Waterloo's *Armageddon* (1898), in which an American inventor produces a dirigible bomber capable of destroying the navies of an evil European coalition massed against the "good" Anglo-American forces. Even more blatantly than in Serviss's novel, it is assumed that the home team is morally justified in using any weapons available, however destructive. The American scientist, "working hard to perfect a deadly machine, destructive beyond all others invented," enunciates in effect the doctrine of mutual assured destruction, which was to become a central linchpin of U.S. policy during the cold war. "To have a world at peace there must be massed in the controlling nations such power of destruction as may not even be questioned. So we shall build our appliances of destruction, calling to our aid every discovery and achievement of science. . . . when it [war] means death to all or the vast majority of all who participate in it, there will be peace."[19]

The dangers of this nineteenth-century, saber-rattling philosophy were to become fully apparent half a century later when these same sentiments, dressed up as the doctrine of deterrence, continued to influence public opinion and were enthusiastically embraced as self-evident justification for the full-scale involvement of science in the arms race. It is illuminating to compare the opinions concerning world politics held by many of the scientists involved in "Star Wars" research of the 1980s[20] with those voiced by Waterloo's American victors at the conference of nations: "We consider ourselves the approved of Providence in directing most of the affairs of the world" (242). From this position of self-righteousness it was only a small step to the assumption that possession of ultimate power justified world domination by the scientists in order to keep the peace. Most of these scientifically based world states rely on the development of some new form of energy, which, it seems, has first to be used in a destructive mode in order to clear the ground for the good society. Two of the most popular of these utopian adventures are enough to suggest the common features of a large group of such stories.

In Simon Newcomb's *His Wisdom, the Defender* (1900), Campbell, a Harvard professor of molecular physics, discovers an anti-gravitational substance "etherine" and a "thermic engine" (the forerunner of a nuclear power plant) which powers cars on minimal fuel, thereby precipitating a new industrial revolution. Such a monopoly of the source of power was frequently used by evil fictional scientists to hold the world to ransom, but Newcomb, himself a distinguished astronomer and professor of mathematics in the U.S. Navy, assumes that his physicist will necessarily be unfailingly benign and paternal. He also believes that absolute authority, vested in a noble scientist somewhat resembling himself, is the only way to ensure that the world is kept in order. Determined to enforce peace upon the world, Campbell, with no suggestion of authorial irony, announces to the nations, "It became evident to me that if I could retain in my own hands the power to guide the revolution, I could bring about its benefits without its attendant evils. To do this my power must be absolute."[21]

Having used his unbeatable force against each of the European nations in order to push through his plan for a federated world state under his own sovereignty, Campbell proclaims himself Defender of the Peace and proceeds to abolish war and colonialism. In this sense, his world federation is utopian and in the forefront of many similar treatments in which scientists not only furnish the vision and the means to usher in utopia but also proceed to govern it with wisdom and benevolence. Like Wells, Newcomb found it necessary to resort to force in order to instigate his utopia but assumed that once it was set up, the scientific lion would lie down happily enough with the democratic lamb in a state of permanent efficiency and bliss.

No such alliance is made, however, in J. Stewart Barney's popular novel *L.P.M.: The End of the Great War* (1915). Like Newcomb's Defender, his scientist protagonist Edestone, a thinly disguised Edison, perfects an ultimate weapon and thereby forces the nations of the world to the conference table. However, his world state has no place for democracy, which he ridicules for its inefficiency. Instead he sets up a limited Aristocracy of Intelligence, with an absolute dictator at its head. The apparent incompatibility of democracy, perceived as inefficient, with the goals of science had troubled H. G. Wells, but it is clear that Barney, at least, had no qualms about his autocratic stance.

In numerous stories and novels of the first decades of the century a world crisis, often precipitated by a "mad" scientist (usually a central European who discovers ultimate powers of destruction and holds the nations to ransom), is averted by a noble scientist, who outwits his evil counterpart. Hollis Godfrey's novel *The Man Who Ended War* (1908) is typical of these in featuring the almost obligatory "mad" scientist who develops weapons of mass destruction and starts a world war that bids fair to end life on the planet until a heroic American scientist intervenes just in time.

The much publicized approach of Halley's comet in 1910 inspired a flood of catastrophe novels and stories, some of which made use of scientists to save the world, or part of it, from imminent natural disasters. In Garrett P. Serviss's "The Second Deluge" (1912) a scientific genius, modestly named Cosmos Versal, predicts that a watery nebula will engulf the Earth and builds an ark to save a select band who will repopulate the Earth after the danger has passed. Versal, of course, has sole choice of the people he will save, and there is no intentional irony implied in his statement that he will "begin with the men of science. They are the true leaders." Versal remains the leader of the new society, to which he "taught the principles of eugenics and implanted deep the germs of science."[22]

Only occasionally are the evil scientist and the noble hero combined in the one person, as in the German novelist Bernhard Kellermann's *Der Tunnel* (1913), an interesting novel that juxtaposes with the heroic scientist many of the elements of the *fin-de-siècle* skepticism about science examined in chapter 10. Kellermann's protagonist is an American engineer, Mac-Allan, who idealistically determines to build a sub-Atlantic tunnel in order to foster worldwide understanding. Although his ruthless obsession with the project takes an enormous toll of money and lives, including those of his wife and daughter, MacAllan persists, and the tunnel, once completed, is hailed as the greatest human feat of all times.[23] The Faustian elements of MacAllan, his determination, arrogance, and willingness to sacrifice others to his project, are finally subsumed in his triumph as "der Odysseus der modernen Technik."

During the First World War the heroic scientist was pressed into service for the war effort on the home side, while his evil counterpart was invariably located in the enemy camp. Thus the age-old tournament between good and evil was briefly entrusted to scientific champions. But with the cessation of war, the scientist was immediately recast as the architect of a new and peaceful society. This had already been foreshadowed in many of Wells's utopian stories. A common Wellsian scenario is one in which scientists end the wars started by others and usher in a reign of uncontested peace; the grateful public is then only too happy to embrace the principles of pure science, rendering a military autocracy redundant. Indeed, in *The World Set Free* (1914) Wells even went to the unprecedented lengths of having his scientist Holsten, who had just worked out the mechanism of a nuclear chain reaction, decide to suppress the knowledge in case it was used for "atomic bombs" (a phrase coined by Wells). Despite his efforts, a world war does break out in the fictional world of the novel (set forty years ahead, in the 1950s), and atomic bombs dropped from planes lead to the near extinction of civilization.[24] Holsten, however, survives to advise the few remaining rulers of the nations about the setting up of a utopian world state based on the principles of science and fueled by the same energy source that had produced the bombs. The message was unequivocal: scientists alone could turn the potentially evil power to good.

In *The Man Who Rocked the Earth* (1915), Arthur Train and Robert Williams Wood also have recourse to a peace-loving inventor, who signs himself Pax and uses his power to stop the world war that, appropriately enough for the date of publication, is raging at the beginning of the novel. Pax, who has discovered how to concentrate and direct nuclear power of such magnitude that, as a mere test case, he can flatten the Atlas Mountains, divert the Mediterranean into the Sahara Desert, and shift the earth's axis of rotation, is depicted as being inspired only by a noble motive—to stop the war. Predictably there is a strong nationalist bias on the part of the authors: the Germans are seen as untrustworthy, reneging on the armistice in order to test their own new weapon, while the Americans are unfailingly honorable and are selected by Pax to arbitrate on the future of the world.[25] On Pax's death, another American scientist is appointed Dictator of Human Destiny to cooperate with the U.S. president in preserving world peace. There is no suggestion that he or his successors might prove unequal to the task of being morally responsible for the world.

The interwar period also produced a spate of novels in which scientists, either individually or in collaboration, were depicted as establishing lasting world peace. These noble scientific heroes have an epic stature little short of godlike as they systematically father a new race and become the acknowledged saviors of mankind. Few people recalled that every one of these attributes had been ascribed to Frankenstein. The means of effecting world peace range from a show of lethal weapons, with which the noble scientists smartly bring the rest of the world under control,[26] to plans for reforming governments by the setting up of an international brain trust of scientific advisers.[27]

In line with this latter action, British and American scientists after the First World War were quick to point to their contribution in ending the war successfully and to capitalize on their strategic importance. Along with their counterparts on the Continent, they endeavored to persuade their governments that it was vital to endow scientific research if they were to remain invulnerable in any subsequent hostilities. The governments of the major powers, even while paying lip service to the League of Nations charter, funded military-oriented science to a degree hitherto unknown. This practice found vigorous support in large numbers of novels and stories of the period. It has been calculated that whereas before 1914 two-thirds of fictional apocalypses were assigned to natural causes, after that time two-thirds were attributed to humans, and of these, three-quarters came in the form of world wars involving scientific weapons.[28]

Not surprisingly, in most of these stories written between 1915 and the early 1930s the villains are terrestrial and the danger is a world war, but during the 1930s, before it was apparent that another major war was looming, the dangers from which scientist-saviors were required to save the earth were often extraterrestrial in origin, either physical disasters or evil aliens. One of the most widely known was Alex Raymond's comic strip *Flash Gordon*, which featured the stereotypical eccentric scientist Hans Zarkov, Flash's confederate in fighting the evil Ming the Merciless from the planet Mongo.[29]

In two classic novels of the 1930s by the science fiction team Edwin Balmer and Philip Wylie scientists are shown as not only predicting a natural catastrophe but preserving at least a section of mankind from destruction. *When Worlds Collide* (1932) begins with the discovery by an astronomer, Bronson, of a dead sun, Bronson Alpha, which, together with its planet Bronson Beta, is about to plow through the solar system, Bronson Alpha being on a collision course with Earth. Although there is no chance of saving the planet Earth, a group of scientists, the League of the Last Days, led by "the world's greatest physicist," Cole Hendron, plans to save a remnant of the human race by transferring five hundred of the best brains and fittest bodies—scientists feature largely in the former category, if not the second—to Bronson Beta as it passes close to Earth. Despite natural disasters and attacks by the millions of desperate people not selected for salvation, Hendron coolly guides his two rocket-powered spaceships to safety on Bronson Beta, having witnessed en route the catastrophic collision of Bronson Alpha and the Earth. The sequel, *After Worlds Collide* (1934), describes the setting up of a stable society on Bronson Beta governed by wise and incorruptible scientists.

It is apparent from these two representative novels and from the spate of pulp fiction featuring a similar scenario that the same personal qualities that had been judged so harshly by the Romantic writers are presented a century later as undeniable grounds for approval. The heroic scientists are characterized by their imperturbability in the face of impending disaster and especially by their refusal to be swayed from rational decisions, made on statistical grounds, by emotional considerations. This suggests that because, at this period, only scientists were perceived as being able to confront and modify impending natural disasters (apparently even disasters on an astronomical scale), the characteristics associated in the popular mind with their ability to do this (namely, their rationality and their cool-headed, even ruthless, decision-making abilities) were reassessed as noble, elevated qualities, while emotional involvement and humane considerations were correspondingly downplayed. This was a dangerous message to impart to the impressionable teenage readers who would become the powerful scientists of the next generation. Some of its repercussions may be seen in the attitudes of the nuclear club, the scientists who worked on the Manhattan Project during World War II and played with the production of a weapon that had a

10 percent chance (considered acceptable) of exploding the earth's atmosphere. William Broad's interviews with the predominantly young scientists involved in "Star Wars" research indicated the perpetuation of these attitudes into the next generation of atomic scientists.[30]

An interesting and more humane variation on the world-savior model is to be found in Rudolf Daumann's *Protuberanzen* (Prominences) (1940). Here the impending fictional disaster (emanating from the ongoing anxiety about increasing entropy) is a terrestrial-based one, the onset of another ice age, which threatens to engulf all existing civilizations. Albin Hegar, Daumann's chemist-hero, predicts the imminent catastrophe but is disbelieved by his fellow scientists. He therefore has to persevere against prejudice and ridicule before he can carry out his plan to melt the glaciers. Daumann presents his hero as a new Galileo taking his righteous stand against reactionary self-interest and limited understanding and declaring that the judgment about the rightness or wrongness of his proposals can rest only with an "authority higher than that of men, Nature itself."[31] Hegar is thus the noble scientist who accepts the responsibility conferred by his wisdom and finally succeeds in saving the world by means of his cool-energy system, seemingly a precursor of "cold fusion."

During the 1930s a new source of potential heroism became available to the scientist—atomic power. This seemed to offer a means of combating not only the plots of evil maniacs but even the cosmic threat of increasing entropy. In these early years there was no suspicion that the benefits might be flawed. On the contrary, the extravagant miracles with which atomic power was credited were endorsed by many of the most respected scientists, including the Nobel laureates Robert Millikan[32] and Frédéric Joliot-Curie, so it is hardly surprising that many novels and stories of the period reflect this euphoria. "Atomic Power" (1934), by Don A. Stuart (a pseudonym for John Campbell, Jr., editor of the most successful science fiction series, *Astounding Stories*), epitomizes this optimism. In this story the Earth is undergoing entropic death, expanding into quiescence, with consequent panic on the part of its inhabitants. Enter a brilliant physicist who discovers atomic power and recharges the energy of the planet, restoring life, health, and happiness to its frozen peoples.[33]

Malcolm Jameson's novel *The Giant Atom* (1944) is possibly even more naive in its combination of the scientist as world savior and intrepid space traveler. When the General Atomic Corporation takes over his research and recklessly constructs a "reintegrator," which functions as a kind of black hole, devouring everything in its path, Jameson's hero Bennion builds an atomic-powered ark, the *Star of Hope*, to convey the chosen few to another planet. Luckily, this Noah escapade proves unnecessary, because Bennion removes the "flood" by transporting the reintegrator to outer space in an atomic-powered rocket. He returns to suitable acclaim as the savior of the world on a grand scale, embodying the power of the atom for good over evil.

By the 1950s, however, the development of atomic power was viewed with more skepticism. The side effects of radioactivity were becoming more widely known, and the unambiguous benevolence of nuclear physicists in promoting nuclear power was being questioned. What is more important, there was Hiroshima. After the bombing of Hiroshima and Nagasaki, it became increasingly difficult to portray scientists as necessarily having both the power and the morality to become world saviors and lead humanity on to a glorious future. In America, moreover, there was the morally awkward fact that former German scientists, some of whom had been prominent Nazis, were welcomed into the national pantheon of missiles-research scientists to boost the U.S. cold war effort. Wernher von Braun, who during the 1950s played an important role in developing American rocket technology, had been a leader in the group that had developed V-2 rockets for Hitler at Peenemünde. The moral implications of this did not bear too much scrutiny and were conveniently glossed over in the media; but although some scientists, notably the Chicago group, which included Leo Szilard, deliberately campaigned to reinstate scientists in the public confidence, arguing that they alone possessed the formula for universal brotherhood, they were never to recover more than a pragmatic respect.[34]

It is significant that at least two of the (few) post–Second World War scientist-saviors in literature reject the reductionist procedures of scientific materialism and instead have recourse to spiritual assistance. In *A Case of Conscience* (1958), set in the twenty-first century, James Blish takes as his protagonist a Jesuit biologist, Ramon Ruiz-Sanchez, who is sent with three other scientists to report on the civilization of the planet Lithia. First impressions suggest that the highly intelligent Lithians have developed a totally rational utopian society in which everyone is both free *and* good, a state of rational harmony far superior to any on Earth.[35] Unlike his fellow scientists, however, the priest soon realizes that the Lithian social system, based on pure reason, implies the perfectibility of humanity without reference to God and concludes that Lithia is no accident but a contrivance of the Devil to destroy mankind. Reprimanded by the scientific establishment for his adverse report and unsound reasoning, he appeals to the pope, who counsels him to perform the ritual of exorcism on the whole planet of Lithia in order to save humanity from such a blasphemy. Ruiz-Sanchez does so at the very moment when another scientist is performing an experiment in nuclear fission. The planet vanishes, apparently because of the fission explosion. "An error in Equation Sixteen," says a mathematician observer, but Ruiz-Sanchez and the author believe otherwise. What makes this novel unusual is the combination of scientist and priest in the one protagonist and the use of a scientist to save the world from a moral and spiritual danger rather than from a merely physical one. Ruiz-Sanchez is, in fact, a twenty-first century

Van Helsing, using the combined arts of religion and science to outwit a satanic Dracula and show up the inadequacies of his fellow scientists, rather like C. S. Lewis's good Christian Ransom confronted by the evil rationalist Weston.[36]

Even more startling and unusual than the scientist-as-priest is the scientist-as-pope, but Oskar Maria Graf's novel *Die Erben des Untergangs. Roman einer Zukunft* (The heirs of death: a novel of the future) (1959) features just such a protagonist, who resurrects the world after the devastation of war. Again, the miracles of science are shown to be consistent with the miracles of faith, and Graf's brilliant scientist-pope, whose research into strains of wheat solves the problem of world hunger (the first reference in fiction to the Green Revolution), remains a humble, idealized figure able to establish peace and harmony in his utopian society. Although this may seem an idealized fantasy, Graf himself stressed that he considered the attitudes and the proposed solution realistic; indeed, Einstein regarded Graf's novel very highly, as suggesting a possible path to world peace and prosperity.[37]

Increasingly, though, after Hiroshima, heroism came to be associated with the suppression, rather than the use, of new knowledge. Many writers suggested that even if the scientists themselves were ethical, they could not avoid being manipulated by power-hungry governments and the military machine. In such a situation, heroism consists in concealing potentially harmful knowledge from the authorities, a theme that was to provide the basis for the large number of novels and plays to be examined in chapter 16, in which the ethical scientist is in conflict with social forces and usually fails to protect his discovery from misuse. One example, in which the scientist is triumphant over the military-industrial complex, is relevant here.

In *The Genesis Machine* (1978), set in the year 2005, James P. Hogan develops a scientist hero who has to battle against the government-military machine. Forced to work on "defence," his physicist, Dr. Bradley Clifford, appears to capitulate and outlines a proposal for a superbomb, code-named "Jericho," or the "J-bomb," which, he assures his superiors, will annihilate any target instantaneously while being itself undetectable and hence unassailable. However, in a clever, logical twist to the story, Clifford reveals that he has programmed the Brunnermont machine (which produces the J-bombs) so that militarily it can be used only in defense, never for offense. Moreover, its surveillance system has a watchdog function to destroy any J-machine that might be produced by another power. Clifford has thus rendered all weapons of mass destruction harmless. The U.S. president remarks, "Doomsday Machines are supposed to guarantee the end of the world. I'd say that this does the exact opposite. . . . it is . . . A Genesis Machine."[38]

In all these stories and novels that feature a scientist-savior, the propagandist element dominates the characterization. In nearly every case the author's concern is to reaffirm the moral status of the hero, but rarely is there evidence of a moral struggle within the protagonist. As with so much science fiction, the interest of the story lies in the intellectual game of inventing a new idea, and the only struggle depicted is the external one of individual heroes battling an evil monolithic authority.

The Scientist as Detective

With the increased importance in judicial enquiry of forensic evidence there emerged a new social role for science, and the scientist-detective who solved a crime baffling to the police (a crime often perpetrated by an evil scientist) became a popular literary hero. The prototype was, of course, Sir Arthur Conan Doyle's best-known character, Sherlock Holmes. It is worth considering this problematic hero in some detail, for Doyle included in his personality many aspects of the traditional villain. Thus Holmes is characterized not only by a dedication to analysis and objectivity but also by his coolness and lack of social involvement.

In the first Holmes story, "A Study in Scarlet" (1888), Holmes is introduced as numbering, amongst his other accomplishments, a comprehensive knowledge of chemistry and as having discovered an infallible test for microscopic amounts of blood. Doyle repeatedly stresses, however, that it is not merely the knowledge and application of chemistry that constitutes scientific enquiry, but the logical processes of analysis and deduction. Along with his analytical cast of mind, Holmes epitomizes the distanced observer, coldly surveying humanity's weakness and stupidity; indeed, although the reader is intended to be dazzled by his intellectual brilliance, it requires all the warmth and unfailing admiration of his assistant Dr. Watson to make him even tolerably human. It is significant that the only person whose intellect Holmes respects is his opponent, Professor Moriarty, a stereotype of the evil scientist, yet described in terms that suggest Holmes's alter ego. Indeed, Holmes's enthusiasm for engaging in battles of wit with Moriarty suggests that his pursuit of criminals is motivated more by an intellectual interest in the chase than by any moral considerations: "He is the Napoleon of crime, Watson. . . . He is a genius, a philosopher, an abstract thinker. He has a brain of the first order. . . . I was forced to confess that I had at last met an antagonist who was my intellectual equal. My horror at his crimes was lost in my admiration at his skill."[39]

Richard Austin Freeman's popular detective, Dr. Thorndyke, a doctor of medicine as well as a lawyer, was another forensic scientist who resolved hitherto impenetrable mysteries by means of the same skills Doyle had popularized; and again, although Thorndyke is a less chilling personality than Holmes, the retrospective explanations in terms of logical deduction formed an important part of the story and the characterization. Stuart

White's novels *The Mystery* (1907) and *The Sign at Six* (1912) are also of this kind, but White offers three scientist types—the cold rationalist who fails to resolve the puzzle; the evil inventor who has discovered how to interfere with electricity; and the seemingly indolent, even effeminate hero, a latter-day Scarlet Pimpernel, who solves the mystery and, in stark contrast to Holmes, frankly attributes his success to his nonrational qualities. "A man must have imagination and human sympathy to get next to this sort of thing."[40]

Another interesting development of the forensic scientist in literature was the character Luther Trant, a psychologist who, in order to track down and convict criminals, allegedly developed the lie-detector. The authors' preface stressed the reality of the hero's methods: "The methods which the fictitious Trant . . . here uses to solve the mysteries which present themselves to him, are real methods; the tests he employs are real tests. . . . they are precisely such as are being used daily in the psychological laboratories of the great universities . . . by means of which modern men of science are at last disclosing and defining the workings of the oldest world-mysteries . . . the human mind."[41]

Trant's successor, Craig Kennedy, forensic-scientist hero of Arthur Reeve's series in *Cosmopolitan* from 1911 to 1915, made use of a whole series of gadgets to track down his criminals, who could "never hope to beat the modern scientific detective"; but generally, there is a minimal attempt to characterize the forensic scientists of literature beyond the pattern developed by Doyle. The connection with laboratory scientists usually resides in their recourse to deductive logic (which is tacitly assumed to be identical with scientific method) and their use of some strategic chemical tests, but overall there is little to distinguish these forensic scientists from routine crime-solvers such as Agatha Christie's Hercule Poirot. Nevertheless, they represent one of the few remaining means whereby an individual scientist can realistically function as a social benefactor, promoting law and order and ensuring that evil does not triumph.

The Scientist as Utopian Ruler

An extension of the scientist who saves the world from crime or disaster is the scientist who ushers in a new society where such evils could never exist. Despite the questioning of technological progress by the writers discussed in chapter 10, there were still those optimists who envisaged a benevolent technology leading to a happy future when all the current social and physical problems of mankind would be solved by the application of science. The genetic improvement of the race, the cessation of war, a limitless source of power, the socialization of science, and world economic planning to eliminate poverty bulk large among the achievements of these heroes.

One of the most influential writers of such technological utopias around the turn of the century was the German Kurd Lasswitz,[42] who argued, both in fiction and in two essays "Über Zukunftsträume I u. II" (Dreams of the future I and II) (1899), for unlimited confidence in technology and the overcoming of all social problems by means of natural science.[43]

The basis for many of these utopias is the development of some new source of power from which all other social benefits are alleged to flow, and in this sense they are not dissimilar to the utopias of his English contemporary Bulwer Lytton, who invoked the power of *vril* or some equivalent.[44] All these writers are reflecting the optimism that attended the discovery, first, of electricity and, later, of atomic power. Unlike the authors of the inventors considered earlier in this chapter, these writers are less interested in the inventions per se and more concerned to describe the resulting utopian society, seen as offering ample riches and leisure for all. It is characteristic of such utopias that scientists are depicted as being directly responsible for introducing a better quality of life, both physical and social. The desired system is usually conceived as being a world state governed by an intellectual-scientific oligarchy rather than by a dictator; there is thus an implicit suggestion that the military-based nationalism favored in the scientist-as-conqueror stories is inimical to a rational society. Despite superficial similarities, these scientist utopian leaders are also essentially different from the more complex idealistic scientists (see chapter 16 below), who suffer inner turmoil and, in most cases, outward defeat. For one thing, they are presented, for the most part, as instruments of technological progress rather than as individuals who embark on their ideological campaigns at great personal cost. Further, these heroic leaders are acclaimed by their society and invariably succeed in implementing their political systems to provide a universally benevolent society.

One of the earliest of these power-based utopias was Kurd Lasswitz's depiction of a future society controlled by benevolent scientists and technicians in *Gegen das Weltgesetz: Erzählung aus dem Jahre 3877* (Against world law: a story of the year 3877) (1878). The scientists and technicians have discovered how to feed the world's population by means of synthetic food production and have developed techniques for speed learning by thought transfer, thereby overcoming the problem of coping with the information explosion. Lasswitz's tales are almost diagrammatic in their propagandist zeal, and the idealized characters—Cotyledo the botanist, Funktionata the mathematician, Atom the physicist—represent particular disciplines rather than realistic individuals.[45]

Despite the literary deficiencies of most of this fiction, the image of the scientist as an exemplar of social responsibility, accepting the scientific equivalent of the white man's burden and ushering in a new society that would render war anachronistic, retained its popularity until the end of the

Second World War. These benevolent scientists, committed to technological progress in the service of mankind, are never tempted to use their power for selfish or evil ends, because the interest of the writers is in social theory rather than psychological exploration. Their simple didactic message is that a world governed by scientists would be as free from greed, inequality, and other moral imperfections as it would be from hunger, war, and pestilence. These stories are essentially variations on Bellamy's technological utopia in *Looking Backwards*, transplanted even further into the future.

Ironically, the best-known creator of science-based utopias was H. G. Wells, who, as we saw in chapter 10, was also one of the most trenchant critics of scientism during the last decades of the nineteenth century. Yet in 1901, as though to welcome the nascent century with a new message of hope, Wells published *Anticipations*, an enormously popular social prophecy about the changing lifestyles to be expected in the century just beginning. With this book Wells effectively became, for the British public, "Mr. Future," and much of his subsequent prolific utopian writing may have resulted from an attempt to live up to this reputation. *Anticipations* marked a turning point in Wells's thought, for although its ostensible method is one of extrapolation from existing trends (and it was this illusion that accounted for its plausibility), the success of his social system in fact hinges on a significant new factor, namely, the emergence of an altruistic scientific elite, the New Republicans, who, while understanding the dangers of technology, are capable of mastering them and who exercise a benevolent paternalistic control over the masses.

Because he believed that he now understood the means for introducing utopia, Wells discarded his former pessimism and embraced the regimen of Bacon's House of Salomon, suitably modified for the twentieth century, as the basis for a revised system of government, education, and social order. This new optimism about the benefits of science and technology was to inform nearly all Wells's subsequent fiction, both the utopias and the realistic novels, requiring him to recast his former morally dubious scientists as champions against the destructive, reactionary forces of disorder, inefficiency, and waste. He thus became the literary spokesperson for those contemporary scientists who were crusading vigorously either for scientists to have greater political influence or for all politicians to have scientific qualifications.[46]

The first fictional example of Wells's changed perspective is his short story "The Land Ironclads" (1903), which is concerned with an alleged military clash between English anti-intellectualism and non-English scientific expertise. The latter is embodied in a group of intelligent and efficient young engineers who, more in sorrow than in anger, speedily defeat the disorganized English patriots. The parallel with the Selenite society of *The First Men in the Moon* is clear, but the authorial position has shifted markedly.

Technological progress is now shown as inevitable, and Wells's contemporaries are counseled to accept it—out of necessity, if for no other reason. Wells was here providing fictional support for the major political campaign being waged by many contemporary scientists (including Wells's own hero, T. H. Huxley) who exploited the threat, both economic and military, posed by a technologically superior Germany in order to gain political influence and research funding.[47]

A year later, in *The Food of the Gods* (1904), technological change, symbolized as "bigness," is represented as not merely strategic but wholly desirable. Here, as in the subsequent series of blueprints for a better society, thinly disguised as novels, Wells endowed his scientist rulers with moral as well as technical supremacy. Like Bacon, he assumes that they will be permanently ennobled by the qualities both writers regarded as intrinsic to the pursuit of science—honesty, internationalism, altruism, and compassion—but Wells adds a new factor, efficiency, which he effectively invests with moral virtue. *A Modern Utopia* (1908), in many ways the prototype of all his utopian novels, contains Wells's manifesto on the primacy of technological efficiency in solving the world's problems:

> The plain message which physical science has for the world at large is this, that were our political and social and moral devices only as well contrived to their ends as the linotype machine, an antiseptic operating plant or an electric tramcar, there need now at the present moment be no appreciable toil in the world and only the smallest fraction of the pain, the fear and the anxiety that now make human life so doubtful in its value. . . . Science stands, a too-competent servant behind her wrangling under-bred masters, holding out resources, devices and remedies they are too stupid to use.[48]

Ironically, this cult of efficiency embodied in his ruling scientific elite, whom he calls Samurai, is the same criterion that Wells had viewed so skeptically in *The Invisible Man* and *The First Men in the Moon*.[49] Gradually, through his repeated insistence that the principles of science were the only universally accepted beliefs, Wells came to see science not merely as rationalism and efficiency but in quasi-religious terms, as a mystical power capable of uniting all mankind, as "something that floats about us and through us. It is that common impersonal will and sense of necessity of which Science is the best understood and most typical aspect. It is the mind of the race."[50] As Wells's prophecies became more mystical, his scientists became shadowy, remote figures, high priests of the new religion and increasingly remote from reality.

Wells was not the only writer of the period to pin his faith on scientists as rulers. A similar science-based utopian future was envisaged by the German writer Martin Atlas. His novel *Die Befreiung* (Liberation) (1910) describes a technological utopia, Penon, ruled by the benevolent, peace-loving scientist Siler, who has, of course, discovered the obligatory limitless source of power. Siler repeatedly stresses that social reforms can emanate only from the wisdom and ability of the noble scientists, who, like Wells's Samurai, constitute the intellectual aristocracy of the world state. Significantly, and without any sarcasm on the author's part, the Penonite motto is *Scientia redemptor mundi* (Science, redeemer of the world), a motto that now has a strangely ironic ring.

Indeed, despite the involvement of scientists during the First World War in developing ever more sophisticated means of destruction, including chemical weapons, several German writers of the interwar period continued to pin their faith on scientific progress as the one force that could hasten the arrival of utopia. Significantly, all these utopias were based on the assumption of a new energy source that would solve the world's economic problems and, by extension, all others as well. Hans Dominik's popular science fiction novel *Die Macht der Drei* (The power of three) (1922) discusses the matter almost diagrammatically, with three scientists representing three different attitudes towards the source of power they have discovered in the *Strahler,* a precursor of the laser. Dominik, himself an engineer and well aware of technical developments of his time, also predicted the widespread use of atomic energy, which he regarded as inherently beneficial, ushering in a new Wellsian age "in which the old Earth will become too small for Man and he will direct his steps towards other stars."[51] Of the three scientists in his novel, one is at first highly idealistic and wants to use the discovery only for the benefit of society; however, his warnings about the devastation that the *Strahler* would cause in a war are ignored, and he becomes vindictive, obsessed with power, and finally mad. A second figure is the typically disengaged scientist, immersed in science for its own sake and unconcerned about the consequences. Dominik's spokesperson, the third member of the trio, is an Asian philosopher-scientist who believes in a future when scientists will be powerful but socially responsible leaders, uncorrupted by the power they wield.

In his later work *Atomgewicht 500* (Atomic weight 500) (1935), a novel strongly reminiscent of Wells's *World Set Free* (1914), Dominik went on to elaborate on what might be involved in such a utopian society. Again the impetus for social revolution comes from the discovery of a new energy source, this time a new heavy element of atomic weight 500, which has the potential to be used either benevolently or as a means of mass destruction. Realizing this and determined to initiate a new age of beneficial technology, the discoverer, Wandel, steadfastly refuses to be intimidated by his scientific peers or manipulated by financial interests, and eventually he succeeds in gaining international cooperation for the production of energy for peaceful purposes.

Ironically, in these technological utopias, instigated and governed by scientists, the very features that were most enthusiastically propounded were to form the basis of the highly critical anti-utopias that characterized the following decades. The authors of *We* (1924), *Brave New World* (1932), and *1984* (1949) regarded with horror the devices and systems that had been hailed so optimistically as the hope of the future, because they saw efficiency and individuality as mutually exclusive and evaluated them in inverse order to their predecessors. Yet there was one more attempt to promulgate the Wellsian-style utopia based on efficiency, this time induced by psychological conditioning, namely, Burrhus Frederic Skinner's *Walden Two* (1948), written specifically to counter the stringent critique of behaviorist practices in *Brave New World*. Skinner, himself a psychologist and one of the main proponents of behaviorism, had a frankly propagandist motive in writing his fictional account of an ideal, nonpolitical society where behaviorist principles ensured that all its members would work for the happiness and fulfillment of the group and achieve those rewards for themselves in the process. In the standard pattern of utopian fiction, Skinner's narrator, Burris (an obvious echo of Burrhus), leads a party to inspect the community founded by a former academic, Frazier, another convinced behaviorist. Against his will, Burris is forced to assent to the superiority of Walden Two over all other societies, to admire the orderly but varied and stimulating life possible when every citizen voluntarily accepts and acts for the overall good of the community. Not surprisingly, Frazier is one of the few psychologists in literature to carry authorial approval. He claims the right to be judged as a scientist, but he is also the most powerful ruler imaginable, since his control of each individual's behavior is largely unrecognized. Although inevitably he remains a flat character as a result of the weight of propaganda he bears, Frazier represents an interesting attempt, unique in the twentieth century, to justify intellectually the possession of ultimate power, the right to decide what is "the best course for mankind forever." "I've admitted neither power nor despotism. But you're quite right in saying that I've exerted an influence and in one sense will continue to exert it forever. . . . I did plan Walden Two—not as an architect plans a building, but as a scientist plans a long-term experiment, uncertain of the conditions he will meet but knowing how he will deal with them when they arise."[52]

Like *Brave New World*, *Walden Two* is based on a denial of the concept of personal freedom, but unlike Huxley, Skinner is prepared to justify this. Frazier asserts, "We can't leave mankind to an accidental or biased control. But by using the principle of positive reinforcement—carefully avoiding force or the threat of force—we can preserve a personal sense of freedom" (220). Notwithstanding the vigorous defense of his theories, Skinner fails to answer the objections raised by Huxley, and Frazier bears a closer resemblance to such doubtful scientists as Giacomo Zapparoni in Ernst Jünger's satire *The Glass Bees* (1957)[53] than to the idealistic rulers mentioned earlier in this chapter. Furthermore, it is finally disclosed that Skinner's primary goal is, not to make people happy and productive, but to create a large-scale laboratory for the study of human behavior. "We can study [these things] only in a living culture, and yet a culture which is under experimental control. Nothing short of Walden Two will suffice. It must be a real world, this laboratory of ours and no foundation can buy a slice of it" (242). Thus, despite his overt role as idealist and world leader, Frazier represents a reincarnation of the Romantic stereotype of the scientist who subordinates humanity to his obsession for science. In this sense the name Walden Two, intended to suggest continuity with the ideals of the recluse naturalist Henry David Thoreau, carries an extra level of irony, of which Skinner was presumably unaware.

In all these high-technology utopias nature is regarded much as Bacon had regarded it, as an endless source of material and energy for exploitation. As this assumption has come to be questioned and the disastrous results of such disregard for environmental integrity have been recognized, the utopias based on technological domination over nature have come to seem naive. In the first quarter of this century, however, there was little hint of planetary fragility, and these utopian schemes were regarded uncritically as pointing the way to the future. Even when scientists themselves fell under a cloud, the necessity of progress through the divine right of technology remained unquestioned for much longer.

From these representative examples it becomes apparent that in British and European literature the disenchantment with the heroic scientist occurred, with few exceptions, almost immediately after Hiroshima. There are two principal reasons for this. First, in the face of nuclear war, unlike any other war, heroism becomes irrelevant. None of the traditional attributes of heroism—courage, self-sacrifice, even cunning—has any effect on an ICBM. Retaliation, even if it proves possible, is at best only a Pyrrhic victory: there are no winners of a nuclear war. Thus in modern warfare the scientist hero is rendered redundant. The real action is carried out by machines, not human beings, and machines are ultimately controlled, if at all, by politicians, not scientists. The best that scientists can do is attempt to outwit the military-industrial complex by recourse to its own strategies of military deterrence, bluff, and deception. Hogan's hero of *The Genesis Machine* (1978), discussed above, has recourse to this option in his "fake" J-machine. Second, and what is perhaps more interesting, scientists came increasingly to be seen as responsible for the nuclear menace. Robert Oppenheimer's memorable statement that "the scientists have known sin" was to echo through the corridors of fiction for decades.

In America the idealism persisted longer than in Europe. As the physical chemist Eugene Rabinowitch wrote in 1956, "In 1945 after the revelation of the atomic bomb, American scientists enjoyed a brief spell of popular respect and acclaim. It was they, most people realised, who had dramatically ended the war in the Far East; it was they, many believed, who held the keys to the future power and prosperity of America. Some went so far as to look up to scientists as men destined to rule the world in the 'atomic age.'"[54]

Among American science fiction writers, in particular, there was a mood of self-congratulation over the apparent fulfillment of their prophecies that atomic weapons would bring the nations to their senses and usher in a reign of peace and unbelievable prosperity based on atomic power. This chimed in well with the generally euphoric mood of a country that saw itself as the real victor of the war, having ended the conflict when its European allies had failed to do so and having emerged not only unaffected by the privations of war but as the richest nation in the world. The defeat of Japan, perceived in the United States as an evil, militaristic power, was taken as a mandate for noble American scientists to recreate the world in their own image. Atomic power, provided it remained in the safekeeping of Americans, would never be used for evil purposes, but only to promote universal happiness and prosperity.

In the United States this euphoria persisted until well into the 1950s. Then, with the launching of Sputniks I and II in 1957, Americans were shocked by the realization that the Russians not only had caught up with them in the development of atomic weapons but had actually preceded them into space.[55] Suddenly scientists, and in particular physicists, far from being perceived as sensitive, honest people racked by ethical torments about the misuse to which their theories were being put, were regarded as morally suspect. Either they were seen as deficient in national loyalty and even likely, in an excess of misguided internationalism, to give away their secrets to the Soviet bloc, as the atom spies had done, or they were retrospectively blamed both for their own part in the development of the weapons and for being unable to render the enemy's weapons harmless. In popular Western films, the role of the heroic savior in matters nuclear passed from the scientists to government agents (a curious but strategic inversion of traditional moral attitudes towards spying). Only another spy, it seemed, could be trusted to outwit the atom spies.[56] Scientists, by contrast, were increasingly portrayed as either irresponsible, caring only for science, regardless of social consequences, or helpless to put back in the bottle the evil genie they had so carelessly released.

CHAPTER TWELVE

MAD, BAD, AND DANGEROUS TO KNOW: REALITY OVERTAKES FICTION

> I'm a scientist . . . who's given his life to pure knowledge. . . . I
> have a chance of performing the last and greatest experiment
> known to science. To release the earth's energy to destroy—I
> hope in a flash—the life on it. . . . one last triumphant stroke . . .
> leaving the mindless cosmos to its own damned dance of blind
> energies for ever.
> —J. B. Priestley

The scientists' reign as heroes of twentieth-century literature was always open to question, and even the promise of a limitless power source was not sufficient to prolong the idea beyond the 1940s. But long before this, and indeed concurrently with much of the scientist-as-hero literature considered in chapter 11, mad and evil scientists holding the human race to ransom had become a commonplace of pulp fiction, horror films, and cartoons and not infrequently were found in mainstream novels and plays. Probably no other profession has provided modern literature and films with so many villains. From his extensive survey of the horror films produced between 1931 and 1960 Andrew Tudor has concluded that one-third posit science as the main cause of disaster, and of that third a large proportion present the scientists behind the science as evil.[1] Certainly mad scientists feature as one of the three most common sources of terror in horror films, the only other contenders being the supernatural and psychic disorder. Why, in a century where technology has been so highly regarded, should the Frankenstein revival have been a paramount feature of popular literature and film?

One clear reason is that the speculative fantasies concerning the evil scientist that we have traced from the medieval alchemist, and that existed in oral tradition long before that in the figure of the sorcerer or prehistoric shaman, had been overtaken by reality. By the first decades of the twentieth century the influence wielded by actual scientists as a result of their technological expertise was perceived as equivalent to, if not surpassing, whatever supernatural efficacy had been attributed to their magic-dependent fictional forebears. The writers to be considered in this chapter expressed the fears of their contemporaries that science and its products had the indisputable physical power to crush individuals and whole societies, even the entire human race. Partly because it still retained an aura of mystery, this power was presented as having an unstoppable momentum, beyond the control of ordinary citizens, and thus it inevitably raised the specter of what would happen if it should fall into evil hands. However, the reasons why it was more commonly represented at this time as being used for malevolent rather than benign purposes are more complex.

At the most basic level, evil characters in fiction are simply more interesting than morally impeccable ones; but a network of other reasons became particularly pertinent at this time. As we saw in chapter 8, one common pattern of response to the immensity of space and time revealed by nineteenth-century astronomy and geology and popular accounts of the heat death of the universe was a sense of powerlessness, loss of religious faith, even despair. As the century drew to a close, however, and the next generation grew familiar with these ideas, such responses came to be regarded as extreme, even naive. They were most often replaced by a cynicism that, discarding ethical perspectives along with religion, resorted to exploiting the here and now without fear of future retribution.

The frequent association of the evil scientists of this period with acts of terrorism whereby they hold a society to ransom suggests an implicit connection between the power of wealthy industrialists and the science-derived technology that enabled them to amass their wealth. Many of the mad and evil scientists in the first decades of the century are represented as extortionists demanding money in return for the use (or nonuse) of their inventions. These characters, who reenact on a world scale the tyranny practiced by H. G. Wells's Invisible Man, are the laboratory counterparts of ruthless capitalists exploiting their workers to gain yet more wealth, confident that they would not be disciplined. The literature in which such characters feature comprises, for the most part, suspense tales in which the interest revolves less around the development of character than around the means whereby some savior, often another scientist, saves society from the evil machinations of the maniac. Although these stories appear to suggest that society pays, in one form or another, for its own obsession with wealth and the technology that makes it possible, it is noteworthy that they characteristically have a happy ending. In this period writers and audience alike retain a residual belief in the long-term benefits of science and technology and the ultimate survival of humanity.

Rather more interesting than these power-obsessed characters are those whose malevolence and determination to destroy the world are alleged to arise from their frustration and despair that the universe is apparently nothing more than a chance eventuality or a flawed mechanism. When such characters have access to sources of apparently limitless power, their potential for destruction offers scope for scenarios of chilling horror.

Also emerging in the literature of this period is the first explicit connection between scientists and the machinery of war. Previously, evil scientists were depicted as individuals bent solely on their own advancement, but by the early twentieth century, by linking technology with the new and more effective military weapons, writers are signaling the increasing reliance of society on its scientists for the nationalistic posturings that were to lead to two world wars and the cold war. Some also refer obliquely to the manipulation of the war machine by scientists in order to obtain funding for their research by arguing that it is of strategic significance to defense. Thus many of the writers considered in this chapter foreshadow, albeit often in simplistic form, the later, more complex discussions regarding the changing relationship between science and the state.

Not surprisingly, the rise of the evil scientists in fiction is directly contingent on some new technology, usually a source of physical power, over which they hold a monopoly. Dynamite had been invented by Alfred Nobel in 1866, Wilhelm Röntgen had discovered X-rays in 1895, and by the turn of the century electricity not only was finding an increasing number of useful applications but was rumored to be capable of annihilating a whole city from a distance. Such a source of power was seen as synonymous with wealth and progress, the twin goals of Western society, but it was also recognized as having equal potential for destruction. Moreover, just as electricity, the "magic" force of the nineteenth century, was being demystified, another source of power, radioactivity, more strange than any before, was being publicized. Henri Becquerel had discovered radioactivity in radium in 1896, and two years later Marie and Pierre Curie had observed the same effect in radium and polonium. Everything that was popularly known about radioactivity seemed mysterious, uncanny, and closely related to the whole alchemist tradition. It glowed in the dark, it was associated with the transmutation of elements, it was capable of killing animals (Pierre Curie had killed a mouse with a dab of radium and told reporters that he would not wish to share a room with a kilogram of the substance) and it released power that, it was rumored, would far surpass that of electricity. Contemporary scientists themselves did nothing to dispel this aura of mystery surrounding radioactivity. In 1903 Sir William Crookes, a British physicist who understood very well how to exploit nationalistic fervor and the power of the press, chose to describe the energy locked in one gram of radium as being able to "blow the British navy sky high."[2] This graphic image, combining both power and destruction, made it seem highly desirable to have a monopoly on such energy. In his Nobel prize address in 1905 Pierre Curie even hinted briefly at the possibility that it might be better not to know the secret

of such power: "One may suppose how radium could become very dangerous in criminal hands, and here we might ask ourselves if it is to mankind's
advantage to know the secrets of nature, if we are mature enough to profit by
them, or if that knowledge will harm us." But like most scientists of the
period, he immediately rejected the pessimistic alternative: "I am one of
those who believe with Nobel that mankind will derive more good than
harm from such discoveries."[3]

The mysterious power of radium was soon overshadowed in the popular mind by the awesome implications of the one equation that everyone,
even the least mathematical, could remember: $E = mc^2$. The possibility of
converting matter into energy was adapted to military uses in popular
fiction long before it became a reality, the first example being in H. G. Wells's
novel *The World Set Free* (1915). Again, the scientists, for reasons as diverse as
the impulse to divest themselves of sole responsibility for the outcome of
their work and the need for research funding, were quick to publicize the
enormous power of their discovery. Frederick Soddy, the British chemist
who had worked with Rutherford on radioactive decay, wrote of scientists
in 1932, "If ever they are in a position to transmute the elements at will—it
would put into the hands of men physical powers as much greater than
those they now possess as these are greater than the forces at the command
of the savage," adding, "but of this particular contingency there is, perhaps
fortunately, as yet no sign whatever."[4] This ambivalent message of warning
and optimism was still current in the 1930s, when Sir Ernest Rutherford,
who had formerly delighted in downplaying his discovery, called his popular book on atomic physics *The Newer Alchemy* and initiated the cliché that
the problem of the alchemists had now been solved, since elements could
now be transmuted into others. It is not surprising that American movie
serials featuring Gene Autry entering an atom-powered underground city
to do battle with an evil prince armed with radium or Flash Gordon sabotag-
ing an "atom furnace" to outwit his archenemy Ming, who boasted that
"Radioactivity will make me Emperor of the Universe," achieved worldwide circulation as emblems of this potential.

Radioactivity and its possible effects, both known and speculated
upon, were not the only source of power and fear to be associated with
science. During the First World War, scientists were pressed into service in
the laboratory to demonstrate their patriotism by devising newer and more
lethal military hardware and chemical weapons. Fighter aircraft, new explosives, bombs, submarines and tanks, and the use of chlorine and mustard
gases to kill thousands at a time turned even the most sinister predictions of
Wells's *War in the Air* and *War of the Worlds* into reality. After the war, these
scientists eagerly publicized their contribution in inventing these weapons
of destruction as part of a strategy to gain more government research funds.
In this immediate aim some were successful,[5] but the long-term cost to their

public image was less fortunate. It became apparent that the power of science could be used for mass evil as well as for good, and the optimistic ideas put forward by H. G. Wells, J.B.S. Haldane, and others that scientists could be safely entrusted with such potential received a severe setback.

As we saw in chapter 11, between the wars scientists reemerged briefly in some fiction as the hope for a new society; and even during the Second World War they were sometimes characterized as battling for good against the military machine and upholding the "scientific" values of openness and internationalism against the secrecy and repressive nationalism of governments invoking emergency powers.[6] But the dropping of the atomic bomb on Hiroshima and on Nagasaki implicated scientists, in the popular mind, in a conspiracy of such moral enormity that they have never wholly extricated themselves. Oppenheimer's memorable statement that "the scientists have known sin" merely revivified and amplified the longstanding archetype of the evil scientist from Dr. Faustus onwards. Spencer Weart remarks, "When the United States dropped atomic bombs on Hiroshima and Nagasaki, the news had no immediate impact on the shape of the stereotype of nuclear scientists. Rather the stereotype already formed leapt into new prominence, its emotional force redoubled. . . . the name of Frankenstein was invoked everywhere from street corners to the US Senate."[7] To many contemporary observers it did indeed seem that power-crazed physicists capable of destroying humanity were walking the earth, respected, courted, and jealously guarded by their governments as a national defense weapon.

The "Frankenstein" tag referred not only to the horror engendered by the bomb itself but to the perceived intellectual arrogance of the scientists, who appeared to believe that the pursuit of science and the power it afforded were all that mattered, denying any social responsibility for the consequences of their research. The American physicist Freeman Dyson, who remarked, "The Faustian bargain is when you sell your soul to the devil in exchange for knowledge and power. That, of course, in a way is what Oppenheimer did," also acknowledged in himself the terrible attraction of such power:

> The glitter of nuclear weapons. It is irresistible if you come to them as a scientist. To feel it's there in your hands—to release this energy that fuels the stars, to let it do your bidding. To perform these miracles—to lift a million tons of rock into the sky. It is something that gives people the illusion of illimitable power and it is, in some ways, responsible for all our troubles, I would say—this, what you would call technical arrogance that overcomes people when they see what they can do with their minds.[8]

The spate of novels and plays dealing explicitly with the scientists working at Los Alamos on the development of the bomb indicates the fascination such a theme has had for the twentieth century. The question why apparently sane and intelligent scientists devote their lives and minds to the development of weapons of mass destruction tantalizes as much as it repels.

Because of their potential for instigating suspense and possible disaster on a cataclysmic, even planetwide scale, evil scientists quickly superseded the scientist hero as a staple of films and comic strips. These mad, bad, and dangerous scientists fall into two main categories: (1) the power maniacs, usually physicists, obsessed with the potential of some new energy source and bent on world domination or destruction on the grand scale; and (2) those whose basic philosophy is inherently evil in some more subtle and sinister way. The latter group usually includes a large component of biologists, perceived, according to the perennial Faustian stereotype, as attempting to usurp the province of the Creator by creating, changing, or extending life. These characters were to prove no less prophetic than the atom-smashing destroyers; they are the fictional precursors of today's researchers into genetic engineering, in vitro fertilization, and ecological manipulation.

Power Maniacs and World Destroyers

As we saw in chapter 10, even by the end of the nineteenth century there was a groundswell of reaction against the claims of that century's scientism. Despite the material benefits the new technology conferred, at least on the middle class, there was a growing unease that technological progress was fast outstripping humanity's moral development and hence its ability to control and direct such power. Such a realization clearly invited a restatement of the Faustus myth. The moral apprehension was accentuated by the general malaise of *fin-de-siècle* pessimism and preoccupation with extinction, both of humanity and of the planet. In part this can be traced to the popular conception of the inevitable increase in entropy, which, combined with the implications of Darwinian theory, emphasized the ruthlessness of nature; but the translation of such long-term disasters into material suitable for fiction required some means of foreshortening the time scale. The belief that scientists were on the brink of discovering a new form of energy incomparably more powerful than anything hitherto known suggested a source of imminent disaster that lent itself readily to literary treatment and led to a flood of stories depicting a mad scientist unleashing his new-found powers of destruction upon the world. Like the Social Darwinists, these paranoiacs often appeal to their catastrophic reading of nature (as revealed by geology and astronomy) as justification for their own obsession with destruction, and it is apparent that all of them actively hate life for one reason or another. A number of such scenarios appeared in simple black-and-white terms at the turn of the century, and they continued to form a large component of the offerings in the pulp science fiction magazines until

1945, when such fictitious horrors were overtaken and rendered obsolete by accounts of the actual effects of the atomic bomb.

Inevitably, the earliest treatments are nonspecific about the details of the threat posed by the mad, bad scientist. He has acquired, usually through physics but sometimes through chemistry, some means of destroying the human race, but the author is usually more preoccupied with the problems of creating the requisite suspense and then averting the imminent disaster than with the psychological motives of the destroyer. Where there is some attempt to account for the scientist's motivation, Darwinism is frequently invoked, either on the grounds that from the evolutionary perspective the extinction of humanity is of little significance, or because, like Doctor Moreau, the evil scientists have adopted what they interpret as the ruthless ethics of nature. In most of the stories dealing with such events, the character is entirely subservient to the plot, but "The Case of Summerfield" and *The Crack of Doom* provide two early and interesting examples where the motivation of the scientist is taken seriously by the author.

In W. H. Rhodes's story "The Case of Summerfield" (1871), the mad chemist Summerfield threatens that unless the citizens of San Francisco pay him the sum of one million dollars, he will throw into the Pacific Ocean a pill that will instantly convert it, and subsequently all the oceans of the world, into a blazing inferno, destroying every living thing in one universal catastrophe.[9] Rhodes insists that Summerfield is both insane ("I thought I could detect in his eye the gleam of madness" [19]) and brilliant (his facial features resemble those of both Newton and Alexander von Humboldt!), and he suggests by way of explanation that his great intellect has been achieved at the expense of ethical development. From his post-Darwinian perspective, Summerfield sees only evil and destruction in nature. Like Swithin St. Cleeve in Thomas Hardy's *Two on a Tower,* he points to the history of catastrophes throughout the universe, quoting von Humboldt's account of the destruction of a star in Cassiopeia and the deterioration of Sirius. That is, he appeals to the violent and inanimate aspects of the universe and uses a nonhuman time scale to measure events, thereby rendering ethical criteria irrelevant.

For Summerfield the inducement underlying the threats is solely hope of profit. A more interesting study is provided by Robert Cromie's classic novel, *The Crack of Doom* (1895), in which the mad scientist, Herbert Brande, is motivated by a warped idealism. Believing, as Summerfield does, that "the universe is a mistake" and animate nature a regression, he determines to reduce the solar system, or at least the planet Earth, to "pure elemental ether." Post-Darwinian nature is the villain of his universe. After citing a catalog of examples of its disorder, cruelty, and waste, he concludes: "She has no system . . . she is not wise. . . . The theory of evolution—her gospel—reeks with ruffianism, nature-patented and promoted. The whole scheme of the universe, all material existence . . . is founded upon and begotten of a system of everlasting suffering. . . . Wholesale murder is Nature's first law" (85–86).[10]

Given these premises and his newly discovered ability to disintegrate the atom, Brande's proposal to undo creation is entirely logical. He also voices a doctrine that was to become almost a cliché in post–Second World War discussions of science, namely, that science *cannot* be stopped. Brande affirms, "No man can say to science, 'thus far and no farther.' No man ever has been able to do so. No man ever shall!" (20). Brande provides, therefore, an early example of that combination of fanatical dedication to an abstract ideal, total faith in the benefits of science, and a lack of feeling for humanity that was to characterize many of the evil scientists in twentieth-century literature.

Because these early writers had little or no understanding of what the immediate agents of destruction might be or how they could be activated, their descriptions of the proposed destruction are necessarily vague. After the publication of Rutherford's *Radioactivity* in 1904, however, a new source of power, and with it new and sinister aspects of behavior, became available for literary embellishment, and a whole spate of novels appeared envisaging the misuse to which the release of such enormous reserves of energy might be put. Typically, the fictional scientists who discover a new energy source are corrupted by its potential for wealth and power, although the fact that they usually have to be overthrown by other "good" scientists indicates a residual faith on the part of these authors in the ultimate benefits of science.

Among the early atom-smashing devices, miraculous rays are almost de rigueur, and nationalist prejudices decide which devices are morally good and which are evil. In one of the earliest of such treatments, George Griffith's *The Lord of Labour* (written in 1906 and published in 1911), a characteristic example of contemporary "invasion" literature, the Germans invent a ray that can "demagnetize" metal, so that it crumbles to dust and the British fleet is destroyed by rays from the enemy's wooden ships, an almost direct transposing into fiction of Sir William Crookes's provocative statement quoted above. However, the British retaliate with the products of their "good" science—helium-radium bullets—which soon demolish the enemy. Similarly, in Arthur Conan Doyle's story "The Disintegration Machine" (1929), which also purports to derive from Rutherford's work, a Latvian scientist (in the British political climate of that time almost any European scientist was regarded as being of doubtful morality), Theodor Nemor, has invented a machine "capable of disintegrating any object placed within its sphere of influence . . . [into] its molecular or atomic condition."[11] Predictably, the machine's potential as a weapon is the first aspect to be seized upon; even more predictably, the amoral Nemor turns out to be intent only on selling his invention to the highest bidder. Happily, he is outwitted by the British scientific hero in the person of the incorruptible Professor Challenger, who persuades Nemor to demonstrate his own machine and be disintegrated—permanently.[12]

Apart from these nationalistic treatments, in which the "good" side invariably triumphs, there were others that were less optimistic about the possibility of controlling the evils. In 1907 Upton Sinclair wrote a play, later published as the novel *The Millennium: A Comedy of the Year 2000*, in which a new radioactive element, radiumite, is let loose by a mad professor and kills all life on Earth except for eleven plane passengers happily flying above the contamination. The following year Anatole France was exploiting the possibilities of atomic rays in his novel *Penguin Island* (1908), in which terrorists destroy whole cities with pocket-sized atomic explosives.

The Czech writer Karel Capek, better known for his play *R.U.R.* (see chapter 14 below), was one of the first to draw the connection between destructiveness on the grand scale and the irresponsible pursuit of what is scientifically interesting, regardless of the consequences. In his novel *Krakatit* (1923), an engineer, Prokop, has invented an explosive, krakatit, and believes that there is within the earth an incalculable force of explosiveness (atomic fission). At first Prokop cares only about the excitement of discovery, refusing to think about the social consequences of his work, but unlike most of the proponents of "value-free" science, he is gradually converted to a sense of social responsibility. His final position is unusual for this period in that his "solution" does not involve the rejection of his discovery, but optimistically looks forward to the peaceful use of the new power. In an earlier, parodic novel, *The Absolute at Large* (1922), Capek postulates an "atomic boiler," which, presumably making use of relativity theory, turns matter into energy. Physics merges into metaphysics when it is discovered that once the matter is destroyed, the Absolute remains behind. "What is left behind is pure God." Religion breaks out everywhere as nations try to monopolize this reified God, and the apocalyptic finale suggests that such powerful inventions are best left alone, since they lead only to global chaos.

Murray Leinster's two ingenious stories "The Storm That Had to be Stopped" and "The Man Who Put Out the Sun" (both 1930), also revolve around the efforts of a mad, power-crazed scientist, Preston, to make himself a world dictator.[13] Preston holds the world to ransom by his ability to interfere with the transmission of the sun's rays through the atmosphere, leading to catastrophic climatic changes with all the symptoms of nuclear winter. Happily, his evil plans are thwarted by another scientist, Schaaf, with the added bonus that Preston's discovery is put to benign use, supplying the world with accident-free solar power—yet another example of the Janus face of science. As in most of these suspense-centered stories, however, the rescuing scientist is a mere deus ex machina, and the authors invariably fail to explain why it is safe to assume that future scientists will not exploit the same discovery for their own malign purposes.

Leinster's two stories typify a large number of similar plots to be found in the pulp magazines of the period.[14] Underlying all of these is the premise that scientists have access to an unprecedented level of technological power, far beyond that of any other group in society, and are therefore capable of affecting the whole planet, if not the solar system. While, as we saw in chapter 11, it was sometimes postulated that this power would be used for the good of mankind to abolish war and usher in an era of plenty for all, twentieth-century literature more frequently expresses the suspicion that scientists, despite their intellectual brilliance, are as subject to the temptations of power as other people and thus are no more to be trusted with the outcome of their discoveries than any other group. Indeed, their habitual preoccupation with abstract thought renders them less humane than most and more ready to sacrifice whole populations if this should appear necessary for furthering their research. With varying degrees of subtlety it is suggested that their brains have developed at the expense, not only of the emotions, but also of their ethical perceptions. Thus many of the negative characteristics attributed to nineteenth-century scientists by the Romantics reappear in an updated form attached to their twentieth-century counterparts, especially to physicists, who epitomize the monopoly of power.

With the development by scientists of increasingly exotic weapons of mass destruction, the speculative attempts by writers who were essentially drawing upon fantasy for their plot devices gave way to factual presentations more horrifying than the products of imagination. Georg Kaiser's popular play *Gas I* (1918) was one of the first to draw on the mass fear of gas attacks during the war in order to intensify the picture of a ruthless scientist/engineer determined to develop his technological expertise in the production of industrial gas, whatever the cost to the workers. "Our gas feeds the industry of the entire world" he proclaims.[15] When a dangerous fault in the process is discovered, the engineer, denying that there could be any flaw in his calculations, refuses to give up the project. Even after an explosion occurs, devastating the factory and killing hundreds of workers, he urges the survivors on with promises of wealth and power in terms reminiscent of the wartime rhetoric of German imperialism: "There is your rule, your mastery—the empire you have established" (623). Kaiser's engineer thus emerges as a personification of the dehumanization of the industrial age, in which the desire for material wealth and the power of scientific calculation outweigh all humane considerations. It is not surprising that in the sequel, *Gas II* (1920), in which a world war breaks out, the same engineer modifies the gas factory to produce nerve gas. "The gas that kills!" he screams maniacally, holding aloft a glass "bomb" of deadly gas in order to rally the workers. "Ours the power! Ours the world!"[16] Unlike many of the other authors discussed here, Kaiser, himself a witness of the horrors of actual gas warfare, makes no attempt to devise a happy ending; his play ends with the extermination of the entire population.

By the mid-1930s, as news of the race to achieve the distinction of producing the first atomic weapon leaked out from behind the security barriers, the atomic bomb superseded all other scientific horrors, both in the public imagination and in literature and film. The first horror movie to refer to the nuclear menace was *The Invisible Ray* (1936), which featured a scientific genius, Dr. Rukh, played, significantly, by Boris Karloff, fresh from his triumph as Frankenstein's Monster. Rukh builds a radium-ray projector, capable both of destroying whole cities and of saving lives; but he himself is as ambivalent as his invention. Infected with the contagious radioactivity, he becomes unstable and sets out to avenge himself on the world. The 1940 horror film *Dr. Cyclops* presents an initially idealistic scientist who intends to improve the world using radium rays, even though his colleagues point out that he is "tampering with powers reserved for God." When his secret weapon is discovered, Cyclops ruthlessly turns it on the interlopers, shrinking them to the size of mice. Again the suggestion is strong that scientists, even if they begin with high ideals, are as unstable and fallible as their political counterparts and are not to be trusted with such extraordinary power. None of these writers, however, proffer suggestions as to how the power, once discovered, should be managed. The implied and fundamentally pessimistic "solution" is little more than a pious wish that such resources had not been unearthed.

More realistic, though certainly no less pessimistic, is J. B. Priestley's novel *The Doomsday Men* (1938), which focuses on a plot to explode an atomic device in the Mojave Desert. The three MacMichael brothers, who are responsible for the project, are mad fanatics who, for diverse reasons, desire to end the world. The novel was to acquire a grim retrospective irony in that seven years later a group of apparently sane men was to carry out just such a project in another American desert at Los Alamos. Of the three brothers in Priestley's novel, only one, Paul, is a scientist, but his reasons for engaging in this attempt at ultimate destruction are the most interesting. Priestley explains his fanaticism as a combination of pride in his own preeminence as an atomic physicist and despair, like that of Cromie's character Brande, at the centrality of chance and accident in the universe. Priestley implies that for physicists obsessed with discovering order in the universe, the realization that chance and accident predominate is likely to induce insanity.[17] Paul MacMichael explains to another scientist his plan to set off an atomic chain reaction that will peel the skin off the Earth, "like an orange, only faster!":

> I'm a scientist. A good one, . . . who's given his life to pure knowledge. . . .
> I have a chance of performing the last and greatest experiment known to

science. To release the earth's energy to destroy—I hope in a flash—the life on it. That life, in my opinion, was an accident. . . . What we call life is matter so arranged that it begins to think and feel. . . . Man or any being like him is doomed from the start. He can't possibly find a lasting place for himself in this universe, which if it has plans are not plans for us. Out of the eternal dance and changing patterns of light and energy . . . mind has somehow emerged, to acquire knowledge but also to understand its own noble despair. But it can still use that knowledge for one last triumphant stroke, one supreme act of defiance, . . . grandly destroying itself, leaving the mindless cosmos to its own damned dance of blind energies for ever.[18]

Given his premise, MacMichael's conclusion, like Brande's in *The Crack of Doom*, is entirely logical.

The actual deployment of atomic bombs on Hiroshima and Nagasaki, at first widely accepted by the Allies as a necessary means of ending the horrors of war, was soon to be reassessed by writers who perceived that such weapons represented not an end but a beginning. One of the earliest treatments of this theme was F. H. Rose's novel *The Maniac's Dream* (1946), which effectively recasts Conrad's problematic character Kurtz, from *Heart of Darkness* (1902), as a maniacal physicist who forces his natives to construct the ultimate weapon. In his characterization of the physicist who sees himself as the agent of vengeance to destroy the destroyers, Rose presciently suggests both the endless proliferation of weapons of mass destruction that the cold war was to produce and the insanity of believing that the chain of destruction can be halted at the will of the victorious party.

The bombs which the Americans made for the destruction of their enemies in Japan, are as nothing compared with the developments that have followed. . . . They—the American and British Governments—thought themselves superior in the science of wholesale death and destruction. But it has often happened in history that the destroyer of others has himself been destroyed by his own device. America and Britain, now that they have unloosed upon humanity this awful power, talk glib nonsense about controlling the uncontrollable . Fools and delusionists, all![19]

The subsequent cold war produced its own particular brand of fictional scientists, who by humanitarian criteria must be considered either evil or deranged in their obsession with testing the latest nuclear weapons and their use of every subterfuge to escalate the arms race and effect a confrontation between the superpowers. The new and more dreadful feature of these evil characters is their apparent normality. They are perceived by their colleagues as dedicated and patriotic Americans or else as motivated by essentially normal psychological reactions. L. Sprague De Camp's story "Judgement Day" (1955) focuses on such an atomic scientist working at Los

Alamos, whose indifference to humanity is traced back to his treatment at school. There, in response to being ostracized for his cleverness, he developed a disaffection for people that has gradually reached psychopathic proportions. At the moment of his research triumph he reflects: "I find myself getting more and more indifferent to everything but physics, and even that is becoming a bore." Once he has the opportunity of destroying most of humanity, including his former tormentors, he scarcely hesitates: "I'd kill them by slow torture if I could. If I can't, blowing up the earth will do. I shall write my report."[20] The British film *Seven Days to Noon* (1950), the story of a formerly naive and compliant physicist, Dr. John Willingdon, who has been involved in the manufacture of atomic bombs, explored this question further. Eventually, Willingdon's mind cracks when he grasps the enormity of what he has done. He steals a bomb and announces that he will blow up himself and the whole of London unless the British government promises to renounce nuclear weapons.

The most influential of all nuclear war films featuring the maniacal scientist was Stanley Kubrick's *Dr. Strangelove, or How I Learned to Stop Worrying and Love the Bomb* (1964), based loosely on the novel *Two Hours to Doom* (1958), by British author Peter George. The central figure, Dr. Strangelove, a scientist-strategist in a motorized wheelchair, is, on the one hand, a reissue of the Romantic stereotype of the emotionless, mechanized scientist who has lost his humanity (the wheelchair suggests a parallel with another dehumanized character, D. H. Lawrence's Clifford Chatterley). But as a composite figure, comprising elements of Otto Hahn, Edward Teller, and Henry Kissinger, he also represents a macabre combination of mad doctor and state scientist.[21] The exaggerated symbolism of medieval sorcery and Frankenstein references, alongside the emphasis on Strangelove's sinister debility (Strangelove, portrayed as a former Nazi, is crippled and has a withered right hand, hidden in a black glove),[22] from which he miraculously "recovers" at the climactic ending, shocked audiences into considering the complex reasons behind the arms race. Not least among these was the motivation of the scientists themselves, especially the physicists, whose exceptional intellectual talents were employed by the military-industrial complex in producing ever more ingenious weapons of mass destruction. The film raised such questions as, Who else would employ these specialist scientists in such numbers, on such funding levels? and was it not, therefore, in the physicists' own interest to further the escalation of the cold war?

A later and more sophisticated treatment of the mad scientist seeking to destroy the world is Philip Dick's allegorical novel *Dr. Bloodmoney* (1965), allegedly written in response to *Dr. Strangelove*. It depicts a dystopian world after a nuclear holocaust, set off when the paranoiac physicist Dr. Bluthgeld incorrectly certified that a high-atmosphere bomb test would be quite safe. Bluthgeld believes that he personally precipitated the destruction by calling down nuclear thunderbolts through his own psychic powers, because in his clinical state of deluded reference he confuses secondary and final causes. Bluthgeld, therefore, represents quite literally the mad scientist; he has even sought treatment from a psychiatrist. On the one hand, his delusions of grandeur lead him to believe that he is the chosen instrument of God for the punishment of mankind: "I am the center. God willed it to be that way."[23] On the other hand, he also, in his way, regrets this (165–66). Thus Bluthgeld stands as the parodic personification of the complex characteristics Dick attributed to the mad scientist archetype—paranoia, delusions of grandeur, obsessive behavior resulting from failure to understand what is happening in the real world, and belief that he is the instrument of God.

Although we at first assume that Bluthgeld's belief in his ability to effect a nuclear cataclysm through psychic powers is further evidence of his insanity, Dick progressively undermines any distinction between "real," or external, events and those that occur within the psyche. That is, he implies that there is no clear demarcation between cause and effect, or in the scientist's case, between discovery and the responsibility for the consequences of that discovery. If Bluthgeld invented the atomic bomb, he *is* responsible, as he himself believes, for the 1972 fallout and for the imminent Third World War; whether the actual means are psychic powers or a causal chain of events is, in Dick's analysis, ultimately irrelevant.

Bluthgeld is finally removed from the action by an equally mad technologist, Hoppy, but such a character, endowed (as a result of mutation caused by radioactive fallout) with increased psychic powers, proves no less dangerous than Bluthgeld. Dick would appear to be suggesting, in response to the escalation of the arms race, that no one can be trusted with the power necessary to arbitrate and control those already in power.[24] If we compare this no-win situation with the easy solutions proposed by writers earlier in the century, involving some allegedly benevolent dictator who enforces world peace, we can see the full extent of Dick's pessimism, which is, in turn, the articulation of a general cynicism in the late-twentieth century about the coexistence of power and benevolence.

Philosophies of Evil: Biologists and Psychologists

More subtle but no less dangerous on the small scale than the insane and evil physicists seeking to destroy the world are those scientists intent on exploiting their influence over living beings to gain wealth, fame, or power. The early years of the century produced a variety of such characters in literature, many of them modeled closely on the Frankenstein archetype, and they too were quickly incorporated into films and pulp fiction. With few exceptions, these sinister characters were biologists, with, later, a sprinkling of psychologists; indeed, until the onset of the cold war these disciplines surpassed physics as the nursery of scientific mania. Biologists, it seemed, incurred the particular occupational hazard of falling prey not only to insanity but also to blasphemy. So frequently were these two conditions depicted together that they were implicitly understood to result from the same cause, a mechanistic view of living things. While a reductionist methodology was accepted as inevitable in the physical sciences, such an approach in biology was censured not only on the Romantic grounds that it led to emotional retardation on the part of the scientist but because it was believed to engender a desire to abrogate divine power over life. The indictment of psychologists in literature later in the century sprang from the same uneasy feeling that they treated even the mind and the personality, seemingly the last bastion of spiritual and moral consciousness, as a mechanism and assumed profane control over them. This latter indictment was not one that had commonly attached to the evil physicists, which suggests that in the popular view there was at least an implicit vitalist belief that living beings, and especially the human mind, were more closely connected to a divine source. Again, the Frankenstein prototype is prominent: charges of intellectual pride and emotional deficiency are most often leveled against biologists, who arrogantly try to apply the mechanistic laws of physics to living systems and are therefore seen as usurping the Creator's role. There was good reason for this association. Darwin's theory was still widely regarded as being at odds with religion, and the public utterances of eminent biologists such as T. H. Huxley and John Tyndall had done little to discourage this idea. Tyndall had asserted that although there was as yet no clear proof of a link between consciousness and molecular activity, nevertheless, all the mysteries of nature would ultimately be explained in mechanistic terms,[25] and Huxley had claimed that consciousness was reducible to a mere reflection of molecular movement, psychic events in the mind being caused solely by physical events in the nervous system. The immoral biologists of fiction invariably adopt this reductionist stance.[26]

An early twentieth-century example of this stereotype occurs in Maurice Renard's popular novel *Le docteur Lerne, sous-dieu* (1908), a horror story concerning an evil German scientist, Klotz (nationalist prejudices resulting from the Franco-Prussian War are clearly in operation), who has "killed" a kindly French doctor, Frédéric Lerne, and, even worse, transplanted his own brain (and hence his immoral personality) into Lerne's body. Klotz has then embarked upon a series of grafting and transmutation experiments clearly based on those of Doctor Moreau (Renard dedicated the novel to Wells), intending eventually to sell new, virile bodies to geriatric millionaires (hence the title of the English translation, *New Bodies for Old*) and ultimately to render himself immortal by transferring his own soul to a piece of machinery. Klotz thus reenacts the alchemist's search for the elixir of life, albeit now pursued through surgery, and is accordingly decked out in all the trappings of the alchemist—clandestine activities, soul transference, a desire for godlike powers, and, predictably, a concomitant lust for gold— but his bizarre belief that immortality resides in machinery makes explicit the connection with the more modern "evil," reductionist biology.

In the same year Somerset Maugham also revived the evil alchemist model, in *The Magician* (1908), a novel strangely prescient of in vitro fertilization and embryo culture. Like his medieval precursors, Maugham's young scientist, Oliver Haddo, is obsessed with a desire to vie with God by creating homunculi and thus studies occult lore, including, significantly, Arabic and Hebrew formulas, in order to create his embryos. When it appears that these require virgin blood for their nourishment, Haddo has no scruples about sacrificing his new wife Margaret for the purpose, a detail that would hardly have surprised medieval readers but proved less convincing in the twentieth century.

Klotz and Haddo are fair representatives of many other characters of the period who are censured for their hubris in attempting to improve upon or reverse the course of nature. The fact that these early twentieth-century writers, being ignorant of the actual procedures of contemporary research, have recourse to vague alchemical details should not discount their expressions of fear and mistrust in relation to biologists. André Gide also chose an amateur biologist as the type of the inhumane scientist. Before his miraculous conversion to Catholicism, Anthime Armand-Dubois features in *Les Caves du Vatican* (1914) as an "unbeliever and freemason," intent on gibing at friends, family, and children. As evidence of his inhumanity, he conducts experiments on rats, blinding, deafening, castrating, skinning, and starving them "with the purpose of wringing from the helpless animal the acknowledgement of its own simplicity." "Anthime's modest pretension, before going on to deal with human beings, was merely to reduce all the animal activities he had under observation to what he termed 'tropisms' . . . an entire category of psychologists would admit nothing in the world but tropisms."[27] Gide takes care to show that Anthime's procedures in his work for the Académie des Sciences are typical of the science fraternity of his time. His rat mazes, with their many variations to keep the animals from their food, are described as "diabolical instruments, which in a later age became the rage in Germany under the name of *Vexierkasten*, and were of great use in helping the new school of psycho-physiologists to take another step forward in the path of unbelief" (6). Such an attitude was not uncommon at the time and suggests that the angst engendered by evolution theory, and associated specifically with biology, was by no means laid to rest. In addition, while the image of the evil physicist was tempered by the American cult of the inventor, biology had no such "champions" and continued to languish under the imputation of godlessness.

In most science fiction of this period there is a strong (albeit simplistic)

moral message, reinforced by the retribution meted out to the presumptuous biologist. Thus in A. Hyatt Verrill's representative story "The Plague
of the Living Dead" (1927) a biologist, Farnham, develops a serum that
inhibits the aging process and even proves capable of restoring dead creatures to life. However, the revived animals and erstwhile dead people be-
have "immorally," and even Farnham is made to see the blasphemy of his
"interference with the laws of a most wise and divine Creator."[28] A similar
line is followed in "The Ultra-Elixir of Youth" (1927), another moralistic
story by the prolific Verrill, in which a scientist attempts to reverse the effects
of aging and the author concludes that this "has brought home to us the
terrible consequence of attempting to interfere with the plan of the Creator."[29] Perhaps the most extended moralizing of this type occurs in Alexander Snyder's "Blasphemers' Plateau" (1926), in which Dr. Santurn, another decadent biologist, determines to prove that the so-called spirit is "merely
the vibrations which stimulate the electrons in their orbits."[30] Predictably,
Santurn proves to be as ruthless and evil as he is irreligious, and he meets a
suitably Faustean, if logically confusing, end.[31]

Such clear-cut moral statements, combined with the exaggerated descriptions of research on living organisms, made these characters eminently
suitable material for the horror movie. The advent of this genre can be dated
from the first version of *Frankenstein* (1931), in which Boris Karloff rocketed
to fame as the nameless Monster produced, as the introduction clearly
states, by the "man of science who sought to create a man after his own
image, without reckoning upon God." The huge success of *Frankenstein* says
much about the climate of popular paranoia at the time. It was followed in
the same year by the equally successful *Dr. X*, which made use of a similar
formula, a group of scientists working in an isolated old Gothic mansion to
create synthetic flesh. These mad, morally doubtful, and certainly irreligious
scientists were to set the trend for the next fifty years, for in the celluloid
world scientists are far more often mad than sane. *Frankenstein* also spawned
Son of Frankenstein (1939), in which the new Baron Frankenstein, seeing that
someone has chalked on his father's tomb "Maker of Monsters," amends it
to read "Maker of Men." In due course there appeared *Bride of Frankenstein*
and *Curse of Frankenstein*, as well as the many cloned versions of *Frankenstein*
itself.[32] All focus upon the image of the isolated, obsessed biologist cutting
himself off from the natural affections of the living to create artificial life
from the materials of death. The series of paradoxes implicit in these films
reflect the ambivalence attaching to the scientists of the period, who were
courted but not loved, feared but not reverenced. Yet, interestingly, science
itself, as distinct from its practitioners, is rarely denigrated in these films,
and even the evil scientist is sometimes presented as more sinned against
than sinning. *Dr. Jekyll and Mr. Hyde* (1931), *Dr. X* (1931), *The Murders in the
Rue Morgue* (1932), *The Vampire Bat* (1933), *The Walking Dead* (1936), and *Dr.*

Renault's Secret (1946) are all concerned with biologists who have gone too far in their attempts to create or restore life, thereby playing God; but there remains a residual, if vague, admiration for science itself. The films can thus be seen as contemporary documents of Western society's own confused cost-benefit analysis of science, its terrors, and its promises.

The attempt to come to terms with the potential for social development in the light of the failings of individual scientists is explored in some depth in Alfred Döblin's *Berge, Meere und Giganten* (1924), which anticipates the adverse ecological consequences of scientists' meddling with nature.[33] In this vast novel, spanning several centuries in the future, industrial scientists and research groups have effectively assumed total power over society, converting it into a huge technological experiment about which no questions may be asked. Insisting on scientific progress as an end in itself, Döblin's scientists, having already devised machines to do all the necessary labor, turn to the production of devices for which there are as yet no applications. In a remarkable prefiguring of the escalation of the cold war, they perfect more and more sophisticated weapons, which in turn require wars for their justification, so that the scientists are portrayed as being directly responsible for the fomenting of another war.[34]

One of the most interesting of Döblin's scientists, and one of the few women scientists in the literature of this period, is the biologist Alice Layard, leader of the North American *Frauenkamaraderie*. Originally motivated by the desire to make a contribution to society, she announces the discovery of a synthetic food whereby entire populations can be fed without the need for arable land or even sunshine. This discovery immediately becomes a political weapon that the women seek to keep for themselves (even lip service to the principle of open research has long since been discarded), and Layard deliberately contaminates the food, causing an epidemic, in order to discredit her (genuine) discovery rather than surrender it. With considerable perception, Döblin thus shows how research can change from an altruistic activity to a means for acquiring prestige and power, until, paradoxically, the research itself is sacrificed to the latter end.

Still obsessed with circumventing nature and striving for the seemingly impossible, Döblin's Promethean scientists embark on another well-intentioned project, which becomes a major disaster: the de-icing of Greenland. In the process, primitive animals and plants, long extinct, are thawed out, producing grotesque monsters; and the scientists, in self-defense, propose to assume similar proportions, a procedure reminiscent of Boom Food in H. G. Wells's *Food of the Gods*. Here, however, the inflated size of the scientists is symbolic of their pretensions and their hubris rather than, as in Wells's treatment, of their moral stature. Eventually all the giants die off, leaving a happy, pastoral, anti-technological society with the motto *Wissend und demütig* (Wise and humble).[35] Thus Döblin's implicit counsel, like

Ruskin's, is a reactionary one: return to a preindustrial system. Scientists are either too ignorant of consequences or too morally flawed to be entrusted with power; hence their attempts to transform nature are doomed to disaster. The enormous box office success of the film *Jurassic Park* (1993) suggests that this view is still widely held.

A more sophisticated philosophical attack on immoral biologists is to be found in the very different premises of Aldous Huxley and C. S. Lewis. Huxley's attitudes towards science and scientists varied considerably during his long life as a writer. In *Along the Road* (1925), written early in his career, Aldous Huxley, grandson of the eminent Thomas Henry Huxley and brother of another distinguished biologist Julian, reflected with regret on his own failure to become a scientist, concluding, "Even if I could be Shakespeare, I think I should still choose to be Faraday." Again at the end of his life, as is evident from his last novel, *Island* (1962), he came to see in science possibilities both for solving the world's ecological problems and for helping humanity to achieve its spiritual potential. However, in the period between, he was intensely critical of the limitations of science, regarding it, first, as inadequate for the evaluation of ethics and aesthetics and inimical to the real purposes of man and, second, as leading to a loss of creativity, individuality, and liberty by its insistence on a narrow range of objective criteria. Already in his early nonfictional work *Jesting Pilate* (1926) Huxley, like the Romantics, blamed scientific paradigms for the disappearance of values from Western civilization, arguing that scientific materialism is a closed intellectual system that precludes the relevance, or even the possibility, of a value system: "If men have doubted the real existence of values, that is because they have not trusted their own immediate and intuitive conviction. They have required an intellectual, a logical and 'scientific' proof of their existence. . . . when you start your argumentation from the premises laid down by scientific materialism, it simply cannot be discovered. Indeed, any argument starting from these premises must infallibly end in a denial of the real existence of values."[36]

Huxley not only saw many of the alleged benefits of technology as illusory but also came to regard science as a root cause of both the loss of individual liberty and the rise of nationalistic totalitarianism. In *Science, Liberty, and Peace* (1946) he outlined the reasons why scientists must bear a large degree of political responsibility. First, by helping to create the powerful weapons of modern warfare, scientists have endowed political leaders with the means of holding the masses in fear and subjection; second, they have acquiesced in the destructive uses to which their discoveries have been put.

It was out of this largely negative assessment of science in the real world that Huxley created his fictional scientists. The best-known expression of his belief that science stunts man's development and insulates him from coming to terms with the reality of his existence is, of course, *Brave New World* (1932), with its description of a society founded exclusively on the principles that Huxley associated with technology: efficiency, pragmatism, standardization, restricted thought, and the consequent dehumanization of the individual. Along with E. M. Forster's "The Machine Stops" (1909),[37] Yevgeny Zamyatin's *We* (1920), and George Orwell's *1984* (1949), it stands as a deliberate and radical rejection of the Wellsian utopias grounded in an unquestioning faith in the beneficence of science.[38] The more immediate cause for Huxley and Orwell's grim picture of a technologically determined society was their reaction against the dehumanizing theories of the behaviorist school; hence research into psychological conditioning is depicted as the most insidious and sinister aspect of science, surpassing even the destructive power unleashed by the physical sciences and the pretensions of biologists peddling eternal life. For both these writers the psychologists are the consummate villains of the whole scientific pantheon.[39]

Huxley presents no individual scientists in *Brave New World* (since loss of individuality is part of the point he is making), but en masse they subvert and parody the scientist-as-hero stereotype. Indeed, the whole ethos of this post-Fordian society, with its complex system of genetic engineering and its redefining of personal relations and emotions as obscene, is a reductio ad absurdum of what Huxley perceived as the inevitable implications of the scientific method linked to the desire for assembly-line efficiency. *Brave New World* became a cult novel, eliciting many derivative treatments of the dystopian theme. Gerald Heard's *Doppelgängers* (1948), for example, is set in 1997, after the Psychological Revolution, when Earth is ruled by a benevolent dictator who uses popular science to keep the masses unthinking and happy. Others follow Huxley less closely, but the message is much the same—rejection of a mechanistic society where humanity has been reduced to an accepting automaton status.[40]

The Christian apologist C. S. Lewis, who from his position as a conservative classicist engaged in a prolonged public controversy with the socialist biologist J.B.S. Haldane over the degree of influence science should exert in society,[41] also produced a popular trilogy; masquerading as science fiction, it was a carefully contrived propagandist attack on scientism. In Lewis's neo-Platonist view, the cultivation of scientific goals and methods must inevitably lead to the demise of religious values. He regards the two basic propositions of modern science—reductionist materialism and chance—as more insidious and dangerous than any overt destruction wreaked by mad, bad scientists, because while the latter will inevitably be abhorrent to society and elicit a counterattack, the rationale of science appears unexceptional, even attractive, and can thus subvert a society's ethical values without resistance. Lewis sets out to denounce this philosophy, embodied in the stereotyped character of Weston (Western man), a physicist, who wholeheartedly embraces accepted scientific values and who is therefore, in Lewis's treatment, both ruthless in his attempt to exploit society, and indeed the universe, and also limited in his ability to understand the complexity of that universe.

Since his main concern is to trounce materialism, Lewis, unlike the majority of science fiction writers, has little interest in the technical side of his story. In the first volume, *Out of the Silent Planet* (1938), we are given no details about the spaceship in which Weston and his companions travel to Malacandra (Mars). Indeed, Lewis decries pulp science fiction of the *Amazing Stories* kind for its "scientification" and describes the Machiavellian Weston as "a man obsessed with the idea which at this moment was circulating all over our planet in obscure works of 'scientification' in little Interplanetary Societies and Rocketry Clubs, and between the covers of monstrous magazines."[42] Being a materialist, Weston is, Lewis assumes, necessarily amoral and therefore has no scruples about dabbling in vivisection, supporting eugenics, or experimenting by force on a hapless local village boy, since in his view the boy is "incapable of serving humanity and only too likely to propagate idiocy. He was the sort of boy who in a civilised community would automatically be handed over to a state laboratory for experimental purposes. . . . The boy was really almost a—a preparation."[43]

To oppose this view, Lewis introduces his champion, Ransom, a philologist and Christian humanist whom Weston regards as "insufferably narrow and individualistic," since he fails to subscribe to the doctrine that the ends justify the means. In practice, this means that Weston adopts the long-term, objective view, sacrificing the individual to the race and the present generation to the hypothetical future, claiming that "all educated opinion—for I do not call classics and history and such trash education—is entirely on my side" (30). He goes on to affirm: "Life is greater than any system of morality; her claims are absolute. It is not by tribal taboos and copy-book maxims that she has pursued her relentless march from the amoeba to man and from man to civilisation" (154).

Unlike most of the mad, bad scientists reviewed in this chapter, Weston is at one level an idealist in that he is interested not in personal gain but in a cause outside himself. His apparent altruism is therefore more dangerous than obvious self-interest. Lewis spends considerable time at the end of the novel making Weston appear morally stupid through the standard satirical device of having him explain his position to someone of a completely different moral perspective, in this case the all-wise Oyarsa of Malacandra.[44] Weston affirms his Darwinian and racist beliefs in the innate superiority of the human species over the individual and, plunging deeper into the moral mire at every step, deduces from this a mandate for man to spread to other worlds, killing the inhabitants if necessary, in order to survive the death of this and other planets. "Humanity, having now sufficiently corrupted the planet where it arose, must at all costs contrive to seed itself over a larger area. . . . planet after planet, system after system, in the end galaxy after galaxy, can be forced to sustain, everywhere and for ever, the sort of life which is contained in the loins of our own species."[45] This is clearly an intertextual allusion to H. G. Wells's frequent statements about "the beings now latent in our thoughts and hidden in our loins [who] shall . . . reach out their hands amidst the stars."[46]

In *Perelandra* (1943), the second volume of the trilogy, Weston's philosophy is modified somewhat in order to allow Lewis to attack another "heresy" of science, the pantheistic view he associated with Olaf Stapledon, namely, that the movement of all life is towards a purposeful life force, which has directed evolution. Such a doctrine, while superficially more compatible with religious tenets, actually runs directly counter to the orthodox belief in original sin, since it is based on what Lewis calls "chronological snobbery," the belief that the new is, ipso facto, superior. Such a view is, of course, diametrically opposed to Lewis's classical humanism, which locates the golden age in the past, and to his affirmation of the Christian doctrine of the Fall. For good measure, Weston also espouses the Baconian ideal of scientific utilitarianism and makes his textbook statement on this point too in order to be knocked down by Lewis's champion Ransom.

In the third novel of the trilogy, *That Hideous Strength* (1945), Lewis turns his attack upon psychologists, in particular the behaviorists. Their spokesperson announces Lewis's version of their intentions: "If science is really given a free hand, it can now take over the human race and recondition it; make man a really efficient animal."[47] The organization set up to implement this program of efficiency is the National Institute of Co-ordinated Experiments (N.I.C.E.), a sinister scientific foundation run by power-seeking bureaucrats who employ techniques of media manipulation to brainwash individuals and eventually society. Its goals include the fusion of state and applied scientific research, control of the environment, eugenics, and the remedial treatment of criminals, all of which Lewis identified with the subjection of nature and the surrender of individual freedom. The goals of N.I.C.E. are listed as "quite simple and obvious things, at first sterilisation of the unfit, liquidation of backward races (we don't want any dead weights), selective breeding. Then real education, including pre-natal education. . . . Of course it'll have to be mainly psychological at first. But we'll get on to biochemical conditioning in the end and direct manipulation of the brain" (47). Underlying this program is clearly the same rationale enunciated by Weston in the first two books—materialism, utilitarianism, the belief that the ends justify the means, the sacrifice of the individual for the species, and so on—in fact all the values that Lewis ascribed to science and scientists and associated with intellectual arrogance and ruthlessness.[48] Lewis's characterization of Weston forms an interesting contrast to the other evil scientists in this chapter because of the detail and ferocity of the attack and the author's intense personal involvement in the argument, which, ironically, renders his case as biased as the one he is bent on destroying.

In this analysis of evil scientists in twentieth-century literature it is apparent that many of their characteristics are identical with those depicted in previous centuries—arrogance, desire to usurp divine authority (particularly as creator), materialism, reductionism, lust for political power. The essentially new element is the resentment voiced by some of these fictional scientists that the classical Newtonian picture of a predictable, orderly system has been overturned by the revelation that chance, accident, and disorder prevail in nature. In several of the works discussed in this chapter this resentment leads to a manic obsession with removing life from the system altogether in order to purge it of its disorder, which suggests that their authors had not yet come to terms with the neo-Gothic horror raised by nineteenth-century astronomy and biology. Most of these writers were still unaware that contemporary physics was positing a parallel model of chance and unpredictability at the subatomic level.

All the characters explored in this chapter are the creations of male writers, whose diagnostic methodology involved searching for particular and individual causes for the violence or obsession they depicted. Later in the century, however, feminist writers and critics were to posit an alternative interpretation. They explain the characteristics of these mad, evil, and dangerous scientists as an extension of the values of their society, and these, in turn, they explain as consequent on the masculine role models perpetrated by that society. Christa Wolf, one of the most outspoken women writers of the (former) Eastern bloc countries, expressed this view pointedly in 1981, at the height of the cold war.

> Rockets and bombs are after all not by-products of this culture; they are the consistent manifestation of thousands of years of expansionist behaviour; they are the inevitable embodiments of the syndrome of industrial societies which, with their more! faster! more accurately! more efficiently! have subordinated all other values, many of which were measured by more human norms. . . . I am plagued by the thought that our culture, which could only attain what it calls "progress" through force, through domestic repression, through the annihilation and exploitation of foreign cultures, which has narrowed its sense of reality by pursuing its material interests, which has become instrumental and efficient—that such a culture necessarily has to reach the point it has now reached. . . . For three thousand years . . . women have not counted and do not count. Half the population of a culture has *by its very nature* no part in those phenomena

through which that culture recognises itself. . . . it occurs to me that they [women] therefore have no part in the experiments in thought and production that concern its destruction.[49]

In Wolf's analysis, then, the mad, bad, and dangerous scientists represent merely a consequence of the excessive concentration on reason, efficiency, and objectivity that a patriarchal hegemony extols. The scientists to be discussed in the next two chapters represent different aspects of this impersonal approach. Their authors are almost invariably critical of such an attitude; hence the fictional presentations are essentially neo-Romantic in their condemnation of the emotionless, inhumane, and amoral scientist. But these analyses are based on a more sophisticated understanding of cosmology, make use of different metaphors (the computer features largely), and perceive the problem as much more pervasive throughout the whole cultural ethos than their nineteenth-century predecessors could have envisaged.

CHAPTER THIRTEEN

THE IMPERSONAL
SCIENTIST

> I have only one relationship and that's with my work . . . I don't
> want relationships, I don't want involvements.
> —Stephen Sewell

> . . . he lived by measuring things
> And died like a recurring decimal
> Run off the page, refusing to be curtailed;
> —Louis Macneice

Of all the charges against the scientist in literature, none has been leveled
more frequently or more vigorously than that of aloofness and emotional
deficiency. This was not an attribute associated with the early representatives of science—the alchemists, the Faustus stereotype, or the Restoration
virtuosi. On the contrary, these precursors of the modern scientist were
depicted as passionate, inspired, even religious in their zeal. Emotional
retardation appeared as a characteristic of fictional scientists only after Newton's celestial mechanics eliminated much of the awe and mystery from
man's perception of the universe and Enlightenment philosophy elevated
the cultivation of objectivity as the precondition of scientific method. The
English Romantics certainly ascribed this diminished respect for the emotions and subconscious states to the cult of scientific rationalism, and although this condemnation was more subdued in the latter half of the nineteenth century, it remained latent, ready to emerge as yet another stick with
which to beat the scientists. Twentieth-century writers have continued to
build on and extend this Romantic archetype of the scientist who has sold
his emotional soul for scientific prestige.

While most of the overtly evil scientists considered in chapter 12 were
drawn from the ranks of physicists, biologists, and psychologists, it is the
mathematicians and computer scientists who bulk large among the literary
portraits of those who have reneged on personal relationships. This suggests that whereas physics and biology are popularly identified with power
and activity, mathematics and its more recent offshoot, computer science,

are associated with abstraction and dehumanization. Why is this?

The exponential rate of expansion of the computer into so many areas of communication appears, at one level, to have transformed Descartes's dream of the mathematization of the world into reality.[1] Although the computer has by no means solved all of humanity's problems, it has significantly changed the prevailing view of what is possible and the value systems espoused by technologically advanced societies. The criteria for evaluating computing systems are primarily those of efficiency, consistency, and simplicity, that is, the qualities most conducive to the reification of experience, and it is these criteria that are, increasingly, being adduced for the evaluation of success in both the social and the personal spheres.[2] The old-fashioned fear that computers would take over the world was seen by some writers as being realized in a more insidious way through this change in values.

In technological societies, communication too has been largely reduced to the transmission of facts through impersonal media—the fax machine, the electronic mail system—which standardize the message in the interests of speed of transmission. The ultimate impersonal message is the binary code of the computer; growing numbers of computer-literate people regard the world, both animate and inanimate, as a series of problems soluble by algorithms. So many daily activities are already based on numbers and computation that we hardly notice the insidious implications of reducing human beings to numbers, a practice that, before the advent of vast computer networks, was reserved for the army and the prison, where the right to individuality was deemed to have been forfeited. Even the social sciences, in order to compete with the physical sciences for funding and academic status, have had recourse to a statistics-based methodology for the assessment of everything from intelligence to social change.

In the broader community, this trend has not gone unchallenged. In the last three decades there have been various attacks on the edifice of scientific materialism. The anti-intellectual movements, which have included the hippie cult of the 1960s, a renewal of interest in the occult, in Eastern religions, meditation, astrology and various forms of mysticism, and alternative medicine, are a protest against the perceived sterility and inhumanity of utilitarian rationalism presided over by science. Writers too have registered their dissent. Kitchen-sink drama, the preoccupation with mental stress and nervous breakdown as a result of social pressures and alienation, and the boom in fantasy fiction and horror films are all aspects of the pervasive discontent with a society dominated by the rationalist and materialist values associated with science.

Another element of this protest is the frequent depiction in literature of inhuman scientist characters, especially psychologists, mathematicians, and computer scientists, for their disciplines are perceived as having displaced eighteenth-century physics and nineteenth-century biology as the cutting edge of reductionism. These impersonal fictional scientists are presented not merely as emotionally deficient individuals but as the archetypal representatives of a deeply depersonalized society. Sometimes they are less vilified than pitied as the primary victims of the system or of their own limitations, but more frequently they themselves are too obtuse or too deeply flawed to realize their own deficiencies.

Representatives of the Depersonalized Society

Scientists are depicted as the fitting representatives of the Western technological society insofar as they embody the values perceived to be most successful in the competition to produce that state-of-the-art technology that accumulates wealth and hence power. These values can be identified as follows:

1. Cultivation of rationalist skills and a corresponding suppression of the emotions
2. Elevation of an objective perspective as being essentially closer to reality than a subjective viewpoint
3. Efficiency as an intrinsically moral value
4. Reduction of individuals to statistical units
5. Integration of technological and economic systems so that the former receives further justification, because it secures wealth, and hence political dominance, for the society that possesses such expertise

Most of these criteria were included to varying degrees in the Romantic attack on science and scientists, but more recently feminist critics have introduced a new perspective by identifying the above factors with culturally induced suppression of male emotions in accordance with the inherent values of a patriarchal society. Their analysis has led in the literature of the 1980s to alleged correlations between male-dominated hegemonic systems and the criteria for evaluation of behavior in those societies. After generations of acceptance, these criteria may appear unbiased, when in fact they are highly eclectic and prejudiced against the so-called feminine values. This subversive perception has produced some of the most innovative studies of the scientist character in recent literature.

One of the least damning depictions of an impersonal scientist is the portrait of the mathematician in Louis Macneice's poem "The Kingdom" (1943), but even here it is implied that the scientist, who is, significantly, nameless, has lost his humanity and substituted his graphs for a normal human relationship with a living child.

> . . . he lived by measuring things
> And died like a recurring decimal
> Run off the page, refusing to be curtailed;
> Died as they say in harness, still believing
> In science, reason, progress,
>
> . . . .
>
> . . . plotting points
> On graph paper he felt the emerging curve
> Like the first flutterings of an embryo
> In somebody's first pregnancy; resembled
> A pregnant woman too in that his logic
> Yet made that hidden child the centre of the world
> And almost a messiah; . . .
>
> . . . .
>
> . . . Patiently
> As Stone Age man he flaked himself away
> By blocked-out pattern on a core of flint
> So that the core which was himself diminished
> Until his friends complained that he had lost
> Something of charm or interest.[3]

This degree of compassion for the depersonalized scientist is rare, however. As in the case of the Romantics, the earliest criticisms of scientific materialism linked it with industrialism, always a topic for disapproval in literature. The most outspoken writer on this theme in the early twentieth century was D. H. Lawrence, whose neo-Luddite attitude towards industrialism and its deleterious effects on both the English countryside and society could best be described as late Romantic. In the main his criticism was focused on characters he had observed, which did not include scientists, but in the character of Winifred Inger, in *The Rainbow* (1915), the connection between science and the machine is explicitly argued.

Lawrence uses Winifred, qua science teacher, primarily to represent both the new independent woman and the broader industrial aspects of his society that he most abhorred, the suppression of instinctive emotions and the consequent perversion of personality. Winifred, the product of a "scientific education," is tainted with many of the same faults as Lawrence's industrialist characters, Clifford Chatterley and Gerald Crich, namely, the subjection of human emotions to the machine as the material equivalent of abstraction. "The real mistress of Winifred was the machine. She too, Winifred, worshipped the impure abstraction, the mechanisms of matter. There, there in the machine, in the service of the machine, was she free from the clog and degradation of human feeling. There, in the monstrous mechanism that held all matter, living or dead, in its service, did she achieve her consummation and her perfect unison, her immortality."[4] Lawrence's reference to Winifred is brief; she represents only a hypothetical option for his main character, Ursula, who rejects her. But before long the complex connections between scientific reductionism, a depersonalized society, and their political equivalent, totalitarianism, were explored in the three great dystopias of the twentieth century—Yevgeny Zamyatin's *We*, Aldous Huxley's *Brave New World*, and George Orwell's *1984*. All three writers conclude that conformity is accepted by populations as the price paid for security.

Inspired by Wells's ambiguous dystopian novel *When the Sleeper Wakes* (1899), the Russian writer Yevgeny Zamyatin, who had been trained as an engineer, created in *We* (1924) the paradigmatic scientific dystopia.[5] In *We* a whole society, the United State, is managed, down to the most intimate details, on wholly mechanistic principles. Wells had depicted a future totalitarian society in which the workers were ruthlessly suppressed, but the villain of his novel was capitalism, not science. For capitalism, Zamyatin (who was himself persecuted for nonconformity)[6] substitutes a scientific obsession with efficiency, showing how this reduces the individual to a nameless number, a robotlike unit in the pattern, and forces him to conform to a "mathematically perfect life."[7] "Every morning, with six-wheeled precision, at the same hour, at the same minute, we wake up, millions of us at once. At the very same hour, millions like one, we begin our work, and millions like one, we finish" (13). Every action, every movement, is part of a stylized, "unfree" dance pattern, every part of which is prearranged according to the Tables of Hours. Even sexual desire has been dealt with in a quantified and scientific way by the Lex Sexualis: "A Number may obtain a licence to use any other Number as a sexual product," and the "product" must acquiesce. Dreams, emotions, imagination, and spontaneity are strictly suppressed.

The narrator, number D-503, is an engineer, Chief Builder of a new spaceship, the *Integral*, which is designed to bring "mathematically faultless happiness," by force if necessary, to aliens in the farthest reaches of the universe. He therefore begins a diary, a guide to life in his society, which he intends to send as part of this galactic crusade, for at first he is in complete accord with the values of the state. Indeed, his conditioning in rational thought has been almost flawless: the great terror of his childhood was his first confrontation with irrational numbers. "I wept and banged the table with my fist and cried, 'I do not want that square root of minus one . . .' This irrational root grew into me as something strange, foreign, terrible; . . . it could not be defeated because it was beyond reason" (37).[8] It is this cultivated fear of the irrational that induces D-503 and the rest of his society to conform: there is security in a system where everyone is an element in the pattern, a configuration in Euclidean geometry.

In the process of writing his diary, however, D-503 looks afresh at his world and becomes increasingly critical of its enforced harmony. This process is initiated by his disturbing encounter with I-330, an experience he regards as a mathematical problem to be solved: "$L = f(D)$, love is the function of death" and should therefore be avoided. In his mind she is associated with the childhood terror: "The woman had a disagreeable effect upon me, like an irrational component of an equation which you cannot eliminate" (8, 10). I-330 turns out to be a member of a secret society, the Mefi, which is planning a revolution, and under her guidance D-503 visits such subversive places as the Ancient House, with its disturbing, nongeometrical shapes, and the Green Wall, beyond which lies the jungle world of the uncivilized, the irrational realm of emotions and the subconscious, characterized by "terrible noise, cawing, stumps, yelling, branches, tree trunks, wings, leaves, whistling" (144).[9] Although D-503 becomes temporarily attracted to the Mefi, he suspects that they are using him in order to gain control of the *Integral;* moreover, he is unable to repudiate completely the safe and conditioned mathematical framework of his life. The revolution is crushed by the state, which executes most of the insurgents but graciously permits D-503 to undergo frontal lobotomy instead, in order to remove any other incipient fancies. This reeducation program is so successful that D-503, who now feels as if "a splinter has been taken out of my head," is able to watch with equanimity while I-330 is tortured to death.

The indictment of scientism in *We* is subtle but relentless, more clearly expressed than in the two other famous dystopias, by Huxley and Orwell, which followed it. By the recurrent use of scientific logic and particularly mathematical language (the "numbers" of the state are designated as plus or minus according to their usefulness), Zamyatin emphasizes the central issue, namely, that the basis of dystopia is a total capitulation to safe, rational, mechanical, and hence inhuman patterns of thought and behavior, to the exclusion of the emotional and creative life of the individual.

In Aldous Huxley's *Brave New World* (1932) the external uniformity of Zamyatin's *We* is extended to the realm of psychology, producing identical components of the social machine, conditioned as fetuses not only to perform the tasks allotted to them but to enjoy them. Like Wells and Zamyatin, Huxley depicts scientific rationalism as closely associated with a totalitarian regime, since philosophically both require the subjection of the individual to the system, and both Zamyatin and Huxley perceive scientists as the dehumanized products of the totalitarian state. In the postwar United States, however, the "system" that subdues individuals is more frequently identified by writers with the military-industrial complex. An early example of this conflation is Algis Budrys's novel *Who?* (1958), in which the symbol of the nonhuman scientist is rendered almost literally as a new form of robot.[10] When Lucas Martino, a brilliant physicist working in West Germany on a military device called K-88, is seriously injured by an explosion in his laboratory, situated near the East German border, a Soviet medical team quickly whisks him off to a hospital. In due course a bionically engineered "person," half resembling Martino and half metal, returns across the border. The FBI spends many weeks trying to establish whether the returned man is the original Martino, a Soviet-convert Martino (the product of brainwashing by his captors), or a Soviet impersonator. Ironically, the FBI is never satisfied that it has discovered the truth, for the bionic Martino, having no personal fallibilities to be manipulated, now represents not only the ideal spy but the least accessible secret agent. The FBI plays safe. In order to preclude any possibility of the secret of K-88 being passed to the Soviets, Martino is removed from the project, even though this means the end of K-88 altogether. Ironically, Martino, inside his metal case, remains loyal, having resisted the Soviet attempts to brainwash him. He thus represents, among his other allegorical aspects, the impotence of science brought about by the very military security systems intended to safeguard it.

At least as interesting as the political intrigue, however, is the characterization of Martino himself. We find that as a student Martino had believed that the universe was "constructed of perfectly fitted parts" (37), and in a series of flashbacks we see that he was never able to form meaningful human relationships, because physics always came first for him. After graduating, he progressively lost what humanity he had, becoming the property of a succession of virtually identical military research establishments. Half inanimate, literally as well as symbolically, this image of the scientist represents a threat to society, partly because of his superior knowledge and power in some areas but also because of his "otherness," as he remains permanently isolated from the human fulfillment he seeks. When Martino tries to renew former acquaintanceships, people flee in terror from his well-intentioned efforts as though from Frankenstein's Monster. Martino is thus a symbol of the depersonalization of the scientist resulting both from his own inclination and from manipulation by others.

The American novelist Thomas Pynchon was a technical writer for Boeing at the time when that company secured the prime contract for the U.S. Apollo *Saturn V* rocket. A year later he left Boeing to write his first novel, *V* (1963), about a fictional company, Yoyodyne, Inc., which grows into a postwar aerospace giant, making gyroscopes for aircraft and missiles and becoming intimately involved in military-industrial collaborations. Three years later Pynchon published *The Crying of Lot 49* (1966), in which the same firm's Galactronics Division extends its operations to the missile and spacecraft business. Pynchon relates the figure of the inhuman scientist to a vast social and philosophical system that, by its very nature, defies coherent description and analysis. His classic novel *Gravity's Rainbow* (1973) seeks to account for the meteoric rise of the American Rocket State, in which unimaginable sums of money were heaped on the National Aeronautics and

Space Administration (NASA) in its joint role as the national flagship of U.S. international prestige and the cardinal element of the cold war.

Gravity's Rainbow explores a world where the previous historical "explanations"—Newtonian mechanism and statistical physics—which offered a measure of certainty about the theoretical rationality of the universe, have been replaced by the irreducible uncertainty of quantum physics. Pynchon shows how the various systems of science and technology have all created their own servants with a mechanical, reductionist explanation of the world, a statistical perception, or a fragmented picture of "reality." Set in Europe near the end of the Second World War, *Gravity's Rainbow* uses the contemporary sociopolitical breakdown as a symbol of the fragmentation that characterizes twentieth-century thought. The symbol of the new order is the V-2 rocket, which was fired on London and, traveling faster than sound, crashed *before* the sound of its approach could be heard, a violation of the classical system of causality.

Pynchon's characters represent the various traditional systems of scientific thought. Pavlov's experiments with conditioned responses in dogs were essentially an attempt to extend Newton's principles into the biological area, and Pointsman, the Pavlovian character, still espouses a clockwork view of the animate as well as the inanimate world. Seeing man himself as a machine, he assumes that human behavior can be predicted in the same way that the engineers plot the path of the rocket using the principles of Newtonian physics. The German rocket engineer Franz Pökler, whose career closely parallels that of Wernher von Braun,[11] is another Newtonian figure, as is Marcel, the mechanical chess player, who is even more symbolic of this reductionist view of the universe, since he becomes indistinguishable from a human being. Yet another character, Roger Mexico, works with statistical models to try and overcome the small perturbations and unforeseen forces. By contrast, the engineer Beláustegui is "a prophet of science" who accepts that every moment has its own value, unconnected with any other, and that chance is the only law. Anti-paranoia, "where nothing is connected to anything, a condition not many of us can bear for long,"[12] is the last desperate response in this bleak contemporary world, which represents the antithesis of classical mechanics. Thus the GI Tyrone Slothrop, engaged in trying to find out how the rocket is assembled, himself becomes disassembled. Pynchon does not develop his characters in depth as individuals; rather they contribute to a panoramic picture of the major thought systems whereby scientists from the seventeenth to the twentieth centuries have attempted to make sense of the universe. All are essentially impersonal in their basic assumptions; hence all regard man as something less than human—as a machine, a set of statistical predictions, or a random array of chance molecules. In Pynchon's view, both extremes—the classical belief in order and determinism on the one hand and the belief in disorder on the other—are equally sterile for humanity.

Dehumanized Individual Scientists

Whereas the characters considered in the preceding section were presented as the more or less inevitable consequences of their society's preoccupation with scientific materialism, many of the more individualized scientist characters of twentieth-century literature have been depicted as active agents in the depersonalization of Western society. Rather than merely gravitating towards a career that rewards their particular strengths, these characters often deliberately exploit their own personality deficiencies in a discipline that empowers them because of their technical qualifications.

Some of the most vituperative attacks on the evils perpetrated by the impersonal, unemotional scientist were made by Aldous Huxley, who had ample opportunity through his family connections alone to observe real-life scientists. His scientist characters, mostly biologists, are presented as being all the more dangerous because they pass for normal in their society, even acquiring honors for their intellectual prowess while they are, in fact, wreaking misery, degradation, and death on those around them. In his novels, the personal failure of these scientist characters is located within a relatively narrow causal range, but the depiction of such characters in successive books became increasingly vicious as Huxley perceived the far-reaching effects on others of the socially accepted peccadilloes permitted to scientists.

Huxley emphasizes two particular aspects in his attack. First, he emphasizes the effect reductionist methodology has on the individual scientist's relations with others. Thus Shearwater, the physiologist of *Antic Hay* (1923), modeled on the biologist J.B.S. Haldane, is so involved in his experiments with one part of the human body, the kidneys, that (like Shadwell's Sir Nicholas Gimcrack, and for similar reasons) he is quite oblivious of his wife's infidelity. A more sinister example of such reductionism is provided by Dr. Sigmund Obispo, the biologist of *After Many a Summer Dies the Swan* (1939), whose particular form of sensuality, considered by himself as intellectually refined, involves tormenting the object of his lust in order to study her reactions, as though she were one of his laboratory animals. Obispo's first name is an obvious reference to Freudian psychology, which Huxley regarded as demeaning of human beings in its reductionist stance. Significantly, Obispo's research into artificially increasing the human life span unearths results very different from his glorious expectations: longevity is found to be inextricably associated with regression, with a senile degradation to the savagery of nonhuman ancestors.[13]

Second, Huxley emphasizes scientists' retardation in all areas other than the intellectual. Lord Edward Tantamount, the eminent biologist of *Point Counter Point* (1928), a fictional portrait of J. B. S. Haldane's father, J. S. Haldane, professor of physiology at Oxford, is described as being "in all but intellect a kind of child." He immerses himself in his laboratory because he is unable to cope with the world of living people and relationships. "In the laboratory, at his desk, he was as old as science itself. But his feelings, his intuitions, his instincts were those of a little boy. Unexercised, the greater part of his spiritual being had never developed."[14]

Huxley's most detailed depiction of an intellectually brilliant but emotionally underdeveloped scientist was Henry Maartens, of *The Genius and the Goddess* (1955). In the world of science he is preeminent and inspired—"working with him was like having your own intelligence raised to a higher power"—but outside his laboratory he is "a kind of high-class monster," says Rivers, the narrator and physicist who works with him.[15] In personal relationships he is a child, selfishly dependent on his mother-figure wife Katy. His chronic hypochondria becomes a weapon to enforce total devotion from her, for Henry is adept at inducing a bout of pneumonia with the most extreme symptoms, ensuring her return from her dying mother's bedside. Totally immersed in himself, Henry is unable to communicate with others, even his wife and children. He lives "in a state of the most profound voluntary ignorance . . . abounding in preconceived opinions about everything." He speaks with authority on education, and indeed he has all the theories on the subject stored away in his filing-cupboard mind, but he can never see their relevance, because he has no interest in actual people.

> He had read Piaget, he had read Dewey, he had read Montessori, he had read the psycho-analysts. It was all there in his cerebral filing cabinet, classified, categorised, instantly available. But when it came to doing something for Ruth and Timmy, he was either hopelessly incompetent or, more often, he just faded out of the picture. For of course they bored him. All children bored him. So did the overwhelming majority of adults. How could it be otherwise? . . . humanity was something in which poor Henry was incapable, congenitally, of taking an interest. (70–71)

Thus he is incapable of communicating with his own children even about their mathematics homework: "Henry had lived so long in the world of Higher Mathematics that he had forgotten how to do sums" (38).

Unlike his apparent predecessor, Sir Nicholas Gimcrack, however, Henry is more than a figure of comedy. In Huxley's view, he represents an insidious evil in society, first, through his emotional blackmail of others and his failure to fulfill his obligations in human relationships, and second, because his complete divorce of theory and practice, and his lack of interest in the latter, mean that he allows his research to be used for destructive ends.

Huxley makes brief but specific reference to the scientists who worked on the atomic bomb during the war. Later in his life Henry makes a series of tape recordings reminiscing "about those exciting war years when he was working on the A-Bomb! Of his gaily apocalyptic speculations about the bigger and better Infernal Machines of the future!" (127).

Yet Huxley also presents Henry as a pathetic, less-than-human figure, an empty shell as bereft of contact with himself as with others. "Between the worlds of quantum theory and epistemology at one end of the spectrum and of sex and pain at the other, there was a kind of limbo peopled only by ghosts. And among those ghosts was about seventy-five per cent of himself" (71). Of the tapes, Rivers remarks: "You could have sworn that it was a real human being who was talking. But gradually, as you went on listening, you began to realise that there was nobody at home. The tapes were being reeled off automatically, it was the *vox et praeterea nihil*—the voice of Henry Maartens without his presence" (127).

Nearly all Huxley's individual scientists are characterized by an intellectual development at the expense of all other aspects of personality and are hence retarded as human beings.[16] They therefore constitute a danger to themselves and to those near to them, as well as to the community at large, insofar as they receive a mandate to promulgate research that is fundamentally dangerous to the integrity of the individual. Extended to a whole society, this attitude is the basis of *Brave New World*.

If Huxley had a rich field of observation to draw upon in his depiction of scientists, Charles Percy Snow had an even more intimate knowledge of the procedures of the research world, and his depictions of scientists in the laboratory must carry particular weight for their realism and revelation. Even in his first novel, *Death Under Sail* (1932), Snow suggests that the single-minded pursuit of science betokens a failure to grow past the childhood desire for the secure, the familiar. One of the characters summarizes the limitations of another, William Garnett, who, though practical and efficient in matters mechanical and set for a brilliant future in science, is emotionally retarded: "He's emotionally underdeveloped in lots of ways. . . . Like a good many scientists, he's still aged fifteen except in just the things that his science encourages. . . . I mean he's at home with *things*, at home with anything where he can use his scientific mind—but very frightened of emotions and people, utterly lost in all the sides of life which seem worthwhile to most of us. He can run a society . . . but he could never run a love affair."[17]

The characters of Snow's novel *The New Men* (1954), working at the fictional equivalent of The Atomic Energy Research Establishment at Harwell on the development of the atomic bomb, will be discussed as examples of the amoral scientist in chapter 14, but his most authentic and partly autobiographical study of an individual scientist lacking emotional qualities is the character of Arthur Miles, whose career is traced in *The Search* (1934).

As a child, Miles was first introduced to the world of science by means of a telescope, often used in literature as a symbol of being distanced from the subject; later he is encouraged to make a career in physics, focusing all his energies on research. When he first finds himself attracted to a girl, he thinks: "No human being has really mattered up to now. . . . The only thing outside myself has been my work. Is this girl going to upset *that?*"[18] It soon becomes evident that the answer to this question is no. After remaining on the periphery of his life as he doggedly pursues a research career at Cambridge during the years of Rutherford's work on the disintegration of the atom, the woman finally rejects his proposal on the grounds that "It'd be as bad as marrying a man with a faith. It *would* be marrying a man with a faith. . . . You oughtn't to marry. Perhaps you ought to be celibate" (120).[19] Thus far Snow's characterization follows the conventional Romantic pattern of the scientist's rejection of relationships in favor of research, but he then introduces an interesting variation on the theme, for Miles, after some reasonable successes and minor setbacks in his laboratory, loses his enthusiasm for research and seeks instead administrative power as director of a prestigious new scientific institute. Ironically, although he had earlier resisted the temptation to further his career by dishonestly removing a damaging piece of counterevidence from an otherwise convincing set of results, he now misses out on being appointed founding director of the Institute of Biophysical Research because he has failed to check one of his assistants' experimental results before publishing an important paper in *Nature* and is thus responsible for an unintentional mistake in the evidence. Although he accepts the justice of his rejection, he is unwilling to rehabilitate himself in the scientific community through years of hard work in a second-rate post.[20] Like H. G. Wells and Snow himself, Miles turns instead to literature, writing novels on the sociology of science.[21]

In his characterization of Miles, then, Snow reverses the normal causal sequence. Most of the characters discussed in this chapter are shown to be emotionally deficient because of their exclusive commitment to science, but Snow suggests a two-way causality: Miles fails to become excited about science precisely because he is deficient in emotions. In retrospect, then, it is not surprising that Miles has been shown as more interested in manipulating committees and administrative posts than in the hours of slog in the laboratory. The former are concerned with power, the latter with dedication.

If this lack of feeling is considered a limitation in a biochemist, it must inevitably appear more so in a medical scientist, whose research impinges directly on his patients. This is the major theme of Irving Fineman's novel *Doctor Addams*, which focuses on a brilliant medical biophysicist whose research into electrophoresis has been applied successfully in a wide range of medical and industrial problems. Addams's sinister qualities are at first masked by the immediate benefits of his work to his grateful patients, who are unaware of his dismissive attitude towards them. Fineman explicitly subverts the stereotype of the dedicated doctor by insisting that Addams has no interest in medical practice or even the small but necessary clinical component of his research. In his clinic the patients are selected, not on the basis of their need for treatment, but according to whether their complaints are useful to his research. *Doctor Addams* was one of the first novels to raise the specter of artificial interference in the process of conception, and Fineman leaves no doubt that in his view such a pursuit is possible only for an impersonal scientist, who regards his patients as so many objects. When Addams's research proves to have application in both contraception and uterine contractions, he, like Frankenstein, clearly sees himself in the role of the Creator, far above those to whom he dispenses joy or sorrow. Indeed, unlike his father, the traditional, kindly general practitioner, Addams despises humanity and sees no future for it. He writes to his father: "In this pursuit of mine . . . I must admit, frankly, to no concern for such visions, optimistic or otherwise, which you insist that men of good will, and especially scientists, should have. Nor am I particularly concerned for the American dream of democracy. . . . To that tradition [of science] I am indeed so devoted that the idea of any further ameliorative service I might perform seems inconsequential."[22]

For Addams, research "had to do with the persistent breakdown of an infinite and timeless barrier of darkness, against which the daily disorders and confusion of men became insignificant and negligible" (292–93). The clearest condemnation of Addams comes from one of his students, who in the middle of Addams's discourse on the movement of blood cells in an electric field recalls him to the fact that the patient is dying: "It was startling to hear the word 'patient'; he had been thinking of tissues and electric potentials, of a process not of a human being" (48).

For the reasons outlined at the beginning of this chapter, the unfeeling scientist in recent literature is more likely to be a mathematician or computer scientist who has become identified with the tools and criteria of his discipline and obsessed with the desire for mathematical clarity and order in all areas of life. One reason for this fascination with mathematics is the apparently astonishing success of mathematics and physics in imposing an orderly explanation on events, by the relatively simple process of discarding as irrelevant those events that do not fit the statistical prediction. Until the rise to preeminence of chaos theory, which sought to mathematize the anomalies as well as those phenomena that conformed to the rules, events that challenged the mathematical thesis were discarded as experimental error. A further reason, and one that recurs in twentieth-century literature, is the suggestion that pure mathematics offers an escape from the complexity of emotional entanglements.

Max Frisch's drama *Don Juan oder die Liebe zur Geometrie: Eine Komödie in fünf Akten* (1953) presents a mathematician who wishes to escape totally from the changes and chances of this uncertain life in order to study geometry as a pure philosophical abstraction. Unlike the authors considered previously in this chapter, Frisch is not interested in either condemning or applauding his protagonist, but only in explaining him. His choice of the unlikely figure of Don Juan, the traditional libertine, to embody such a character may be intended to indicate some skepticism about the Don's alleged desire to escape from relationships in order to embrace the static absolute of mathematical laws as the ultimate perfection of human experience. Equally, it may be Frisch's attempt to locate the psychological basis for the legend in a restless dissatisfaction with the emotional complexities epitomized by love relationships. Fleeing from his fiancée, Juan tells his friend Roderigo: "I feel freer than I have ever felt before, Roderigo, empty and alert and filled with the masculine need for geometry. . . . Have you never experienced the feeling of sober amazement at a science that is correct? . . . at the nature of a circle, at the purity of a geometrical locus. I long for the pure, my friend, for the sober, the exact. I have a horror of the morass of our emotions. . . . Do you know what a triangle is? It's as inexorable as destiny. . . . No deception and no changing mood affects the issue."[23]

Secure in the absolute certainty of two- and three-dimensional geometry, Juan becomes visibly upset when the bishop, representing a metaphysical view, imports the subject of four-dimensional geometry, for this introduces complexities and uncertainties, the mathematical counterpart of the emotional entanglements from which he is fleeing. In Frisch's treatment, Don Juan's dissatisfaction with women arises because they remind him that he, unlike a geometrical figure, is not complete in himself. "My dislike of Creation, which has divided us into men and women, is more intense than ever. . . . What a monstrous mistake that the individual alone is not a whole!" (150). It is because of his determination to escape from emotional entanglements that Frisch's Juan commits the infidelities and murders associated with the legendary Don. In his postscript to the play Frisch adds: "If he lived in our own day, Don Juan (as I see him) would probably concern himself with atomic physics: in search of ultimate truth. . . . As an atomic physicist too he would sooner or later be faced with the choice: death or capitulation—capitulation of that masculine spirit which obviously, if it remains autocratic, is going to blow up Creation as soon as it possesses the technical ability to do so" (158–59). Frisch might, alternatively, have made his Don a computer scientist. Artificial intelligence epitomizes reason undiluted by other aspects of the human condition, in particular, physical and emotional areas of experience; hence the literary analysis of computers and the responses of scientists to them offers the basis for a modern parable about the unbalanced cultivation of those traits.

Almost without exception, the computer scientist is depicted as wholly absorbed in his (and such characters are universally male, a point that has not escaped feminist critics) computer as the focal icon of his life, an end in itself rather than a means to achieving other goals. This may be seen as an oblique statement of the perceived high-priestly stance of computer experts, who insist on preserving their intimate knowledge of computing as uniquely theirs, inaccessible to the layperson. The resultant irony explored by several writers is that such scientists eventually become subservient to the computer, sacrificing their own human traits in order to become more "acceptable" to it. Thus, although originally they are attracted to the computer because it appears to give them power and control, they end up being controlled by it. It is therefore not surprising that this situation has come to characterize the depersonalized scientist in modern literature. It is, in fact, a more insidious version of the robot-cloned society Capek depicted in *R.U.R.*

In his short story "Die Schalttafel" (The switchboard) (1956) Erich Nossack describes a chemistry student who has mapped out his plan for life on a switchboard, so that at each moment he can calculate the possibilities and probable consequences of any proposed action.[24] His purpose is to choose always the most conformist course. That is, he has effectively programmed himself to become as much like a computer as possible, eliminating freedom and spontaneity.

A similar critique of pure reason underlies Heinz von Cramer's modern parable "Aufzeichnungen eines ordentlichen Menschen" (Notes of an orderly person) (1964), about a mathematician who has constructed for himself a completely regulated life as insulation from the outside world. For him, order has not only a pragmatic value but also the quality of metaphysical truth: "I love order. I am a scientific man, not merely a scientist. That is, my profession has taken possession of me, body and soul. However, order has, for me, also a metaphysical meaning; a threshold on the other side of which it becomes apparent as a universal principle, a principle of creation."[25]

The logical extension of such obsession with order is his progressive identification with a supercomputer, and in von Cramer's story this symbolic truth is actualized. The mathematician comes to consider his computer as a living organism, to which he ascribes emotions and personal needs. When the computer performs operations for which he has not programmed it, he feels subordinate to it and makes an unsuccessful and increasingly frantic attempt to establish that he is really its master.[26] Unable to communicate with the computer through his own language, he resorts to its limited binary language—0/1—thereby effectively reducing himself to its level. He then enters into a personal contest with the computer, progressively seeing himself as a modern Frankenstein at the mercy of his own creation. To his frenzied and hallucinating mind the machine appears to transmit to *him* the commands with which he has programmed *it*, to observe him, analyze him, and rule his life. His last, irrational diary entries indicate his Kafka-like metamorphosis into an inferior computer.[27] Thus von Cramer locates the causes of computer dominance in the minds of the programmers, who endow their inanimate machines with human qualities while relinquishing their own humanity.

The seductive aspects of mathematical order as a refuge from the frightening complexities of life have also been explored by the Austrian writer Hermann Broch, who himself studied mathematics and philosophy. Richard Hieck, the protagonist of Broch's *Die unbekannte Größe* (The unknown quantity) (1933), is a mathematician and astronomer who places all his faith in mathematical certainty and logical positivism. Mathematics seems to be an island of decency in a world of uncertainty and irrationality, and he expects it to provide him with stability and assurance and, ultimately, to lead to the understanding of the whole of life. His intention is that of Descartes, "to grasp all the appearances of Life, to grasp them mathematically and rationally, because through this prolific 'mathematicising' of the world, he would arrive at the totality of his own life."[28]

In his attempts to exorcise from his personality all irrational elements, including emotion, Hieck comes, paradoxically, to ascribe a religious dimension to mathematics, which he now regards as a deliverance from sin and as the source of moral probity. "Within mathematics, something could be accomplished only by those human beings who remain pure from all that is sinful or whatever else one wants to call it."[29] When Hieck invites his girlfriend, also a mathematician, to visit his observatory, he becomes increasingly conscious of the inadequacy of the Einsteinian system he is expounding to account for the emotion he is experiencing. Yet, despite this incident and the warnings of his elderly professor, Weitprecht, that mathematics is no substitute for life, it takes the death of his brother to bring Hieck to the full realization that scientific knowledge expresses only part of a much broader mystical truth.[30]

Broch related this realization to the theory of relativity itself, which he saw as an affirmation of the integral relationship between the observer and his field of observation. Ironically, he regarded many scientists, preoccupied as they were with the search for objectivity, as slow to accept the full implications of this central precept of modern physics.[31] Hence the paradox of Hieck expatiating on the cosmology of relativity theory while trying, in effect, to subvert it by remaining objective and impersonal.

Stephen Sewell's play *Welcome the Bright World* (1982) goes further than exploring the effect of such computer identification on the devotee; it suggests a close connection between the allegiance to inanimate conceptualizations of the world, the denigration of people, the rejection of accountability by the scientists involved, and potential destruction of the world by a nuclear war. *Welcome the Bright World* depicts two nuclear physicists, Max Lewin and Sebastian Ayalti, who have invested all their emotion in trying to understand the nature of the electromagnetic forces that hold the atom together and the radioactive disintegration of nuclear particles. In the notes for the play's premiere, Sewell wrote: "This work is currently moving rapidly ahead. Unfortunately, and tragically, so is the work of those scientists and technologists involved in the development of even more hideous nuclear weapons." Within the play he explores this anomaly by showing how these highly enthusiastic and committed scientists insist on remaining detached in their deteriorating personal relationships and in the worsening political situation of a virtual police state. Early in the play, Sebastian and Max discuss the scientist's duty to explain to the public the dangers of nuclear energy, Max arguing for the view that scientists are beyond accountability:

> MAX: How can you talk about something as complex as nuclear energy in front of ten thousand people?
> SEBAST: People have the right to information.
> MAX: Most people don't count, Sebastian![32]

Later Sebastian too finds it expedient to adopt a cosmic perspective of events, since this permits him to ignore social responsibility on a human time scale. Discussing the concept of a black hole with Rebekah, he tells her that it is ten thousand light years away:

> REBEK: Everything's so big.
> SEBAST: That's not very far. . . . Ten thousand years is nothing—the end of the last Ice Age. Human beings had already sent the mammoth into extinction.
> REBEK: How old is the universe?
> SEBAST: Between fifteen and twenty billion years. (25–26)

As in Hardy's novel about an astronomer, *Two on a Tower*, discussed in chapter 8, this astronomical time scale is seen as having a depersonalizing effect as physics becomes an escape from personal commitments. When Rebekah asks Sebastian what he likes about physics, he answers:

> SEBAST: I always wanted to understand the world. . . . I'm an obsessive.
> REBEK: What are you obsessed with?
> SEBAST: Knowledge and discovery are like drugs. . . . Schopenhauer sought relief from the will in art. . . . contemplating nature gives me the same relief.
> REBEK: From the will?
> SEBAST: From involvement. (51)

Sebastian later proclaims, "I only have one relationship, and that's with my work. . . . I didn't want to get involved; I didn't want to have any scenes; I only want to do my work. . . . I don't want relationships, I don't want involvements!" (62–63).

Mathematics and the particular dehumanizing effect it has on its devotees formed the basis of much of the work of the Austrian writer Robert Musil. In his early work *Die Verwirrungen des Zöglings Törleß* (The confusion of young Törless) (1906), his adolescent schoolboy protagonist is obsessed by a desire to understand life "with the eye of reason," that is, in the simple, objective terms of mathematics, as a means of escaping from the frightening depths of emotion and mysticism that he dimly perceives are also part of life. Impatient with simplistic analogies and the master's patronizing assurances that he will be able to understand it when he is older and has mastered Kant, Törless craves the certainty that mathematics, alone of all subjects, seems to offer. Yet even in the sphere of mathematics, the inexplicable, the mathematical correlative of the mystical, haunts him (as it had D-503 in *We*) in the form of such concepts as infinity and imaginary numbers such as the square root of -1. Simultaneously, Törless is drawn into an intensely emotional situation of brutal cruelty and exploitation, from which he escapes only after a virtual nervous breakdown. Like Broch's mathematician Hieck, Törless thus learns the lesson, foreshadowed in both the *Verwirrungen* (confusion) of the title and the introductory epigraph from Maeterlinck, which proposes that for everything there are always two aspects—reason and mysticism—that cannot be reconciled.[33]

In Musil's later story "Tonka" (1924), in the *Drei Frauen* volume, these two perspectives are embodied in different characters.[34] The young scientist, who, significantly, remains nameless, represents the rational component developed at the expense of other aspects of life. Tonka, the servant with whom he enters into an affair, embodies the mysterious, irrational forces of nature. The sex roles ascribed to these opposites are certainly intentional. For Musil, it is the male mind that reaches its fullest intellectual development in the conscious activities of logic, science, and verbal articulateness, so his scientist gains strength and confidence from his work on a new invention. Tonka's needs, on the other hand, are emotional rather than rational; like nature, she seems merely to *be*. The ensuing conflict between reason and mysticism is tragic for Tonka. When she becomes pregnant, and there is doubt about the time of conception, the scientist demands irrefutable empirical evidence for or against her possible infidelity; he cannot accept a truth that must be based on trust in the girl. He therefore abandons her, although he is equally unsure of her infidelity. The completion of his invention on the day Tonka dies represents the triumph of reason over the emotions and mysticism; yet, paradoxically, her death represents both the only temporary resolution of the problem and the permanent impossibility of any final solution to the uncertainty. Although it is certainly possible to read the story as a vindication of Tonka and the intuitive, emotional perspective she represents, Musil's own evaluation is less clear. Here, as in *Törleß*, he remains ambivalent, for there is also an implicit Nietzschean justification of the strong, intellectual male in vanquishing the weak emotional female.

Musil was unable to leave the topic. In his epic novel *Der Mann ohne Eigenschaften* (The man without qualities) (1930–43), the protagonist, Ulrich, like Musil himself, has tried several careers—the army, mathematical research, engineering, the public service—but he remains throughout a pure mathematician in his noninvolved approach to life. As a representative of scientific method, he analyzes every thought and feeling with disinterested objectivity, ceaselessly dissecting, questioning, and measuring every experience as an abstraction until, as the title suggests, he loses any identity and cohesion as an individual. Ulrich's obsession with order arises out his quest for meaning. At the start of the novel, he has given himself one year to discover meaning in life; otherwise he will commit suicide. Thus, like Törless and the scientist of "Tonka," Ulrich sees the world of abstract mathematical thought as a bulwark against the unpredictable dangers of the emotions and meaninglessness.

Embarking on his "hypothetical life" as on a mathematical experiment, Ulrich remains detached, skeptical, a relativist. While to the other characters in the novel, this stance seems intolerable, impersonal, and destructive of human relationships, Musil suggests that it has its positive side too—Ulrich's lack of prejudice and his openness to the future. He regards himself as a man of possibilities, a *Möglichkeitsmensch*, continually speculating on the options available but never committing himself to any one of them. As long as they are merely conjectures, they may be discarded at any time without regret. "He sensed that this order was not so firm as it pretended to be; no thing, no self, no form, no principle was certain. Everything is caught up in invisible but never-resting metamorphosis; in the unfirm lies more of the future than the firm, and the present is nothing but a hypothesis which one has not yet gotten beyond."[35]

Ulrich undertakes two experiments: a "rational experiment" designed to find an intellectual basis for understanding life and a "mystical experiment." Like Musil, he believes that these alternative, even opposite, roads to truth must be combined into a single creative road and that this is the greatest challenge awaiting modern man. But it is the first experiment that concerns us here.[36] In this phase of his life, Ulrich commits himself to scientific method and the principles of mathematics with the utmost rigor. His friends, predictably, find him cold, hypercritical, lacking in compassion, and, withal, strangely impractical, for it is part of Ulrich's commitment to pure science that it should not be required to be useful. Musil calls modern science "the miracle of the Anti-Christ" (360), and he notes Ulrich's "tendency towards malevolence and hardness of heart" (359).

However, Musil's depiction of Ulrich remains ambiguous. On the one hand, he towers over the other characters, with their limited, egocentric grasp on life; on the other hand, he fails to attain even the partial understanding of life he had expected. Even limited solutions evade the relativist. Yet Musil, writing about the futility and violence of Austrian society in the last days of the Hapsburg monarchy, does not imply that Ulrich's failure to discover a structure and order among the disorder and pluralism of his society is evidence of his own deficiency. On the contrary, the novel suggests that although belief in a firm, universal order is a delusion, disinterested objectivity may be the only appropriate stance in a time of ideological breakdown.

If Musil vacillates about the validity of the impersonal scientist's objective stance, Isaac Asimov has no such reservations.[37] His many stories about robots and the reactions of their scientist-"minders" constitute a striking contrast to the view of most of the writers discussed in this chapter about the effect of "intelligent" machines. Asimov, himself a biochemist and author of a large number of books on popular science, was scornful of what he called the "Frankenstein complex" and fundamentally optimistic about technological progress, which he, like H. G. Wells before him, saw as relieving humanity of "those mental tasks that are dull, repetitive, stultifying and degrading, leaving to human beings themselves the far greater work of creative thought in every field from art and literature to science and ethics."[38]

Asimov saw danger from robots and computers only for those who feared change and had thus, in effect, already abrogated their autonomy in a technological age, effectively becoming like the machines they attacked. This is made apparent in his story "Profession" (1957), set in a future world where most of the inhabitants, fearful of change, have had their brains wired and programmed to act in a routine fashion so as to avoid the agony of decision-making. Like the citizens of E. M. Forster's "The Machine Stops," they have become voluntary appendages of the machine as a result of their reactionary paranoia, for, according to Asimov's treatment, their state of total dependency is of their own making.

Asimov's most popular sequence of stories, beginning with "I, Robot" (1950), was based on an exploration of robots as essentially "benign," being controlled by the Three Laws of Robotics, which Asimov enunciated for the protection of society against a takeover by robots such as Capek depicted.[39] The reader is confidently assured that all will be well in the wonderful world of robots. However, contrary to Asimov's intention, the stories themselves subvert this contention, for not only are the human characters upstaged by the robots but the more "favored" human characters seem, themselves, to be governed by the Three Laws of Robotics. The robots are the real heroes of these stories; where problems in their functioning occur, it is nearly always because of a failure in perception or logic on the part of the humans. The interest of the stories thus centers, not on human qualities and emotions, but on the laws of logic and intellectual wordplay; that is, the plot interest is determined by the rules of the robots and, in the later stories, the computer.

Asimov's most recurrent character in these stories, and the one closest to the robots, is the robopsychologist Susan Calvin, an unusual female version of the unemotional, completely rational scientist stereotype. "Susan Calvin talked about Powell and Donovan with unsmiling amusement, but warmth came into her voice when she mentioned robots."[40] Her only departure from this strict rationality occurs in the story "Liar!" (1941), in which she too falls prey to the emotion of vanity, even acquiescing in the effective death of a favorite robot, Herbie, in order to save face about her mistake. Thus, apart from her impeccable logic and rigidly intellectual approach to the world, Calvin's only other demonstrated traits are far from admirable. On the other hand, in "Evidence" (1946) the most "ethical" character in the robot stories, the politician Byerly, is accused of being a humanoid with a robot's brain, and Calvin concedes, "Actions such as his could come only from a robot, or from a very honourable and decent human being. But you see, you just can't differentiate between a robot and the very best of humans."[41]

The approved stereotype emerging from Asimov's stories is a thoroughgoing materialist and pragmatist who, without a qualm, exploits the solar system in the name of efficiency and human imperialism. In *View from a Height* (1963) Asimov discusses the most efficient means of colonizing the other planets, where existing life could provide an immediate source of food for the prospective Terran colonists; the use of space as a garbage dump for radioactive waste; and an ingenious real-estate scheme for selling off planets broken into asteroids. Indeed, Asimov's heroes bear a striking resemblance to Lewis's archvillain, Weston.

A similar, optimistic view of artificial intelligence informs Frank Herbert's novel *Destination: Void* (1966). His four scientists aboard the spaceship *Earthling*—a psychiatrist, a life-systems engineer, a doctor who specializes in brain chemistry, and a computer scientist—represent the four disciplines most closely allied with the understanding and development of cognitive science. In the critical circumstances that attend their lone journey through space, they come to the realization that their survival depends on developing high-level artificial intelligence. Herbert's view is clearly that machine intelligence in cooperation with human intelligence is our only hope for the future and that scientists are therefore indispensable for the very reasons that led to their vilification by the majority of writers represented in this chapter.

In stark contrast to Asimov and Herbert's optimistic assessment of man-machine symbiosis is the bleak, postholocaust vision of Philip K. Dick, whose mad scientist Dr. Bloodmoney was considered in chapter 12. Like Asimov, Dick insists on the inevitability of machine intelligence, and he consistently refers to both anthropomorphic electronic constructs and humans who have become machinelike in their behavior as "androids" rather than as robots. Dick's two nonfictional statements on this subject, "The Android and the Human" (1972) and "Man, Android, and Machine" (1976), focus on the centrality of the relationship between humans and machines, the question who is really human.[42] In the former he defines the characteristics of the android mind that distinguish it from the fully human as deficiency in feelings, inflexibility, unswerving obedience, and predictability—the qualities ascribed to the figure of the scientist by most of the writers discussed in this chapter. Dick, however, is less sure that the impersonal qualities can be so readily identified with one group. In *Vulcan's Hammer* (1960) the machines act as destructive humans, while in *The Man in the High Castle* (1962) the humans have become destructive machines, and in *Do Androids Dream of Electric Sheep?* (1968) Dick explores the premise that man has two brains in the one skull, approximating to an intellectual, unfeeling personality and an empathizing, intuitive one.[43] Thus in Dick's reading, the impersonal scientist is only a special example of twentieth-century man for whom the most fitting symbol is the android.

An interesting feminist analysis of the inhuman scientist is to be found in Christa Wolf's short story "Selbstversuch. Traktat zu einem Protokoll" ("Self-Experiment: Appendix to a Report") (1973). Wolf's story concerns a female scientist, a physiopsychologist specializing in sex-change research and condescendingly regarded by her professor as "the equal of any male scientist."[44] By means of a drug developed by her professor, she has voluntarily undergone a sex change, which is hailed as highly successful. Yet before a month of her life as a male has passed, she rejects the impersonal role imposed on her by her male colleagues and requests a reversal of the process. As in Musil's writing, the sex roles are identified with two different modes of perception. In assuming a male role, the woman of the story is reinforcing the intellectual, scientific aspect of herself and suppressing her feminine, intuitive self. Her demand to be restored to her female role, a demand that for her male colleagues is as embarrassing as it is incomprehensible, results from her actual experience of the limitations of the purely rational perception. Knowing, as the mythical Tiresias does, both sex roles, she opts unhesitatingly for the more complex, nonscientific awareness that Wolf identifies with the female consciousness. The unfeeling professor, who epitomizes the scientific assumption that he has caught reality "fast in a net of numbers, diagrams, and calculations," is exposed as a figure of ultimate weakness. The woman tells him: "Your ingeniously constructed system of rules, your hopeless addiction to work, all your withdrawal manoeuvres were only an attempt to protect yourself from this discovery: that you cannot love, and know it" (130).

In her short novel *Störfall* (Breakdown) (1987), written in response to the Chernobyl nuclear reactor accident of 1986, Wolf further indicts what she sees as the obsessive fascination of male scientists with technical problems at the expense of personal relationships and social responsibility and links it with their isolation from their emotional sources. Specifically targeting the physicists working on the Star Wars project at the U.S. Livermore Laboratories, Wolf compares their situation to that of boys in a boarding school, deprived of relationships with women or families, and asks whether this is the cause or effect of their intense relationship with their computer. "These were men in an isolation station, without women, without children or friends, without any pleasures other than their work, subject to the most rigorous controls of security and secrecy. . . . I have read that they know neither father nor mother, neither brother nor sister, neither wife nor child (there are no women there . . . ! Is this oppressive fact the reason for the computer-love of the young people or its result?)."[45]

Significantly, Wolf also focuses on what she has read about the girlfriend of one of these young Star Warriors. Josephine Stein, the girlfriend of Peter Hagelstein, is a pacifist and protests publicly against Hagelstein's work in developing the x-ray laser for the Star Wars project. Eventually she leaves him, and Wolf casts her, not as Faust's passive Gretchen, but as a new Lysistrata. This is consistent with Wolf's feminist view that women, being marginalized by the patriarchal hegemony, are better able to perceive its defects and to protest against its inhumane principles. So, too, the female narrator of Störfall, whose preoccupations are with growing plants in her garden, nurturing children and friends, and meditating critically on the implications for humanity of the Chernobyl disaster, is contrasted throughout with her brother, a computer scientist currently undergoing surgery for the removal of a brain tumor. His condition symbolizes Wolf's belief that a brain that is overdeveloped in relation to the rest of the person is in fact diseased. This juxtaposition of the personal and the public breakdowns indicated by the title, reiterates the cause-and-effect question quoted above. Is the uncontrolled technology apparent in the breakdown of the Chernobyl reactor a cause or a result of the disproportionately evolved brain of scientists and technologists, who display a correspondingly stunted emotional development?

Wolf's affirmation of complementarity provides an interesting metaphorical parallel with Heisenberg's uncertainty principle and Bohr's complementarity principle, of which it is a special case.[46] Increased knowledge in one area occurs at the expense of decreased certainty in another. We cannot know accurately both the position and the momentum of a given particle simultaneously; we cannot overly develop our technologically focused rationality without sacrificing our emotional well-being, and with it our social and environmental responsibilities. Viewed in this light, the preoccupation of the scientist characters discussed in this chapter with objectivity, measurement, and accuracy of definition takes on an extra hue of irony.

As treated by most of the authors discussed in this chapter (Asimov is, of course, the notable exception), there is at least a suggested link between the impersonal scientist and the amoral one. This in turn carries the implication that morality is associated not with objectivity and rationality but rather with subjectivity and feelings, with sensibilities nurtured by emotional experiences. This connection is made explicit in Carl Zuckmayer's play *Das kalte Licht* (Cold light) (1955).

Factually, Zuckmayer's play is based on the story of Emil Julius Klaus Fuchs, a naturalized British subject of German origin and communist sympathies who worked in atomic weapons research establishments in both Britain and America and passed on secret information to the Soviet Union.[47] However, Zuckmayer's interest lies not so much in the moral dilemma of the scientist working on weapons of mass destruction—the subject of the works examined in the preceding section—as in the more general problem of alienated modern man, bereft of any system of values. The moral rootlessness of his atom spy, Christof Wolters, is symbolized by his involuntary movements from country to country. An emigrant from Nazi Germany because of his communist sympathies, he has applied for British citizenship, but his application is delayed by the outbreak of war, and he is deported to Canada as a civilian detainee. When his background in physics becomes known, he is quickly reassessed as an asset and dispatched to London to help in atomic weapons research. Subsequently he is sent to work with the American scientists at Los Alamos, then returns to Britain, where he adopts the religion of science: "In mathematics there is something—idea and reality. It can only be proved hypothetically, of course—yet it's the basis of our whole existence."[48]

Resenting the restrictions and secrecy imposed by the war and believing that science should be international property, Wolters agrees to pass on scientific results to the Russians. His perfunctory stipulations that the information must not be abused indicate a troubled conscience, but Wolters believes there is no going back, either in science (where the point of no return is symbolized by the explosion of the atomic bomb at Hiroshima) or in one's private actions. He thus affirms the determinist doctrine that Dürrenmatt and Kipphardt were also to examine. "There's no way out. What's done, can't be undone" (71).

Wolters is depicted by Zuckmayer as a flawed, incomplete individual, preoccupied with abstract thought and having "a conscience that somehow short-circuited and a kind of a fatal split in his ethical code" (113). He is a man lost and split (like the atom). This represents an interesting twentieth-century variation of the Romantic image of the scientist as cold and unfeeling.[49] As the title suggests, coldness is a recurrent image of the play. Although the epigraph (allegedly from a textbook on nuclear physics) relates the title to science—"Cold light is a light which cannot ignite, nor does it generate any heat"—and at one level clearly refers to the atomic bomb,[50] the underlying meaning of the phrase is developed in moral terms through contrast with the warm glow of conscience. In what is effectively the pivotal statement of the play Wolters says: "All my life I've been standing in the beam of a cold light—one that comes from outside and turns you to ice. . . . But one single moment can be a great blaze—transforming everything" (141).

In the face of his moral confusion, Wolters comes to affirm the centrality of conscience, the "inner light," as a guide to action, as "an integral part of our bodies—born with us. . . . It's there to help us when we're lost, and remind us that there's a better and finer—a higher order of things, or quite simply to show us the truth" (129). Eventually he chooses to confess his espionage, even though there is insufficient evidence to convict him, discovering inner peace by doing the duty that lies closest to hand: "Good can only be lived, and like charity, it begins at home!" Thus Wolters effectively discards the public world of social responsibility and returns to the pietistic Protestant ethic of individual integrity, the religious background from which he, like Fuchs himself, had originally come.

The next chapter explores the character of the amoral scientist, the contributory causes postulated by writers for the association of science with amorality, and the effects on society of such a stance.

CHAPTER FOURTEEN

SCIENTIA GRATIA SCIENTIAE:
THE AMORAL SCIENTIST

> We scientists have only one God of free research. And this God
> says *"Fiat scientia pereat mundus"*—let there be knowledge though
> the world perish!
> —Heinrich Schirmbeck

A distinctive subset of the stereotype of the impersonal scientist is that of the amoral scientist. Whereas the scientists discussed in chapter 13 affected mainly, perhaps only, their immediate family and friends, the amoral scientist carries the stance of noncommitment into the broader ethical sphere and thus causes major repercussions for his society. Since these scientists are often powerful people, eminent in government policy-making or acting as advisers to the military-industrial complex, their impact may be pervasive and insidious.

Compared with the mad or evil scientist stereotype, scientists who fit the stereotype of the amoral scientist are less readily identifiable as evil; they do not pursue science for power or wealth, but merely for the apparently modest reward of solving an abstract intellectual problem or serving their country. As with the utopian scientists of chapter 11, there is often a deceptive air of idealism about such dedicated individuals who devote themselves to the pursuit of truth, oblivious of personal comfort or material gain. Compared with the barons of industry, they may even appear to be figures of wisdom or innocence. Shades of the Curies hover around them, and Sir Francis Bacon stands benignly in the background. But although the evil is less obvious, it is no less dangerous. By achieving respectability, even fame, these scientists influence others to discredit ethical concerns as irrelevant, even inimical, to science and hence to the development and prosperity of society. Whether explicitly or implicitly, these characters adopt the principle that doing X because X is possible is a prerequisite of science. Indeed, such an attitude is often accepted, both in the media and in the profession, as a guarantee of the highly motivated scientist's credentials.

In literature, however, amorality has rarely been treated favorably, and where it is associated with science the authorial voice has in most cases come down overwhelmingly against the pursuit of truth for its own sake. It should be noted that this adverse presentation of the amoral scientist is almost entirely a function of the literary perspective rather than a reflection of the attitudes existing in actual research establishments or even in the wider community, where an amoral approach generally has not been considered improper. It is only in the last decade or so that ethics committees, which monitor research programs for experiments likely to infringe upon human or animal rights, have been widely accepted within the scientific community; formerly they would have been regarded as a violation of academic freedom, and even now their presence is tolerated only because it is often a precondition for funding.

There are several reasons for the acceptance, even the cultivation, of amorality within the scientific fraternity. The one most frequently invoked is the assumption that science is value-free, that science in itself is neither good nor bad, only its applications. Many scientists believe that they cannot, and should not, have any influence over the use to which their discoveries are put. They see their role as properly focused only on the pursuit of knowledge for its own sake; indeed, they regard this as an essentially noble, even aesthetic, enterprise, morally superior to the petty self-seeking that is perceived to motivate most human endeavor. Enrico Fermi is quoted as having said in relation to his work on the atomic bomb, "Don't bother me with your conscientious scruples. After all the thing is beautiful physics." This attitude was not unique to Fermi. Shiv Visvanathan cites Robert Jungk's encounter at Los Alamos with a mathematician whose face was "wreathed in smiles of almost angelic beauty. He looked as if his gaze was fixed upon the world of harmonies. But in fact he told me later that he was thinking about a mathematical problem whose solution was essential to the construction of a new type of H-Bomb."[1]

Further support for such a stance of amorality comes from the widespread fatalistic belief by scientists that discoveries will emerge inevitably when the intellectual climate is right. Although at the time of their discovery important breakthroughs in research may seem to be the work of particular brilliant scientists, in hindsight it often appears that they would necessarily have arisen within a short time, facilitated by one research group or another. The fact that many major projects in this century have been worked on simultaneously by several research groups in competition adds support for this view. This assumption of the inevitability of discovery became a particularly significant factor during the Second World War, when research on the atom bomb was fueled largely by the argument that the "other side" was close to making the breakthrough first. It was Albert Einstein's letter of 2 August 1939 to President Roosevelt suggesting that Germany might already have begun such a program that initiated the U.S. project to develop the atomic bomb.[2] The belief that the enemy might "get there first" was, of course, exploited both by the military as war propaganda and by the scientists themselves in order to obtain funding. A parallel argument was later used throughout the cold war for exactly the same purposes. Despite this pragmatic and usually unacknowledged subtext, it is clear that the overt argument for multiple possibilities of discovery appears to absolve scientists from specific responsibility, or even, ultimately, any responsibility, for their work. They can claim that they are at most midwives assisting at a birth that would have occurred without their particular intervention and that to refuse to work on a particular project merely allows another research group to reap the honors and rewards.

A related factor in the peer-group acceptance of the amoral scientist has been the growth during the twentieth century of the research team. As science has become increasingly expensive, it has become virtually impossible to finance it other than through the group project. While there may officially be a team leader, who functions as the figurehead (and who may well receive the public honors for the team's success), the other members usually regard themselves as being equally important in achieving the results. Such a system serves to blur the responsibility for the outcome of the project and even to mask the consequences of the research. It therefore becomes easy for individual scientists to disclaim responsibility. Loyalty to the group and to the project overrides accountability to the rest of society.

To most writers from a humanities background, however, these teams of scientists immured in their vast and often secret research institutions are faceless and irresponsible, extending the impersonal focus of their work to justify a position of ethical minimalism. Their disregard of the human factor in their research is interpreted as evidence of their alienation from their own humanity; thus many of the characters examined here are depicted as emotionally retarded, like the impersonal scientists of chapter 13.

In addition, this amoral and pragmatic stance is perceived by many writers as a fundamental characteristic of Western technological society, bent only on extending the boundaries of knowledge and dominating both the natural and the human world. It is frequently suggested that the scientists' amorality arises from their habit of treating both the inanimate and the animate world as a mechanism to be manipulated (*engineered* in its original meaning) and from the related assumption that rationalism represents the highest human endeavor. These are, of course, the postulates of the Enlightenment transposed into the twentieth century; therefore it is not surprising that for most of the century the literary response has been a recapitulation of the nineteenth-century Romantics' opposition to the Newtonian edifice of their time. These writers often regard themselves as modern Cassandras marginalized in a technological society; they see the danger, but their warnings are largely ignored.

Since the 1970s a new perspective on this amoral, rationalist hegemony has been offered by feminist critics. They have identified it specifically with male-devised parameters emanating from gender-linked proclivities, which they explain as the result of the emotionally deficient conditioning and lack of direction experienced by males in a society where the role of raising children has been almost exclusively assigned to women. The former atomic physicist Brian Easlea has explored this hypothesis in detail, both generally in relation to the perceived reification of nature and more specifically in relation to the cult of nuclear weapons research as a masculine alternative to childbirth.[3]

Symptomatic of the paradigm of amorality in science is the realization in the twentieth century of two of the alchemists' dreams: the creation of mechanical "human" beings and the discovery of a source of almost limitless power (see fig. 23). Both are important symbols in the analysis of the technological society, and both feature largely in the presentation of the scientists discussed in this chapter. Amoral scientists in fiction are drawn largely from the ranks of biologists (especially those concerned with genetic engineering), psychologists, mathematicians, physicists, and, more recently, the cyberneticists, who, representing a hybridization of mathematicians and psychologists, are frequently perceived as combining the worst features of both groups. The reason for the emphasis on these particular disciplines is to be sought in their reductionist methodology, which is popularly regarded, especially in the case of biologists and psychologists, as being even more abhorrent than mechanical modeling of the physical universe (see fig. 24).

Physicists too have provided a major focus for the investigation of amorality through their involvement in the development of atomic weapons. Numerous writers have felt compelled to explore the motives that could lead brilliant scientists, many of whom had at some stage embraced a creed of internationalism, to collaborate in producing weapons of unprecedented mass destruction. In *Night Thoughts of a Classical Physicist* (1982) Russell McCormmach suggests how easily a course of action that is fundamentally inimical to both the moral premises of the individual and the postulates of science can be justified in terms of high-sounding principles. His protagonist, Jakob, an aging German physics professor during the First World War, reflects: "It is no betrayal of our scientific creed for us to yield to the dominant pull of patriotism, despite what scientists abroad may say about us. To work for the fatherland, we temporarily sacrifice what is essential to our work as scientists, but we do not for a moment renounce our desire for a peaceful world that knows no national scientific barriers. . . . in science and only in science, [people] work seriously together, in harmony instead of in conflict, in selfless co-ordination instead of in egoistic isolation."[4]

With few exceptions, the writers discussed in this chapter account for this anomaly by portraying their scientist characters as concentrating on the

FIGURE 23.
*The Energy
Explosion.*
Electricity,
personified as a
provocative woman
with evil intent,
liberates and
enslaves the world
in this illustration
by Albert Robida
from *La Vie
électrique*, 1887.

science as an end in itself and rejecting any social or moral responsibility for the outcome. If they are asked how they can devote their lives to research that may lead to the destruction of large sections of society, if not the extinction of the whole species, they reply, in effect, that they themselves are engaged only in an intellectual exercise and cannot be held responsible for what others may do with their research.

Although the attitudes of the writers vary greatly, from satirical condemnation to pity, the questions raised concerning the social responsibility of scientists revolve around the same few recurrent issues: ethical considerations of particular research projects and whether they are appropriate, given the traditional belief that science is value-free; questions about the degree to which scientists can be held responsible for the use to which their discoveries are put by others; the control they can hope to assume; and the certainty with which possible harmful ramifications can be predicted. Invariably the realization emerges that scientists cannot "take back" a discov-

FIGURE 24. *The Test Tube Baby* as envisaged by Albert Robida in *Le Vingtième Siècle,* 1883.

ery; and thus the scientists whose only concern is for their research often lose control over their discoveries.

Because the power of twentieth-century scientists, albeit indirect, far exceeds that hitherto held by any person or group in the whole violent history of the human race, writers have increasingly turned to such characters in their analysis of Western society; thus the number of novels and plays featuring scientists has increased dramatically, especially since 1945. The works chosen for discussion here by no means exhaust the subject, but have been selected as representative. Although the majority have appeared since 1945, there were some important earlier depictions by authors who foresaw the path that a technologically dedicated society would almost certainly follow and endeavored to warn their contemporaries. These early examples, such as *R.U.R,* *Adam Link's Vengeance,* and *The Heart of a Dog,* pose the issues more starkly and diagrammatically than the later works, because they posited a world that was still unrealized. It is illuminating to see how closely the later, more realistic presentations corroborate these prophecies of doom.

Robots

Robots—here I have used the term generically, for mechanical simulations of life forms, in particular the machines that approximate most closely to human behavior—have been particularly important signifiers in literature of the values and attitudes ascribed to their creators. In addition to typifying the hubris of their creators, robots exemplify the substitution of mechanism for man and hence are used to imply severe limitations in the scientists who have created them in their own image. Robotics engineers, it is suggested, see no problem in creating mechanical men, because they themselves have lost their humanity. Indeed, the ultimate symbol of the amoral intellectual would be not a person but a robot. As presented in fiction, robots are essentially hybrids, combining the intelligence of human beings with the unquestioning functionality and programmed pragmatism of machines. They epitomize the criteria of efficiency, rationality, amorality, objectivity, and the pursuit of a predetermined end regardless of the consequences. The Polish cyberneticist-novelist Stanislaw Lem has suggested that ethically a robot with humanlike intelligence must be regarded as a human.[5] In many of the works discussed in this chapter, robots provide such an important symbol that if they had not been actualized, writers would have had to invent them. Essentially this is what Karel Capek did in *R.U.R.,* giving the world the term *robot.*

In mainstream fiction robots are used almost exclusively to encapsulate the amoral mentality that the authors associate with scientists, engineers, and, in some cases, the general ethos of a technological society. For writers with a humanities background, the authorial voice is invariably critical, usually satirical. On the other hand, the robot stories of writers who have come to fiction from a career in science are usually markedly different in tone. In Isaac Asimov's stories, for example, the robots are the heroes, being "morally" as well as intellectually superior to the flawed human characters they so devotedly serve.[6] Asimov enthusiastically endorses the "values" of the robots, who are free from the meaner human emotions, and most of the incidents in his robot series are designed to demonstrate the inferiority of their human attendants. In order to reassure readers that his robots, unlike the accepted stereotype, were nonthreatening, Asimov devised three laws that amounted to a system of ethics designed invariably to favor human beings (see chapter 13 above). Ironically, these laws were accepted and propagated by later science fiction writers as though they had some inherent necessity. In the broader context, however, Asimov's robots are atypical, for they are so humanized that they blur, if they do not actually deny, the issues being explored by the other writers discussed in this chapter. Asimov's robots have spawned a lucrative progeny of "cute," harmless robot characters, popularized in films—the R_2-D_2 and C-3PO models of *Star Wars* (1977) and the like. Like E.T., these robots are essentially novel pets with just enough initiative to make the games interesting but always, in the long run, deferential to their humans. In some ways, the complacency they generate could be regarded as the most sinister response of all. The writers represented in this section, however, are far from complacent. They regard robots as special kinds of computers whose creators evince all the failings of the computer scientists in chapter 13, with the added flaw of trying to simulate human actions and, in the case of more sophisticated robots and cyborgs, create alternative human beings.

The first in-depth literary study of the impact of robots and the motives of their creators was Karel Capek's famous play *R.U.R.* (1921).[7] Capek was concerned primarily with the effect on society of robots designed to relieve their masters of toil and leave them free to enjoy endless leisure. (The word *robot* comes from the Czech word *robota*, compulsory work or drudgery.) It is this feature—their suggested potential for liberating humanity from the burden of heavy and repetitive tasks—that has contributed to their continuing ambiguous status as being both desired and feared, an ambiguity that, as we have seen, has beset science itself from its origins in alchemy. Capek also explores the motives of the Rossums, father and son, who have created the firm designated by the play's title, Rossum's Universal Robots. Having accidentally discovered a substance that behaved like protoplasm, Rossum Senior, a physiologist, proceeds to experiment with making curious artificial beings. He thus exemplifies the irresponsible hubris that leads scientists to believe that reason (science) is sufficient to create anything required. He is eventually killed by one of his monstrous creatures, but not before he has produced a man. Like Frankenstein and Moreau, Rossum Senior was motivated by a desire to usurp the role of God, an intention about which he was more explicit than his predecessors.[8] He "wanted to become a sort of scientific substitute for God. He was a frightful materialist, and that's why he did it all . . . to prove that God was no longer necessary" (14).

Rossum Junior, eager to exploit the discovery commercially, has an engineer's impatience with the inefficiency of his father's methods and has therefore simplified the design, omitting all aesthetic and emotional elements. The result is not an android but a robot. Domin, the manager of the firm explains: "The Robots are not people. Mechanically they are *more perfect* than we are, they have an enormously developed intelligence, but they have no soul" (17, my italics). Domin himself cherishes a utopian vision of a world without work, but it is apparent that this ideal is also strongly tinged with hubris: "Man shall have no other aim, no other labour, no other care than to perfect himself. . . . He will be Lord of creation" (51–52). Capek is also concerned to show how this philosophy of scientific materialism affects those who espouse it. The other scientists at Rossum's factory are almost indistinguishable from each other and from the robots, for they too have become standardized, as though fresh from the assembly line. They are among the first and most stylized symbols of the anonymity and disappearance of individuality in a scientific, industrialized society, where people are interchangeable, being endowed with meaning only by the tasks assigned to them.⁹ Eventually the robots revolt against their ineffectual masters and kill every human being except one, who is spared only on condition that he work to rediscover the lost formula for creating the robots. The parallel with Frankenstein, begged by the Monster to create a mate for him, is clear as the play evolves to show that the reckless pursuit of science as an end in itself leads inevitably to the actual destruction of humanity, a humanity which, Capek suggests, has been effectively sterile and symbolically dead for years.

R.U.R. quickly became the prototype for a succession of robot stories, in most of which the scientists, like those in Capek's story, lose control over their creations. Usually the robots constitute a threat to humanity, but Eando Binder's story *Adam Link's Vengeance* (1941) has an interesting twist in that the ethical roles are reversed. The robot Adam Link proves to be the "moral" hero pitted against an evil scientist, Hillory, who, by means of remotely controlled motor impulses, directs Adam to rob and murder. Again the reference to Frankenstein is strong, even explicit, in that Hillory, recalling the earlier novel, offers to construct a mate for the conscience-stricken Adam. The metal woman is named Eve, an allusion also to Capek's play, in which Adam and Eve are the two near-human robots remaining at the end.

Another and more contentious variation of the amoral robot-maker is Zapparoni, the scientist-creator of Ernst Jünger's *Gläserne Bienen* (Glass bees) (1957). This Prospero-like figure creates aesthetically perfect glass bees, tiny, sophisticated automatons that, in their functionality and craftsmanship, combine technology and art, prefiguring the evolution during the 1980s of the powerful microchip. Symbolically, Zapparoni creates for himself a complete culture, endowing his creatures not only with physical being but even with the ability to create in their turn. Superficially, Zapparoni's world appears utopian, like the world of marvels, both functional and aesthetic, offered by technology; but through his narrator, Rittmeister Richard, Jünger indicates his reservations about Zapparoni's pose of godlike detachment from his world, living in happy ignorance and refusing to accept accountability for a Brave New World society. In Jünger's analysis, Western society's complacent acceptance of the gifts and controls of technology gives a new and sinister meaning to Bacon's dictum that knowledge is power;¹⁰ thus the serene and amoral Zapparoni may be even more dangerous to humanity than a recognizably evil dictator.

Human Guinea Pigs

While the robot stories provided stark metaphors, they had limited potential for psychological interest and were largely superseded, even in science fiction, by attempts to make recognizable human beings or to alter the personality of existing people. In these cases the obsessed scientists are almost invariably biologists, who share with the physicists and psychologists the doubtful distinction of being the scientific archvillains of twentieth-century literature. When such characters are manifestly evil, such as those discussed in chapter 12, they are easily recognized, but their amoral counterparts are, for the most part, socially accepted and hence more dangerous.

Mikhail Bulgakov's Professor Preobrazhensky, the austere biologist of *The Heart of a Dog* (1925), exemplifies an early twentieth-century image of the individual scientist who sacrifices the human means to the scientific end. Like Hawthorne's Dr. Heidegger, he has long been conducting experiments in rejuvenation, the more plausible, twentieth-century equivalent of the alchemists' promise of eternal youth, endowing his elderly, grotesque patients with beauty and sexual potency in order to observe their vanities and foibles. Thus, although his scientific and creative genius is not disputed, there is clearly a moral ambivalence about his experimental triumphs.[11]

Allegedly "concerned . . . about the improvement of the human race" (109), Preobrazhensky, like his precursor Frankenstein, is sure that his attempts to usurp the role of Creator and change the nature of living things are both justified and certain to succeed. The central events of the novel revolve around his intention to create a human being from a dog, Sharikov, into which he transplants a human pituitary gland and sex organs. It is clear, therefore, that Bulgakov regards the cultivation of amorality in science as closely aligned with an immoral quest for power. During the transplant experiment, Preobrazhensky is described in unmistakably priestly terms: "In the white glare stood the high priest, . . . His smoothed-back gray hair was hidden under a white cap, making him look as if he were dressed up as a patriarch. The divine figure was all in white and over the white, like a stole, he wore a narrow rubber apron. His hands were in black gloves" (51).[12] Although, as in the Frankenstein story, the creation Sharikov turns out to be a monster, Preobrazhensky is more fortunate than his predecessor in being able to reverse the process and return Sharikov to a dog's form and mentality. Nevertheless, Bulgakov stresses that such a reversion is not to be relied upon. Preobrazhensky undergoes a conversion to human values and, renouncing his exotic experiments, opts for a greater reverence for nature.[13]

More recently, another Russian author, Anatoly Dneprov, has contributed several ingenious short stories on this theme of scientists pursuing research into human life without thought for the consequences. "Formula for Immortality" deals with the attempts by two biologists to breed human beings by chemically synthesizing the genetic "formula of man." Despite the moral problems arising from the "failures" of the experiments, a new generation of research students is totally committed to extending the project, justifying it, as their predecessor Frankenstein had, on grounds of social benevolence. "Now couples can have a balanced family. Also, when necessary, say, like in wartime, the government can maintain a balanced population. Neat, eh?"[14] When challenged about the morality of their work, the students blandly assert that whatever has once been discovered will be discovered again, that even if they were to desist from their research, it would soon be carried out by someone else. This has become one of the most frequently recurring arguments of twentieth-century science, in literature as in life.

The Bomb Makers

It is one of the most provocative paradoxes of the twentieth century that although prior to the 1940s fictional characters who invented and proposed to explode bombs were considered, almost without exception, to be both mad and evil or, at the very least, misguided, the actual scientists involved in the construction, testing, and use against a civilian population of weapons of mass destruction have, for the most part, been regarded by their contemporaries as sane, even highly intelligent individuals, their research being funded with an ever-increasing slice of their society's wealth. Only in literature has their humanity or their sanity been rigorously and systematically questioned, as writers have felt compelled to investigate the underlying reasons for the collaboration of seemingly idealistic scientists with governments and the military-industrial complex to produce such weapons. Their fascination with the subject has produced a spate of books set in weapons research establishments, but as few writers have sufficient technical knowledge to make their work detailed or convincing, most have concentrated on the scientists in their relations with their families rather than with other scientists or have resorted to symbolic rather than realistic treatments.

In assessing the motivation of these scientists, some authors have posited mad, evil, or power-seeking characters, but more have had recourse to the stereotype of the amoral scientist, single-mindedly pursuing knowledge for its own sake and denying responsibility for the consequences. If challenged, these characters frequently take refuge in appeals to the inevitability of events that have proved to be beyond their control. Albert Einstein's own explanation for the involvement of physicists in atomic weapons research can be considered a prototype of this "argument":

> Scientists have been driven by the intellectual fascination of complex questions, and on occasion they have found added justification for their work in the hope that their activities might transform the world. They have often offered policy makers undreamed-of opportunities or have prodded the policy makers to set limits to their ambitions. But ultimately the scientists have worked as the servants of policy. Those who would praise or blame the weapons scientists for the consequences brought

about by the products of their work should first direct their attention to the authors of policy.[15]

When we deconstruct this apparently ingenuous statement, we can see the image of the noble scientist presented in its most attractive form—all of the nobility and none of the responsibility. This is undoubtedly the way most weapons scientists would like to see themselves, and it is the self-image of most of characters discussed here.

The authors, however, are less permissive. They have inclined more towards the view put forward by the physicist Freeman Dyson, who was also involved in research on the atomic bomb:

> Up to the year 1982, six countries had overtly acquired nuclear status: the United States, the Soviet Union, Britain, France, China and India. It was certainly true in the first four cases, and possibly true in all six, that scientists rather than generals took the initiative in getting nuclear weapons programs started. In each case of which the world has knowledge, scientists were motivated to build weapons by feelings of professional pride as well as of patriotic duty. The construction of the bomb was a technical challenge that stirred their fiercest competitive instincts. In each case the scientists felt themselves to be in competition with the scientists of some other country. . . . It is no great exaggeration to say that the British and French programs were driven by professional pique of scientists rather than by a careful consideration of strategic needs. The same thing might be said of the American hydrogen-bomb program.[16]

The force of this statement is accentuated when we take into account the fact that it was evident to the American physicists towards the end of 1944 that Germany had abandoned its atomic bomb project and hence that the alleged rationale of needing to perfect the bomb before Hitler could do so was no longer valid. Joseph Rotblat, one of the physicists who left Los Alamos at that point, in discussing why his colleagues remained, asserted that "the most frequent reason given was pure and simple scientific curiosity—the strong urge to find out whether the theoretical calculations and predictions would come true. These scientists felt that only after the test at Alamogordo should they enter into the debate about the use of the bomb."[17] By then, of course, the project had attained its own political momentum. There would always be a new pretext to justify its continuation; if not Hitler, then the Soviets. The seeds of the cold war were already germinating in the warmth of Los Alamos before the end of the Second World War. To a nonscientist, Rotblat's statement about the "urge to find out whether the theoretical calculations and predictions would come true," when the consequences would be the death of thousands of innocent people and the long-term irradiation of the environment, may well seem incredible lunacy or the most inhuman malevolence. Yet this encapsulates precisely the relative values of the amoral scientist in literature and of many highly respected atomic physicists in real life.

Some of the most enduring and effective literary treatments of this theme made no pretense at realism, but dwelt on a symbolic, representative character. One of the earliest was Carl Sandburg's black comic poem (dated, significantly, August 1945) about a seemingly harmless, absent-minded atomic physicist, "Mr. Attila."

They made a myth of you,
  professor,
you of the gentle voice,
the books, the specs,
the furtive rabbit manners
in the mortar-board cap
and the medieval gown.

They didn't think it, eh professor?

On account of you're so absent-minded,
you bumping into the tree and saying,
"Excuse me, I thought you were a tree,"
passing on again blank and absent-
  minded.

Now it's "Mr. Attila, *how* do you do?"
Do you pack wallops of wholesale death?
Are you the practical dynamite son-of-a-gun?
Have you come through with a few abstractions?
Is it you Mr. Attila we hear saying,
"I beg your pardon but we believe we have made
some degree of progress on the residual
qualities of the atom"?[18]

Sandburg clearly implies that cultivated ignorance of consequences is no excuse for the physicist; the role of the comic and harmless absent-minded professor has become untenable. Despite his mild manners and timidity, Mr. Attila has committed mass murder as certainly as his barbaric namesake did.

One of the most influential, though necessarily oblique, treatments of the German physicists working on the bomb was Bertolt Brecht's second version (1947) of his *Life of Galileo*. This modification of his 1938–39 version was written in the United States after the bombing of Hiroshima and Nagasaki. Brecht wrote in his introduction: "Overnight the biography of the founder of the new system of physics read differently. The infernal effect of the Great Bomb placed the conflict between Galileo and the authorities of his day in a new, sharper light."[19] Unlike his earlier Galileo, the courageous rebel against repressive authority, this revised Galileo stands as the prototype of the amoral scientist pursuing research without concern for its consequences. For Brecht, the scientific discoveries do not compensate for the crime against society. He commented: "In the end, Galileo is a promoter of science and a social criminal. . . . the scientific work does not outweigh the social failure."[20] Having begun his career in the service of humanity, Galileo becomes obsessed with knowledge for its own sake and fails to maintain communication with ordinary people. This was another aspect of public accountability that Brecht was concerned to stress in the contemporary debate about the social responsibility of scientists. As he explained in his notes on the play, "Galileo's crime can be regarded as the 'original sin' of modern natural sciences. From the new astronomy, which deeply interested a new class—the bourgeoisie—since it gave an impetus to the revolutionary social current of the time, he made a sharply defined special science which . . . was able to develop comparatively undisturbed. The atom bomb is, both as a technical and as a social phenomenon, the classical end-product of his contribution to science and his failure to society."[21] Brecht's third version of *Galileo* (1953–55) is an even more rigorous criticism of the divorce of scientists from society and their cult of *scientia gratia scientiae*. Galileo becomes the central focus of Brecht's condemnation of the inhumanity of contemporary science and the irresponsibility of scientists.

Given the particular focus of atomic physics at the time, it was scarcely surprising that Brecht should have seen Galileo as a metaphor for the twentieth-century scientist who had betrayed his trust to work for the welfare of humanity; but in charging Galileo with social treason, he has set him in an anachronistic framework, one that certainly did not exist in the seventeenth century. At worst, Galileo's "sin" is one of omission: he has turned his back on the dawning enthusiasm of the bourgeoisie for knowledge and isolated himself in the pursuit of pure science. Compared with the active sins of commission of those who developed nuclear weapons, his "treason" appears almost a peccadillo, a minor ethical failure at a personal level. Yet in another sense Brecht was correct in placing Galileo at a critical moment in the development of science, the moment when, according to Arthur Koestler, the confrontation between Galileo and the church resulted in a progressive divergence between scientific and technological progress on the one hand and the affirmation of a moral and spiritual dimension on the other.[22] Viewed in this light, it is perhaps not too much of an exaggeration to say that Galileo's withdrawal from society, as depicted by Brecht, made possible the behavior of Oppenheimer and his colleagues working on the Manhattan Project in similar isolation.

Brecht's treatment was deliberately stylized and universalized, characteristics that were an important part of his philosophy of the theater.[23] When the bomb makers appear in person as differentiated characters in twentieth-century novels or plays, the presentation, often based on carefully collected biographical detail, is still critical but varies markedly in the degree of condemnation assigned by the authors to individuals. In Pearl Buck's novel *Command the Morning* (1959) the intention is to explore how otherwise well-intentioned individuals can allow themselves to engage in military research, while Dürrenmatt's play *The Physicists* (1962), on the other hand, poses the question how a brilliant physicist can avoid having his ideas misused. In other treatments the focus is satirical, as in Vonnegut's Felix Hoenikker, the physicist of *Cat's Cradle* (1963), or overtly condemnatory, as in Schirmbeck's Richard Tzessar in *Ärgert dich dein rechtes Auge* (1957).

Often these characters are "explained" in terms of moral schizophrenia. In other areas of life they may be sensitive, humane individuals, but they are unable to see any anomaly in working on a project designed to destroy thousands of civilians. Some of the most trenchant criticism of the atomic physicists in this regard comes from German writers of the two decades following the war. The inventor-protagonist of Wolfgang Borchert's *Lesebuchgeschichten* (1946) begins by making home appliances but subsequently, without any sense of immorality, moves on to making bombs. His mind is so rigidly compartmentalized that he can agonize over a wilting flower while regarding society as merely a market for his products—whether appliances or bombs. Borchert is doubtless reworking the often-remarked phenomenon of Nazi concentration camp supervisors enjoying classical music while perpetrating horrors against humanity.

One of the first realistic portrayals of research scientists involved in a political dilemma was C. P. Snow's novel *The New Men* (1954). Snow, a molecular physicist by training, worked during the Second World War in the British civil service, selecting scientific personnel. He was therefore one of the few novelists writing with inside knowledge about a group of research physicists engaged in nuclear fission research during the war. Lewis Eliot, Snow's nonscientist narrator, explicitly comments upon the absence of ethical anguish among these "new men" at Barford (the fictional equivalent of Harwell), their naive acceptance of the view that they must race the Nazis to the production of the atomic bomb—"For those who had a qualm of doubt, that was a complete ethical solvent"[24]—and their equally naive belief that mere possession of the bomb would be sufficient to end the war (and all wars).

Given Snow's own background in both science and the public service, it is significant that in his novels politicians by no means always play villain to the innocent scientists. Bevill, the government minister supervising the project comments: "I used to think scientists were supermen. But they're not supermen, are they? Some of them are brilliant, I grant you that. But . . . a good many of them are like garage hands. Those are the chaps who are going to blow us all up" (69–70). Bevill is also the spokesperson through whom Snow subverts the scientists' complacent assumption that the bomb, once developed, would not be used: "Has there ever been a weapon that someone did not want to let off?" (71). Snow also pinpoints the scientists' traditional allegiance to internationalism as one of the major sources of conflict between them and their government during wartime. Even though "many scientists congratulated themselves on their professional ethic and acted otherwise" (130), they still believed in "free science, without secrets, without much national feeling. . . . the truth was the truth and, in a sensible world, should not be withheld; science belonged to mankind" (130–31). Such an attitude runs entirely counter to the rationale of military and political hegemonies, grounded in the belief that secrecy is a precondition of power.

Insofar as Snow's scientists have a claim to heroism, it depends on their idealism concerning the free, international community of scientists and their revulsion at the idea that the atomic bomb should be used against Japanese civilians. Yet Snow also portrays this idealism as coexisting with a quite amoral sense of elation at the success of the project. "Several men of good will felt above all excitement and wonder. . . . I thought the emotion was awe; a not unpleasurable, a self-congratulatory awe" (166). The real-life model for this ambivalence was almost certainly the sensitive and widely read J. Robert Oppenheimer, whose private fears about the increasing hold by the military over the products of research in atomic physics did not prevent his public expression of delight in the H-bomb as "technically so sweet that you could not argue about that. It was purely the military, the political and the humane problem of what you were going to do about it once you had it."[25] As Philip Stern remarked, "If scientists as sensitive as Oppenheimer can indeed wall off their moral sensibilities so completely and successfully, then technology is an even more fearsome monster than most of us realise."[26]

Unlike those writers who approach the subject from the outside, Snow also presents a broad spectrum of attitudes and motives among his scientists. Two, Martin and Mounteney, regard science as a private satisfaction; content with research as an end in itself, they have no interest in the social relevance of their work. Walter Luke, on the other hand, is a convinced Baconian; he believes that the justification of science is the power it offers over nature, a power that he, like Bacon, never doubts will be used for the betterment of mankind. Other characters are less accepting of the establishment's position. Determined to protest against the use of the bomb before the decision to use it is irrevocable, two members of the Barford group travel to Washington to put their case. Francis Getliffe, Snow's most intelligent and idealized scientist, voices the pragmatic argument that, quite apart from moral considerations, it would be insane to use the bomb, because it would lead to a proliferation of such weapons in other countries. "Non-scientists never understood, he said, for how short a time you could keep a technical lead. Within five years any major country could make these bombs for itself" (174). Here Snow is almost certainly voicing his personal critique of the Baruch Plan of 1948 (supported by some politically naive scientists), which blandly assumed that because only the United States at that time possessed the knowledge necessary to construct the atomic bomb, it would be able to retain honorable control over it and prevent any other nations from duplicating the discovery. Subsequent events have vindicated Snow's skepticism on a scale unimagined by Snow himself or, indeed, anyone else at the time. The majority of Snow's fictional physicists, however, like their real-life counterparts, refuse to believe that the bomb will be used and incline to the view that "the fission bomb was the final product of scientific civilisation; if it were used at once to destroy, neither science nor the civilisation of which science was bone and fibre, would be free from guilt again" (178).

When the second bomb is dropped on Nagasaki, even those of Snow's scientists who had given qualified assent to the first are outraged and cynical: "They all assumed . . . that the plutonium bomb was dropped as an experiment, to measure its 'effectiveness' against the other. 'It had to be dropped in a hurry,' said someone, 'because the war will be over and there won't be another chance'" (201). It becomes clear that the scientists feel that the "perfection" of the atomic bomb has changed the rules of the game; "the relics of liberal humanism had no place there" (300). Ironically, therefore, those who have been most opposed to the dropping of the bomb now see the only hope for the future in making more bombs in Britain. Luke speaks for most of them when he voices the conventional argument for the bargaining power of deterrence: "All I can say is that, if we're going to get any decency back, then first this country must have a bit of power" (203).

One of the many interesting comments arising from Snow's inside knowledge of a research establishment relates to the contrast he draws between physicists and engineers, since it qualifies many of the stereotyped attributes traditionally imputed to scientists in general. Snow characterizes physicists as rebellious, opposing the use of the bomb and refusing to accept government edicts about their work; he sees engineers, on the other hand, as more conservative, less reflective. Snow attributes this disparity to the different kind of thinking involved in their professions:

> The engineers . . . the people who made the hardware, who used existing knowledge to make something go, were in nine cases out of ten, conservatives in politics, acceptant of any regime in which they found themselves, interested in making their machine work, indifferent to long-term social guesses.
>
> Whereas the physicists, whose whole intellectual life was spent in seeking new truths, found it uncongenial to stop seeking when they had a look at society. They were rebellious, questioning, protestant, curious for the future and unable to resist reshaping it. The engineers bucked to their jobs and gave no trouble, in America, in Russia, in Germany; it was not from them, but from the scientists, that came heretics, forerunners, martyrs, traitors. (176)

For most of Snow's scientists, therefore, the greatest temptation is to capitulate to the doctrine that events are too big for individuals to change. As Luke affirms, "It may be so. *But we've got to act as though they're not*" (235). Snow clearly perceives a nexus between the fatalistic belief that events proceed inexorably and the withdrawal to research pursued in isolation; and his intimate knowledge of a weapons research establishment must lend considerable weight to his depiction of the factors motivating such scientists.

Whereas Snow strove for apparent objectivity of presentation, Heinrich Schirmbeck's monolithic novel of some six hundred pages, *Ärgert dich dein rechtes Auge* (1957), translated into English as *The Blinding Light*, is strongly judgmental. Although there is much realistic detail, Schirmbeck's interest, like Brecht's, lies in the universal rather than in the particular; where Brecht used Galileo as an archetype, Schirmbeck employs the mathematician Pascal to suggest the demonic nature of all knowledge. Early in the novel, Schirmbeck's narrator, the young physicist Thomas Grey, imagines Pascal's thoughts upon constructing the first calculator. At first Pascal claims that his motive was solely to free mankind from the burden of mental calculation, but he later confesses that he was moved by his own desire to transcend the limitations of knowledge.

From among Schirmbeck's large cast of scientists, the two most striking representatives of the amoral pursuit of science are the brilliant atomic physicist Richard Tzessar and Colonel Elliot, head of a cybernetics establishment. Tzessar, obviously modeled on the American physicist Edward Teller, is engaged in research on the hydrogen bomb, which he sees as bringing the solar process down to earth. Introduced under the chapter heading "Lucifer" and described as "the new Faust," Tzessar not only embodies the pride of Lucifer but rejects the call for moral responsibility as "Callow sentimentality!" Through an interchange between the still idealistic Grey and Tzessar, Schirmbeck explores the dialectical positions adopted by scientists at the time. Predictably, Tzessar adopts the pragmatic doctrine that science is inherently neither good nor bad, but value-free, and that it is not the scientists' business to become involved in moral issues. "We [scientists] serve the God of free research, the God who says '*Fiat scientia pereat mundus*'—let there be knowledge though the world perish! . . . We have no power to prevent it." Grey, acting as authorial spokesperson, counters: "You're still a child of the eighteenth century, Professor—the Age of Enlightenment. . . . But those voices of scientific idealism are not the voices we need now. Any argument based on them can be nothing but an anachronism, a threadbare travesty with the devil grinning behind the scenes." He then proceeds to compare Tzessar's attitude to that of the firms that during the war took pride in manufacturing high-quality gas-chambers and corpse-ovens, "also in the name of science."[27]

Tzessar also invokes the well-worn argument of inevitability and the helplessness of the individual to change the course of events: "Scientific research is governed by its own laws; it follows an ineluctable course that cannot be affected by the actions of individuals" (342). He even resorts to an implicit categorical imperative for his own purposes: "Anyone who can't bury himself heart and soul in scientific work, letting himself be possessed by it . . . has no right to call himself a scientist" (343). Like Mary Shelley, Schirmbeck links this stance of high-sounding amorality to an unacknowledged lust for power. Tzessar repeats the arguments of Frankenstein's charismatic mentor Waldman in his defense of the right to limitless power: "Only when he [man] commands *every* possibility, including that of self-destruction, will he be truly man, the image of God and the master of his destiny. He will still remain the central figure in the cosmic drama, even if he destroys the biological conditions which permit him to exist. Our job as scientists is to put him in the saddle. It is for him to ride the horse" (343). Thus, despite his acknowledged intellectual power, Tzessar is depicted by Schirmbeck as morally retarded. As one of the other characters remarks, "He's an excellent example of the schizoid nature of the scientist. Science grows in them like a tumour. Their organic balance has been upset" (263).

Nor is Tzessar alone in this debility. As Mary Shelley achieved added force for her characterization of Frankenstein by creating the parallel figure of the explorer Walton, Schirmbeck explores the same amoral philosophy in another scientific discipline, cybernetics, and thereby suggests a universal judgment on the state of contemporary science. Colonel Elliot, head of a cybernetics research institute, is one of the first representatives of this discipline in literature. The institute is located in a hexagonal building, implying that it is a future derivative of the Pentagon; it is also significant that it is actually under the control of the government's Department for Strategic Information and that its head is both a scientist and a military official. Schirmbeck suggests that although the cyberneticists believe they are equal partners in the collaboration and are free to do their own research, they are in fact being subtly manipulated by the military. Elliot, who embodies the worst faults of earlier fictional psychologists and mathematicians, regards cybernetics as an intellectual game that represents the supreme alienation of the scientist from the world of experience and that, at a more sinister level, can be used to control society by rigorous conditioning and propaganda. His position is the sociological equivalent of Tzessar's in physics: essentially, while professing objectivity, he recognizes only power. He explains enthusiastically: "By feeding him [man] with selected data . . . we can direct his thoughts and emotions along specific channels. That is the art of propaganda . . . as social cybernetics. . . . It comes down to the problem of scientifically turning bad citizens into good ones" (381). Elliot, symbolically flanked by his assistants, the "technical eunuch" Norf and a former military officer, Moras, does not define what he means by "good" or "bad" citizens;

but Schirmbeck clearly implies that by its very nature cybernetics subordinates all human and moral considerations to the criteria of efficiency and control.

Schirmbeck provides explicit commentary on the evils of the amoral stance of Tzessar and Elliot through the heroic character of Prince De Bary, based on the French physicist Louis de Broglie. De Bary tells Grey that as a younger scientist he too had been seduced into a life of contemplation focused on the pure beauty of mathematics and theoretical physics. Later he rejects this course as dangerous and immoral insofar as it omits the human factor, the unpredictable, the indescribable, and hence can discover only a part of the truth. Whatever its merits as literature, Schirmbeck's novel remains a tour de force in its formulation of contemporary arguments concerning involvement or disengagement by scientists in relation to the needs of the emerging military-industrial complexes of the 1950s and puts forward the uncomfortable view that the morality of a project may be gauged by reference to its funding body.

Another vigorous indictment of atomic physicists for their determination to test lethal weapons regardless of the consequences is provided by Heinz von Cramer's novel *Die einzige Staatsvernunft in Konzession des Himmels* (The political wisdom for the franchise of the heavens) (1961), set in post–Second World War Australia and intended as a comment on the British nuclear tests in South Australia during the 1950s.[28] Von Cramer's scientists demand freedom to continue their experiments without being answerable to anyone, and given the climate of the cold war, political, military, and even religious bodies condone the bomb tests in the name of expediency, even though the aboriginal population of the area has already suffered terrible injuries attributable to radiation and will continue to do so.

It is noteworthy that despite these sophisticated and, for the most part, damning analyses of the bomb makers by British and European writers, disenchantment with scientists still working on nuclear weapons emerged relatively slowly in the United States. Immediately after the war, Hiroshima and Nagasaki notwithstanding, science was, as American journalist Daniel Lang records, "synonymous with victory, and its practitioners, notably physicists, were regarded as celebrities. . . . they were not only looked up to as the inventors of a successful secret weapon, but they themselves seemed secret weapons—a hitherto undiscovered national resource, a fraternity of geniuses who could bail the country out of any crisis. . . . If they could bring off so formidable a feat as ending a war, then what problem could possibly stump them in the era of peace that was now at hand?"[29]

The explanation for this disparity between attitudes in America and Europe lies partly in the fact that in the United States the atomic bomb was associated with the ending of the war in the Pacific and hence with the saving of American soldiers' lives. Thus the nexus between science and war was viewed by most Americans at this time as a benign cooperation. Moreover, as long as America alone possessed the secret of the bomb (refusing to share it even with her allies Britain and Canada, who had contributed much to America's "success" in the field), nationalist feeling assumed that such knowledge was in good hands (several instances of this assumption appeared in the literature discussed in chapter 11) and would be used only in the defense of democracy.

Two critical factors led to the reversal of this estimate. First, several leading scientists, including Einstein and Oppenheimer, appalled at the power and destruction that had resulted from their work, set forth on a crusade to educate the public about the dangers of unchecked research. They wanted to ensure not only that the bomb would never be used again but that there would be international control of atomic energy. To the popular mind this seemed either naive or, in the face of growing fears about the spread of communism, irresponsible and dangerous. When, in 1949, the Soviet Union exploded its first nuclear weapon, the American reaction was one of panic. Having ignored the scientists' warnings that the knowledge gap was not fixed but an ever-diminishing variable, the public immediately assumed that scientists were not to be trusted with their own discovery, since they might be harboring treacherous internationalist thoughts. The fear and suspicion generated by the Oppenheimer investigation in 1953, which was to culminate in a witch hunt for atom spies, left, in the public mind, a taint on all scientists. If the former leader of America's science team could fall under suspicion as a security risk, what perfidy might not be alleged against his subordinates?

Second, and more spectacular, the successful launching of the Russian spacecraft Sputnik on 4 October 1957 sent a shock wave of incredulity through the United States as it was suddenly perceived that the Soviet Union was the leader in space research. As a consequence, U.S. scientists were berated for losing the knowledge advantage that had existed at the end of the war, and the gathering opprobrium attaching to scientists suddenly crystallized. They were seen as the embodiment of the double-edged sword, beneficent only if kept under the strictest control, otherwise a dangerous resource and not to be trusted.

Even while this view of scientists was being aired in the media, however, literary disapproval tended to focus less on their alleged political disaffection than on the sociological effect of their devotion to research regardless of its consequences. Where the popular press pointed to possible un-American activities on the part of scientists at the onset of the cold war, novelists and playwrights emphasized the inhumanity of the "patriotic" scientists, playing intellectual games while humanity as a whole staggered ever closer to the brink of destruction. This aspect has been emphasized in several films based on the Manhattan Project, including *The Day after Trinity*, in which the physicist Freeman Dyson (quoted above in relation to the manipulatory and strategic power exercised by scientists in weapons research) acknowledges the fascination that such power exerts. "I have felt it myself, the glitter of nuclear weapons. It is irresistible if you come to them as a scientist. To feel it's there in your hands—to release this energy that fuels the stars—to lift a million tons of rock into the sky."[30]

It was in this ambivalent climate that Pearl Buck's novel *Command the Morning* (1959) was written, reviewing the motivation of the scientists who worked on the Los Alamos project.[31] Having spent many days studying the site and interviewing the scientists and their families, Buck integrated real and fictitious characters in her novel, the American counterpart of Snow's *The New Men*. Enrico Fermi features in the cast as the archetype of the scientist who refuses to acknowledge any social responsibility for his research. "'Don't bother me with your conscientious scruples,' Fermi had said. 'After all, the thing is superb physics'" (206).

Although many of Buck's other scientists begin with scruples on various accounts, these are dissolved by the Japanese attack on Pearl Harbor, after which the decision to make the bomb is for them no longer an abstract question but one involving American lives. This, combined with the belief that the Germans have almost perfected an atomic bomb, produces that sense of inevitability about the project that had been depicted by both Schirmbeck and Snow; it no longer seems possible to stop the research. In Buck's novel the American scientists are portrayed as avoiding individual responsibility for the decisions, resorting to opinion polls among the scientific fraternity, seeking safety in numbers, and ultimately assigning to the military the decision whether or not to use the bomb. Buck indicates that such avoidance of responsibility is not, as is frequently claimed, a position of moral neutrality but a positive evil and implies that the traditional claims for a value-free science will inevitably issue in scientific arrogance in the next generation: the younger scientists, having tasted power, no longer have any scruples to overcome. The novel ends with the ambivalent attitude of the project leader, Burton Hall. In his description of this younger generation, a rewriting of Job's capitulation to God, there is both pride and despair: "These kids! We found the divine fire for them and they grabbed it away from us. They're riding into space on the wings of power. . . . I have a hunch that when the first one goes soaring off into space and driving for the moon . . . he'll yell back to the rest of us here . . . 'Yeah, man, I do command the morning!'" (316–17). In this symbol of the scientist challenging, if not playing, God, claiming exemption from the human condition of responsibility, Buck suggests a form of evil all the more insidious for being socially acceptable, even tacitly admired.

Similar unmistakable criticism of the atomic physicists working on the bomb is found in Dexter Masters's novel *The Accident* (1955), which also focuses on the group at Los Alamos, but after the war. Like Buck's novel, it is based on actual incidents and people and thereby signals to readers the urgency of becoming informed about contemporary issues that might otherwise be dismissed as fiction. In 1946 Louis Slotin, one of the nuclear physicists involved in the development of the atomic bomb, fell victim to an accidentally triggered burst of radiation during a test exercise on the critical mass of uranium[238]. Nine days later he died, having suffered terrible physical disintegration, even though the radiation levels received were nowhere near as high as those suffered by the victims of the Hiroshima and Nagasaki explosions. In Masters's novel the life of Louis Saxl, Slotin's fictional counterpart, is reconstructed in detail—the motivation of the scientist, his moral values and scruples, the conflict between the scientific and the social conscience, and the battle of both with military directives. Early in the novel Masters draws a parallel between the foreseen possibility of Saxl's "accident," which was considered an "acceptable risk," and the possibility that had been put forward by the Nobel prize–winning physicist Baillie, that the explosion of the atomic bomb might ignite the whole of the earth's atmosphere. (This possibility was also discussed in Buck's *Command the Morning.*) To the physicists involved, this too was an "acceptable risk," being of the order of only 1 percent, but to the doctor, Pederson, it is incredible that they would "go ahead and explode a bomb that might blow up the world."[32]

In describing the spectrum of the other physicists' responses to Saxl's accident, ranging from intense pity and horror to anger that this accident was "allowed" to happen, Masters emphasizes the unease of the scientists who had stayed on after the war to work on the Bikini tests and who were then forced to witness the consequences for their colleague of the research they were doing.

> "The three conditions of scientific work are the feeling that one ought not to leave his path, the belief that work is not all intellectual but moral as well, and the feeling of human solidarity." So wrote Szent-Gyorgyi, a Nobel Prize winner of 1937, two years before the fission of the atom was discovered. In the spring of 1946 all three of these conditions were lacking at Los Alamos for the scientists who had left their science to take on the building of the bomb, who were having trouble with their beliefs in their work, and whose feelings of human solidarity were the source of the trouble.   (46)

Yet ultimately, whatever their reservations, they all return to the development of more bombs. Saxl becomes a useful set of statistics in the ongoing medical research into radiation effects, and the production of the weapons continues unabated.

Although Buck and Masters evolved their novels from actual events, they had no hesitation in including their own views and interpretation of those events. Heiner Kipphardt, by contrast, used a different and perhaps ultimately more effective method. His play *In the Matter of J. Robert Oppenheimer* (1967), based on the tapes of the enquiry into Oppenheimer's security clearance, took documentary realism to new lengths and gave the impression of being wholly objective, a mere transcription of interviews. However, Kipphardt contrives to give his authorial view obliquely by means of rigorous selection and juxtaposition from a much longer transcript. The character of Oppenheimer himself, leader of the Manhattan Project, is discussed in chapter 15; here the comments of Edward Teller, who appears in person uttering the actual words he spoke at that time before the Personnel Security Board, are pertinent as a measure of how closely the fictional characters discussed above resemble their real-life counterparts. At the enquiry Teller puts forward his faith in military deterrence. "I am convinced that people will learn political common sense only when they are really and truly scared. Only when the bombs are so big that they will destroy everything there is."[33] He thus denies any personal responsibility for the use of the hydrogen bomb he has worked on. When giving his testimony against his former colleague Oppenheimer, Teller is asked if he has ever had any moral scruples about the hydrogen bomb.

> TELLER: No.
> EVANS: How did you manage to solve that problem for yourself?
> TELLER: I never regarded it as *my* problem. . . . It is not a matter of indifference to me, but I cannot possibly foresee the consequences, the full range of practical uses, which are part and parcel of an invention. . . . Discoveries in themselves are neither good nor evil, neither moral nor immoral, but merely factual. They can be used or misused.   (83–84)

Here Teller enunciates the prototypical argument that the scientist is not responsible for the uses to which his discoveries may be put, because he is unable to foresee them. Later in the hearing Teller is asked about his reaction when he heard that twenty-three Japanese fisherman in a trawler had drifted into a radioactive blizzard because the wind had suddenly shifted after the Bikini explosion of the hydrogen bomb. Teller replies: "We set up a commission to study all the effects. And we were able to make considerable improvements in the meteorological forecasts for our tests." It is this that prompts Evans to repeat his former question, "What kind of people are physicists?" (85). Teller, oblivious of the irony, replies: "They need a little bit more imagination and a little bit better brains, for their job. Apart from that, they are just like other people" (85). This same argument was put forward by his fictional counterpart, Tzessar, in Schirmbeck's *Blinding Light*.

Martin Cruz Smith's novel *Stallion Gate* (1986) offers an interesting exploration of the attitudes of Robert Oppenheimer and his group at Los

Alamos against the background of the Mexican Indians' belief in the sacredness of life. Smith presents Oppenheimer as a relatively limited individual who needs the bomb as evidence of his identity at least as much as the development of the bomb needs him. Anna Weiss, a German Jewish refugee working on the project, remarks, "There wouldn't be an Oppy without the bomb. There are other physicists here more brilliant by far. . . . He's quick to finish other people's thoughts, but they're still other people's thoughts." Just before the Trinity test, Oppenheimer is like a man possessed, his life drained from him, existing only as an intellect. "His entire body seemed to maintain a faint existence only to carry the painfully brooding skull."[34] Other members of the group have been enlisted because of the belief that they need to make the bomb before Hitler's scientists do. When it becomes apparent that there is no such danger, the rationale is transferred to the need to end the war in the Pacific and save American lives. Smith suggests how glibly the manipulation takes place: "Oppy's voice became nearly tender. 'But you tell a mother of a young soldier who died on the beach of Japan . . . that you had a bomb that could have ended the war and that you chose not to use it. Tell his wife. Tell his children'" (176).

Anna Weiss retains her sanity throughout the process of modeling the impact of Trinity by refusing to consider the process as anything other than a mathematical exercise. She asserts that the other scientists, too, simply refuse to think about consequences. "No one looks ahead to after the bomb is used. Or asks whether the bomb should be used or, at least, demonstrated to the Japanese first. . . . Every day I kill these thousands and thousands of imaginary people. The only way to do it is to be positive they are purely imaginary, simply numbers" (167). In one intense discussion between Oppenheimer and another physicist, Harvey Pillsbury, the latter argues for an end to the project after the German surrender or, at least, a demonstration of the bomb to the Japanese emperor in a relatively uninhabited area rather than an attack on a major city, but Oppenheimer, ever the pragmatist, replies in terms of cost-effectiveness: "A waste. A waste of a bomb and a waste of the soldiers. . . . There wouldn't be much to see in a camp—not like a city, not like buildings" (175).

Another factually based discussion of the amoral approach to wartime science, namely, the argument that it is of no consequence to the researcher whether his work has a life-giving or a life-destroying outcome, is to be found in Richard Gordon's well-researched historical novel *The Invisible Victory* (1977). Gordon, himself a doctor, traces the discovery of the sulpha drugs by chemists in Nazi Germany and the subsequent medical breakthrough with penicillin in Britain. In 1939 Gerhard Domagk was awarded the Nobel prize for physiology and medicine for his discovery of the efficacy of prontosil, precursor of the sulpha drugs or sulphanilamide compounds.[35] After the war the narrator, revisits the I. G. Farbenindustrie laboratories in

Wuppertal, where the drugs were discovered, and learns that the actual suggestion of introducing the sulphanyl group into the basic azo molecule, thereby turning a chemical dye into a life-saving drug, was made by Professor Hörlein, who was therefore "the true originator of the sulphonamide drugs . . . so opening the eyes of Florey, that he might see the potentialities of penicillin."[36] However, the narrator learns almost immediately that Hörlein is currently on trial for war crimes, charged with mass murder, since he was also on the board of the subsidiary chemical company that made poisonous gases, including Zyklon-B, used by the SS for mass killings in the concentration camps. Pondering on this, Gordon's narrator concludes: "So the man responsible for modern chemotherapy was also responsible for the gas used in genocide. A sickly paradox. But perhaps Hörlein had not seen it as a paradox? Drugs to cure and drugs to kill are still only chemicals. When to do either one or the other is equally laudable, who is the technician to object? Such moral autism was the secret of Nazi power" (265).[37]

In all these representative novels and plays it is clear that the authors themselves perceive the amoral stance of scientists as potentially dangerous to society and, more explicitly, as a basic factor in the development of weapons research. The belief that science is neither good nor bad but value-free is taken by scientists to absolve them from responsibility in an area that would otherwise be too terrible to contemplate. Implicit in this reasoning is a suggestion that the scientists need the military (to take responsibility for the front-line decisions) as much as the military needs the scientists (as the backroom producers of their weapons). It is precisely this cozy symbiosis that is under attack from the writers who deny scientists the right to play Pilate and wash their hands of guilt.

This point is also made in Roland Joffe's film *Shadow Makers*, another exploration of the motives of the physicists working on the Manhattan Project. In analyzing the relationship between General Leslie Groves, the military director of the project, and J. Robert Oppenheimer, the scientific director, Joffe suggests that Groves plays Mephistopheles to Oppenheimer's Faust, to the mutual satisfaction of both. Ultimately, what this Faust desires as much as achieving the "technologically sweet" success of the bomb is the popular acclaim accorded to the great physicist who designed and completed the project, untainted by any blame for the ensuing destruction. For this, a stated belief in the amorality of science is a necessary precondition: Oppenheimer needs Groves to take total moral responsibility for the tactical decisions as to how and when the bomb will be deployed.

In his devastating satire on the American Rocket State, *Gravity's Rainbow* (1973), discussed in chapter 13, Thomas Pynchon focuses on another form of amorality associated with weapons research, the claim that scientists are forced to carry out whatever research is funded. By means of his character Franz Pökler, whose career closely parallels that of Wernher von

Braun, Pynchon links the German project to build the V-2 rockets during the Second World War with the U.S. weapons research program during the cold war and exposes the ease with which German physicists such as Wernher von Braun, who from 1937 to 1944 had held eminent positions at Peenemünde working to develop military rockets for the Third Reich, were quickly recruited by the United States, any questions about political allegiance being conveniently glossed over. Pynchon's character Pökler, after attending the Technische Hochschule in Munich, moves to the Verein für Raumschiffahrt (the German Rocket Society) in Berlin, then to Peenemünde, and finally to Nordhausen, where he constructs the Imipolex "insulation device" for the 00000.[38] Pökler's politics extend no further than the acquiring of research funding; hence he is quite happy to join the winning side after the war and move to a lucrative U.S. research institute, which is equally ready to accept his services against the new enemy, the Soviet Union.

Although his presentation differs markedly from the other examples in this chapter in being superficially comic, the American-born writer Kurt Vonnegut, Jr., who was a U.S. soldier in Dresden at the time of the firebombing, is also deeply cynical about the trustworthiness of scientists who reject the ethical perspective and regard science as amoral play. In *Cat's Cradle* (1963) he presents Dr. Felix Hoenikker, another fictional father of the atomic bomb and another emotionally retarded scientist, repeatedly described as a child in his "innocent" irresponsibility for the consequences of his research.[39] For Hoenikker everything is a game: turtles, a piece of string, atomic weapons—to him all are equally interesting and morally indistinguishable. At one level Hoenikker epitomizes the engaging, absent-minded professor, so impractical that he is dependent on his daughter to get him dressed in the morning; but Vonnegut uses the stonemason who has made Hoenikker's tombstone to deconstruct this picture for us: "I know all about how harmless and gentle and dreamy he was supposed to be, how he'd never hurt a fly, how he didn't care about money and power and fancy clothes and automobiles and things . . . how he was so innocent . . . but how the hell innocent is a man who helps make a thing like the atomic bomb? . . . I never met a man who was less interested in the living" (63).

Besides the doubtful distinction of being a father of the atomic bomb, Hoenikker has also "in his playful way, and *all* his ways were playful," produced a crystalline substance, ice-nine, which, by virtue of having a melting point of 114.4° F, causes all water in contact with it to crystallize as ice-nine. The reaction proceeds exponentially and is thus potentially capable of freezing the entire planet.[40] Instead of instantly suppressing this discovery, Hoenikker has continued to play with it up until his death, teasing his children with little hints as in a game of hide-and-seek. "Come now, stretch your minds a little. . . . What could the explanation be? Think a little. Don't be afraid of stretching your brains. They won't break" (201). When he dies, his three children divide up the lethal crystals of ice-nine as their inheritance ("there was no talk of morals"), walking around unconcernedly with their samples of the substance capable of destroying virtually all life on the planet. Ice-nine is thus a parallel to the nuclear weapons Hoenikker was already working on. As one character comments, "Anything a scientist worked on was sure to wind up as a weapon, one way or another" (27). Nor is Hoenikker unique; Asa Breed, whose name suggests his representative function, the head of the research laboratory where Hoenikker works, evinces much the same values. His platitudinous message to the high school students is characteristic: "The trouble with the world was that people were still superstitious instead of scientific. He said if everybody would study science more, there wouldn't be all the trouble there was" (25). Breed and Hoenikker are contrasted with a convert to the mystic, holistic religion of Bokononism, Dr. von Koenigswald, who says, "I am a very bad scientist. I will do anything to make a human being feel better, even if it's unscientific" (148).[41] Despite the wry humor of his presentation, Vonnegut's condemnation of scientists is fundamentally serious, as, perhaps, is the pseudo-comic reason he suggests for their inhumanity: "Beware of the man who works hard to learn something, learns it, and finds himself no wiser than before. . . . He is full of murderous resentment of people who are ignorant without having come by their ignorance the hard way" (187).

Pragmatic Postwar Scientists

This amoral stance of scientists explored in a military context has found scarcely more favor in literary treatments during peacetime. Writers have pointed to the resultant dangers to society in a wide variety of hypothetical situations, from space research to materials technology and industrial pollution. In these analyses, many of which represent other smaller "wars," the nexus between denial of responsibility and desire for power and transcendence continues to be emphasized.

A number of contemporary writers level the charge of immaturity against their scientist characters. In Edward Albee's play *Who's Afraid of Virginia Woolf?* (1962) the young geneticist Nick, whose ruthless pursuit of a scientific career within the university is contrasted with the ethical rejection of such ambition by George, an aging history lecturer, is presented as having the tastes and judgment of an immature teenager. George's wife Martha finally passes judgment on Nick in terms that suggest a twentieth-century paraphrase of Blake's condemnation of Newton: "Oh, little boy, you got yourself hunched over that microphone [*sic*] of yours . . . you don't see anything do you? You see everything but the goddam mind; you see all the little specks and crap, but you don't see what goes on, do you? . . . You know so little. And you're going to take over the world, hunh?"[42]

Albee indicates, however, that in the pragmatic and shallow university world, his microcosm of American society, Nick will indeed achieve promotion and power. Albee's fictional assessment of Nick as immature is supported by William Broad's *Star Warriors* (1985), a documentary study of the young weapons scientists at the Lawrence Livermore Laboratory "whose labours had helped to inspire the 'Star Wars' speech of President Reagan and were now aimed at bringing that vision to life. . . . At times they seemed to believe that their labours gave them the power to save or destroy the world."[43] After weeks spent interviewing these scientists, Broad characterizes them as precocious teenagers in all but their technical and mathematical brilliance.

Christa Wolf also uses the term Star Warriors in her novel *Störfall* (1987), already referred to in chapter 13. Wolf draws specific parallels between the explosion of the atomic reactor in the nuclear power plant at Chernobyl in 1986, occurring as the result of a "peaceful" use of nuclear power, and the militaristic Star Wars program, which was promoted by the Reagan administration as a path to peace and security. She sees the young scientists at the Livermore Laboratories as modern-day Fausts who, "driven . . . by the hyperactivity of certain centres of their brain, have given themselves not to the devil . . . but to the fascination of a technical problem."[44] As the result of an article she had read about the Star Wars physicists, Wolf was particularly interested in the character of Peter Hagelstein, who had originally wanted to develop an X-ray laser for medical purposes, hoping to win the Nobel prize for such a humanitarian invention. However, unable to obtain research funding for this project, Hagelstein-Faust sold his idealism for the chance to work on an X-ray laser weapon instead. Wolf attributes this subversion of purpose partly to ambition but mainly to the hothouse, rationalist atmosphere of the Livermore Laboratories, where the all-male scientists live isolated from emotional relationships with women, children, or friends. "What they know, these half-children with their highly-trained brains, with the restless left hemisphere of their brains feverishly working day and night—what they know is their machine. Their dearly beloved computer. . . . What they know is the goal of constructing an atom-powered X-ray laser, the core of that fantasy of a totally secure America achieved by transferring future nuclear battles into space" (70–71).[45]

The American Nobel prize novelist Saul Bellow, whose black humor is not unlike Vonnegut's, has also turned his attention to the amoral scientific world-view and its effect on postwar American society. In *Mr. Sammler's Planet* (1970) he extends the discussion of responsibility to include man's proposed excursion into space. Bellow's representative scientist is the Indian physicist Dr. Govinda Lal, who, like H. G. Wells in a previous generation, cherishes an optimistic belief in science as the hope for a utopian future. Discussing with the humanist Mr. Sammler the issue of man's future and responsibility in science, Dr. Lal asserts, albeit with considerable charm, the pragmatic irresponsibility of the scientist, the ethic of "take what you want because it is there": "There is a universe into which we can overflow. Obviously we cannot manage with one single planet. . . . If we could soar out and did not, we would condemn ourselves. . . . There is no duty in biology. There is no sovereign obligation to one's breed. When biological destiny is fulfilled in reproduction the desire is often to die. We please ourselves in extracting ideas of duty from biology. But duty is plain. Duty is hateful— misery oppressive."[46]

Once a disciple of Wells's Open Conspiracy project,[47] Sammler is temporarily attracted to the order and serenity of science, especially in the face of its prevailing opposite, the violence, aggressive individualism, and antirationalist anarchy of modern American society, but he is unable, finally, to accept its "solutions," because they omit too much from human experience. Having observed the atrocities of war at first hand and become more skeptical about human nature, he answers Lal: "As for duty—you are wrong. The pain of duty makes the creature upright, and this uprightness is no negligible thing" (177). Like science itself, Lal (a rare example of an Asian scientist in Western literature) oversimplifies experience in the interests of pragmatism; and it is clear that Bellow himself is aligned with Sammler rather than Lal in insisting on complexity and on a moral frame of reference. The space race, Bellow implies, will not solve man's problems; he will merely take them with him into space.

One very unusual example of the amoral scientist is provided by an American cyberneticist, Sibylle von Koçalski, the protagonist of Jan Hammerström's *Die Abenteur der Sibylle Kyberneta* (1963). This novel is interesting not only because it provides an unusual example of a woman scientist aligning herself with the amoral point of view (generally considered a peculiarly male proclivity) but also because the author, far from being critical about the undue influence of cybernetics and computing, appears optimistically to endorse the heroine's program of worldwide control to wrench society into the cybernetic age.

Hammerström was considerably influenced by the American mathematician Norbert Wiener, whose popular books on cybernetic theory had appeared in the 1950s.[48] Inspired by Wiener's example, Sibylle has already devised tables and statistical computer models to calculate and hence predetermine economic and political decisions, and she looks forward to extending this control over other aspects of society. Although Sibylle has to justify her system for world governance to more conservative cyberneticists and economists, her success is interpreted by Hammerström as auguring a turning point in world relations, substituting rationality and planning for imprecision and inefficiency. It might be Wells the utopian speaking, especially when Sibylle expresses the hope that eventually she will establish mathematical functions for determining social morality. This program accords with Norbert Wiener's outline of a *machine à gouverner* for the rational conduct of human affairs. Wiener had asked: "Can't one even conceive a State apparatus concerning all systems of political decisions either under a regime of many states distributed over the earth or under the apparently much more simple regime of human government on this planet. . . . We may dream of a time when the *machine à gouverner* may come to supply— whether for good or evil—the present obvious inadequacy of the brain when the latter is concerned with the customary machinery of politics."[49] Hammerström is virtually the only mainstream writer to present the amoral stance favorably or even neutrally. Most portrayals carry a strong element of authorial cynicism, if not overt condemnation, and many link it with an unacknowledged desire for power. In such presentations there is only a thin line between the amoral characters and the dangerous scientists considered in chapter 12.

In this and the two previous chapters we have seen a progression in the depiction of the scientist that is roughly chronological: the power maniac seeking to destroy the world; the scientist with an essentially evil philosophy; the scientist who believes that science per se is of more value than its social consequences or fails to think of the latter at all; and the scientist as exemplar of an amoral perspective, either because of intellectual arrogance, which discounts other facets of human experience, or because of his own personal inadequacies in nonrational pursuits. While this trend appears to suggest decreasing overt evil on the part of the scientist, it involves, in fact, an increasing danger to society, because the attitudes become more socially acceptable and are thus permitted greater freedom for maneuver and the exercise of power. In the progression from the insane and overtly evil scientists who featured so prominently in the early years of the century to the "dedicated" scientists, with their impersonal, objective perspective and amoral standards, there is an increasingly insidious (because more socially acceptable) and pervasive danger. The attitudes of these characters are seen as gradually affecting the whole society until it is powerless to break loose from the stranglehold science and technology have upon it.

In recent literature, amoral scientists frequently have been censured for the escalation of events beyond what they themselves could have predicted. This aspect is most often discussed in relation to the responsibility of scientists, particularly chemists, for environmental pollution, a topic that has been addressed chiefly by German writers. In *Die Sintflut: Der graue Regenbogen* (1959), translated as *Dusty Rainbow,* the German writer Stefan Andres reflected critically on the various arguments used by many real-life scientists to exonerate themselves from moral responsibility for this aspect of their work. His character Lorenz, a chemist who has worked in America, returns to postwar Germany to find the country polluted by a noxious gas.

To the assertion that scientists should not be permitted to invent such horrors, Lorenz counters with the standard twentieth-century arguments that the scientists have not invented anything new, but merely discovered what was already there; that if one scientist did not publish his discovery, another would very soon do so and thus a scientist owes it to his country to make sure that such discoveries are not preempted by the enemy. These arguments from inevitability, used so frequently about the development of the atomic bomb and about chemical and biological weapons, are found repeatedly in the literature, nearly always in order to be overthrown, at least implicitly, by the authorial voice.[50] Heinrich Böll's *Fürsorgliche Belagerung* (1979), translated as *The Safety Net;* Monika Maron's *Flugasche* (1981), Günter Grass's *Die Rättin* (1986), translated as *The Rat;* and Christa Wolf's *Störfall* (1987) all address aspects of environmental pollution and firmly assert the responsibility of the scientists for such effects, even though the consequences have far exceeded their intentions. These scientists who have lost control of their research will be examined in chapter 15.

THE SCIENTIST'S SCIENCE
OUT OF CONTROL

What was once thought can never be unthought.
—Friedrich Dürrenmatt

We, the physicists, find that we have never before been of such
consequence and that we have never before been so completely
helpless.
—Heiner Kipphardt

Unlike the characters of the previous two chapters, who were all, to a
greater or lesser degree, judged as morally reprehensible, the scientist figures discussed here neither intend evil to result from their work nor deliberately isolate themselves from their fellow human beings; on the contrary,
most begin with high moral intentions. Nevertheless, they have all, for one
reason or another, opened a Pandora's box of social or environmental disasters, over which they have lost control in either the technological or the
managerial sense. Either they did not foresee the full consequences of their
discoveries and therefore, like their archetypal predecessor, Frankenstein,
are at the mercy of a creation that cannot now be "undiscovered," or else
they thought they would be given a more authoritative voice in its deployment than, in the event, they were permitted.

The theme of the scientist hoist by his own petard has always been a
popular one in anti-rational times; it assumes a pleasing quality of moral
justice for those who regard science as exceeding the limits of what it is
"proper" for humanity to know. In addition, there is often a perceptible
element of *Schadenfreude* on the part of writers who resent the power and
prestige that Western society accords its scientists compared with its evaluation of the humanities. This resentment was already in evidence in much
Romantic writing, but it has emerged with greater urgency in contemporary
fiction, as the arms race, genetic engineering, the fears associated with nuclear power, fifth-generation computers, in vitro fertilization, and large-scale
industrial pollution of the environment have rendered the dividing line
between fiction and nonfiction more difficult to establish. For the first time,

actual events have provided more spectacular and horrific material than the creations of fantasy. Whereas the technical problem for Shelley, Verne, Wells, and most science fiction writers was to make their fantastic creations credible, the difficulty for many contemporary writers is to avoid mere documentation of actual or highly probable events. What more potent symbol of science out of control could be conceived than the significant probability that existed throughout the 1980s of accidental nuclear war caused by failure of computers or communications, the two areas that epitomized the cutting edge of technology? Burdick and Wheeler's pseudo-documentary thriller *Fail-Safe* (1963) and the film of the same name (1964) were the first fictional treatments of such an eventuality, which was soon to be regarded as highly probable.[1]

Second only to nuclear war was the fear of a major accident in a nuclear power station, and the actuality of Three Mile Island and the Chernobyl disaster, which gave a new and literal meaning to the notion of global catastrophe, permanently demolished the previously widespread faith in fail-safe nuclear technology. As radioactive fallout from the exploded reactor at Chernobyl proceeded to contaminate huge tracts of land, turning food into poison and rendering national boundaries and political alliances irrelevant, the technical phrase "chain reaction" began to assume a suggestion of macabre determinism.

A different kind of troubled and inadequate scientist, a peculiarly twentieth-century figure, is the one who, while in fact morally superior to his society in accepting responsibility for his discovery, has lost control over its implementation to some organization, usually the government-military complex or a multinational company. His own moral imperative may even drive him to subvert government authorities, thereby risking the label of traitor. This theme of what could be called "ethical treason" emerged during the Second World War, when scientists in defense work suddenly found themselves involved in escalated research on atomic weapons. It became even more pertinent as the paranoia generated by the cold war continued to provide a similar moral quandary for those caught up in the superpowers' escalating weapons research and outraged by the barriers to free exchange of scientific information, even between political allies. Klaus Fuchs and the other "atom spies," who for diverse reasons passed the secrets of atomic research on to the Soviets, were only the most publicized representatives of a much larger group of scientists who experienced similar moral anguish about their research into weapons of mass destruction. This theme was explored in fiction before any actual examples of nuclear espionage were known.

Runaway Inventions

Mary Shelley's analysis of the factors leading to Frankenstein's subjection to his creation could stand as a prototype of one of the central preoccupations of twentieth-century literature about scientists. The scientist is unable to control the Monster he has so thoughtlessly and optimistically brought to life because, having deliberately cut himself off from his society in order to pursue his research, he, like Frankenstein, has no support network to sustain him when he is confronted with the results of his labors. Deluding himself throughout the period of his research with the wishful belief that *his* creature would necessarily be beautiful, he has failed to observe all the indications to the contrary. Attempting to disguise his real motives (desire for fame and power), he persuades himself that his research is for the good of mankind. This egotism underlies his disgust and horror on seeing the misshapen creature, which he characteristically interprets as a reflection on himself, his subsequent attempts to deny responsibility for it, and his failure to proffer an adequate emotional response. Thus the tragedy follows inexorably from Frankenstein's own deficiencies.

While not all twentieth-century writers emphasize these particular aspects, most do strongly imply that an element of hubris underlies the scientists' otherwise incredible failure to see where their experiment will lead and to take action accordingly. To predict the horror developing in the laboratory might mean the discontinuation of the precious experiment. It is therefore more opportunistic not to notice the Monster, or to hope that it will, after all, prove to be a friendly, biddable monster, entirely under one's control. The earliest twentieth-century examples of the scientist overpowered by his own invention are closest to the Frankenstein archetype, involving the creation of a living human being or some approximation thereto; all have a strongly moral, even religious, lesson to impart. Such characters are clearly culpable and hence closer to the mad, evil, or dangerous scientists of chapter 12, while their successors may even be idealistic in their intentions and fail only in lacking the power to retain control over their discoveries.

Harle Owen Cummins's short story "The Man Who Made a Man" (1902), for example, closely follows the central Frankenstein theme, though without Shelley's subtlety. Professor Aloysius Holbrok (whole broken?), head of a department of synthetic chemistry and presumably a disciple of Moleschott, has reduced the human body to its simplest components and recombined these elements to synthesize the frame of a man. All that remains is to bring it to life, and this the professor recklessly undertakes by means of a high-voltage shock. The experiment is briefly successful. The professor's assistant relates: "Before our staring eyes that thing which was only a mass of chemical compounds *became a man*. A convulsive twitching brought the body almost into a sitting position, then the mouth opened and there burst forth from the lips a groan."[2] At this crucial moment, the assistant faints, "but those wires, with their deadly current, which I tried to tear away as I fainted, must have been directed back by a Higher Hand, for there remained on the slab only a charred and cinder-like mass." The strong moral overtones suggested by this divine intervention, which saves the world from an unholy creation, are amplified in the theme of retribution. The professor, whose effrontery has been emphasized by his repeated comparison of himself to God—"We will do what no one but God has ever done before" (8)[3]—is duly punished for his intellectual pride by becoming insane. "The man who had made a man . . . was crawling about on the floor, counting the nails in the boards and laughing wildly" (10). This "happy" outcome, resulting in the restoration of divine control, is clearly intended to underline the failure of human attempts to control life.

Related to the desire to create a look-alike human being was the perhaps more modest ambition to construct a thinking machine. E. A. Hoffmann's use of the beautiful automaton Olimpia in "Der Sandmann" and the chess-playing machines in Edgar Allen Poe's "Maelzel's Chess-Player" and Ambrose Bierce's "Moxon's Master" were nineteenth-century examples.[4] In all of these there is a degree of retribution contingent upon the creation of these automatons. Two examples are sufficient to indicate the prevailing thought pattern.

Edmond Hamilton's early story "Metal Giants" (1926) makes use of a similar concept in the creation of a metal man endowed with a metal brain and consciousness. Heedless of the obvious potential problems, the scientist-creator Detmold proceeds to supply his creature with arms animated by magnetic fields generated by thought currents in the brain. Worse still, he allows the brain to "think" for itself by means of internal atomic changes, thus making it both self-sufficient and physically empowered to carry out its will. Predictably, the brain takes over its own evolution and constructs more metal giants, which it controls remotely, directing them to embark on a program of world conquest. Detmold is able to destroy the controlling brain, thereby causing the collapse of the giants, but in falling they destroy their creator, who thereby combines the roles of enemy and martyr for the human race.[5]

A similar example of hubris leads to near destruction in Philip Wylie's novel *The Gladiator* (1930), in which a biologist, Abednego Danner, injects his pregnant wife with a chemical intended to endow their child Hugo with superhuman strength. Fortunately, although this infant Hercules unintentionally smashes his cradle before he is six months old, he is morally as well as physically superior to his parents and aware of the potential dangers of his strength. When a professor proposes cloning Hugo to create a superrace in the wilderness, Hugo is doubtful and asks Heaven for a sign. It comes: a bolt of lightning strikes the noble martyr dead, thus subverting the professor's plan. In these and most similar stories from the early decades of the century the disastrous situations resulting from scientific discoveries are redeemed, and control restored, whether by human or supernatural agency. (Chance is usually taken to indicate divine intervention.)[6] But such optimism about the possibility of averting the disasters contingent on technological progress was not to last long.

One of the first "modern" warnings was Karel Capek's play R.U.R. (1921), already mentioned as an exploration of the pride and materialism that fuel technological advances. R.U.R. was even more obviously Capek's analysis of an experiment that backfired. As in the Frankenstein archetype, the very success of the Rossum invention proves the undoing of the company—and of humanity—as the robots rise against their masters throughout the world, demanding the designs to reproduce their kind and proclaiming humanity as their enemy, as parasites who must be eliminated. Again it is stressed by Capek that the warning was there from the start. Significantly, it was voiced explicitly by the nonscientific, female characters, Helena and Nana, who protested against "all these new-fangled things [which] are an offence to the Lord." The complexity of meaning associated with the robots expands the levels of irony in the play. Capek himself commented cynically:

> This terrible machinery [of industrialism] must not stop, for if it does it would destroy the lives of thousands. It must, on the contrary, go on faster and faster, although it destroy in the process thousands and thousands of other existences. Those who think to master industry are themselves mastered by it; Robots must be produced although they are, or rather *because* they are, a war industry. The conception of the human brain has at last escaped from the control of human hands. That is the comedy of science.[7]

R.U.R. provided a new myth for the twentieth century, pointing out the Damoclean sword poised over the developing sciences of robotics and computing. The threat of an alien power enforcing its will on humanity is now located, not in invading Martians, against whom the world's armaments would be instantly directed, but in the development of artificial intelligence, the runaway child of man's own intellect.

C. P. Snow's second novel, *New Lives for Old* (1933), concerns the development by two heedless scientists, Pilgrim and Callan, of a rejuvenation process that will extend the normal life span by some thirty years and restore to their geriatric customers the physical attributes of youth. This apparently beneficial process, however, turns out to be a social catastrophe, for not only do the old retain the follies of youth but the whole experiment precipitates a communist revolution when it is realized that only the rich can indulge their desire to become young again.

By contrast with Snow's bleak vision, Anatoly Dneprov's elegant story

"The S*T*A*P*L*E Farm," based on the paradox of determinism, provides a devastating but humorous attack on the pride of scientists and their cult of efficiency. Dneprov's unnamed biochemist believes that social harmony is possible only through standardization. Having discovered how to clone individuals from a single cell, he does not hesitate to select himself and his wife as prototypes for this cloning process. "Margie and I were outstanding . . . we discovered in ourselves enough wisdom to envisage the standardised man in a monolithic, homogeneous society."[8] Like Frankenstein, only more literally, this scientist sees himself immortalized as the father of a future race. The resulting offspring are brought up in environments identical to those of their parents, but far from achieving complete harmony, the standardized children have not only their parents' intelligence but also their moral failings. All the female descendants will inevitably commit suicide, as Margie did on learning that her husband had once attacked a neighboring girl, while all the male descendants are likewise doomed to reenact this attack. At the time the story was written, this nightmare future "which must go on repeating itself with unswerving inevitability on an ever-increasing scale" (195) seemed merely a humorous exercise on the paradoxes inherent in determinism; today the cloning procedures of the story are almost achievable, yet there are still no legal or moral parameters for dealing with the consequences.

A similar style of trenchant satire masquerading as comedy about hapless scientists who have little idea where their ideas are leading characterizes the work of the Polish writer Stanislaw Lem, whose formidable qualifications include a background in medicine, cybernetics, literature, and philosophy. The cybernetic theme is central to much of Lem's work,[9] and his many attacks on scientists invariably focus on their failure to communicate despite the massive and sophisticated systems of hardware and software designed for this express purpose. Because of this basic deficiency, such characters inevitably lose control and are unable to extricate their society from the disasters they have unintentionally perpetrated.

In *The Futurological Conference* (1954), set at the Hilton Hotel in Costa Rica in the year 2039, a conference of scientists is assembled to save the world from numerous crises that science and technology have unleashed. Characteristically, their approach to the problem is to look for a quick, technological "fix." The agenda for the first day alone includes the world urban crisis, the ecology crisis, the air pollution crisis, the energy crisis, and the food crisis; the technological, military, and political crises are to be dealt with the following day. However, although the scientists have every electronic device to assist them, they are quite unable to communicate meaningfully. Each speaker is given only four minutes, so all reports, with numbered paragraphs, are distributed beforehand. The lecturer speaks only in these numerals. "Stan Hazelton of the US delegation threw the hall into a flurry by emphatically repeating: 4, 6, 11, and therefore 22; 5, 9, hence 22; 3, 7, 2, 11, from which it followed that 22 and only 22! Someone jumped up saying yes but 5, and what about 6, 18, or 4, for that matter; Hazelton countered this objection with the crushing retort that, either way, 22. I turned to the number key in his paper and discovered that 22 meant the end of the world."[10]

The scientists gathered to discuss technology are unable to control their own robots, which are too intelligent to be obedient and, instead, follow Chapulier's Rule: "A smart machine will first consider which is more worth its while: to perform the given task or, instead, to figure out some way out of it. Whichever is easier . . . therefore we have malingerants, fudgerators and drudge-dodgers, not to mention the special phenomenon of simulimbecility or mimicretinism. . . . some get completely out of hand—the dynamoks, the locomotors, the cyberserkers" (84–85). Far from being able to solve the problems on their agenda, the members of the Futurological Association are unable to agree about anything. Their schemes are mutually contradictory, and the conference ends, appropriately, in catastrophe when the hotel is accidentally blown up by Costa Rican armed forces trying to quell the revolutionaries in the streets outside. These scientists, affecting to organize the world, are unable to complete even one day of their conference without falling victim to primitive violence or their own inadequacy.

Lem returned to this theme of scientists' failure to communicate in *His Master's Voice* (1968). In this novel, Professor Hogarth leads a special project (His Master's Voice) to decipher a signal from outer space. The signal, at first thought to be merely random noise, is later found to have a repetitive pattern suggestive of a message.[11] Symbolically, the novel is about communication, including the difficulties of communicating even with other human beings, much less with other civilizations; but it is particularly concerned with scientists who, in this age of electronics, computers, and information technology, have set themselves up as the great communicators. The project has collected together the world's most eminent theoretical scientists, and Lem proceeds to show their personal failings as well as their ineptitude in solving the riddle. Locked up in the concrete city that was formerly (and significantly) the Nevada atomic testing site—for Lem's scientists too produce nothing but destruction—they become increasingly childish; petty feuds waged with graffiti (minimal communication) break out between the biologists and the psychoanalysts. Like their real-life counterparts, Lem's scientists are rigorously supervised by the government, the military, and the police, just in case the message should have military implications. They tolerate this supervision in exchange for research funding. Lem comments that they are like pigs kept to hunt for truffles, in their "well-equipped sties—that is to say the shining laboratories."

More devastating for these theoretical scientists, conditioned to objectivity, is the completely new frame of reference within which they now have to work. Instead of dealing with an impersonal, objective problem, they are suddenly confronted with a personal antagonist, the Someone who sent the message. Here Lem raises fundamental questions about the nature of scientific enquiry and the rigid mind-set of scientists. "Never before had physicists, engineers, chemists, nucleonicists, biologists or information theorists held in their hands an object of research that represented not only a certain material—hence natural—puzzle, but which had intentionally been made by Someone and transmitted, and where the intent must have taken into account the potential addressee. . . . their whole life's training, the whole acquired expertise of their respective fields, worked against that knowledge" (33).

Lem also shows, by means of two subplots, that science may lead to either creation (the Frog Eggs Project) or destruction (the TX Bomb Project, in which the most brilliant of the scientists develop the ultimate bomb).[12] Gradually the project founders as the scientists find themselves unable either to deal with the implications of the evidence available or to agree about possible theories. Hogarth, the eminent mathematician and the doyen of the scientific community, is depicted as essentially evil and ruthless, and to a lesser extent this is true of the scientists as a group. They have gathered idealistically, to discover a new breakthrough in communications, but instead they develop the deadly TX bomb, capable of destroying the Earth. For Hogarth, as for the others, "the source of his greatness—the dispassionate taste for challenges and for pure mathematical/scientific analysis—is also the source of his undoing. He and his colleagues cannot resist the temptation to develop ideas even when they are clearly evil."[13] Thus the scientists are seen as fallible mortals, pretentiously engaged on a project far beyond their capabilities, unaware of their unimportance or even that they are merely being kept as useful animals by their own governments.

The bleak visionary landscape of the American novelist Philip K. Dick provides many more examples of scientists whose inventions have led to even more macabre disasters, and Dr. Bloodmoney has already been discussed as an example of a mad and evil scientist. Dr. Labyrinth, the protagonist in Dick's story "The Preserving Machine" (1953), has no such evil intentions; on the contrary, he hopes to preserve the fine and beautiful things from a civilization on the brink of collapse. To avert the loss of classical music, he builds a machine to process great musical scores into living forms, producing a Mozart bird, a Beethoven beetle, a Wagner animal, and so on. However, unexpected problems arise, and he too finds that he has no control over the result. "It was all out of his hands, subject to some strong, invisible law that had subtly taken over, and this worried him greatly. The creatures were bending, changing before a deep, impersonal force, a force that Labyrinth could neither see nor understand. And it made him afraid."[14]

When the musical animals become brutal, enacting the violent emotions that underlie the order and beauty of the music they represent, Labyrinth tries to reverse his act of creation by returning his creatures to the machine; but alas, when they are translated back into music, the order of the original score has been irretrievably lost, and the result is a terrible cacophony. This situation encapsulates Dick's perception of the future as full of unpredictable possibilities, to which humanity will have, often painfully, to adapt. There is no return to a pretechnological world, nor can there be control over technology; man's survival depends on his transformation of technology, which will, in turn, transform him. There can be no "preserving machine" in which we can selectively enshrine those parts of the world that most appeal to us. Dr. Labyrinth represents, therefore, both the scientist who professes to offer us the technological power to determine our lives and, simultaneously, modern man, who attempts to use technology on his own terms only to find that the results are beyond his control.

C. M. Kornbluth's short story "The Altar at Midnight" (1952) puts forward a similar message about the ignored human casualties of the space race, another topic where the reality has outstripped literary predictions. Kornbluth's space scientist Dr. Francis Bowman, "the man who made space flight a reality," is ultimately responsible for the physical deterioration of the astronauts, or "spacers," whose blood vessels have been ruptured by repeated compression and decompression and whose eyes have been irreparably scarred by radiation. Overtly, Bowman shifts the responsibility to the government, which has pushed the scientists' ideas to their conclusion, but his own compulsive alcoholism suggests acknowledgment of guilt for the human suffering contingent on his work. Thomas Pynchon was to explore these issues with more telling satire in his tour de force *Gravity's Rainbow* (1973) (see chapter 13 above) by modeling his characters on actual luminaries of the U.S. space research program. However, it was to take a real tragedy, the death of seven American astronauts in the space shuttle *Challenger* in 1986, to elicit searching questions from a society that hitherto had funded the space race at the expense of other programs without questioning its long-term costs and benefits.

Compared with the enormities laid at the door of physicists, biologists, and chemists, astronomers have generally managed to retain a "clean" image, untainted by the suspicion of military collaboration and weapons production. However, the former Cambridge theoretical astronomers Fred Hoyle and John Elliot exploded this comforting belief in their science fiction story *A for Andromeda* (1962) and its sequel, *Andromeda Breakthrough* (1964), later combined to make a popular film. Hoyle and Elliot explore the general moral issue of pursuing cutting-edge research that may produce dangerous results; in the process, they also question the belief that there is such a thing as harmless, "pure" research. Hoyle and Elliot's protagonist, John Fleming, provides an interesting example of the scientist who loses control of his experiment to a series of external agents and finally regains autonomy only through brute force. Fleming, an astronomer-cum-computer scientist, slowly comes to realize that his deciphering of an encoded message from the Andromeda nebula is being controlled by an alien intelligence, which directs him to build a new, massive computer with a brain superior to his own. This computer (a fifth-generation model with artificial intelligence) thereupon becomes the real controller of the project, demanding details of human DNA and then directing the researchers to construct progressively more complex organisms, culminating in a woman with the superior intelligence of the aliens who sent the signals from the Andromeda nebula. In due course Fleming acknowledges his instinctive forebodings and tries to terminate the project, but his coworkers, especially the female biologist Dawnay, have become obsessed with the potential breakthrough and are determined to continue the project. Fleming alone realizes that the creature is merely an extension of the computer. He observes, "Dawnay thinks the machine's given her the power to create life; but she's wrong. It's given itself the power."[15]

Ultimately he is able to remove the danger, not by scientific or intelligent means, but only by resorting to primitive violence and smashing the computer with an axe. He also releases Andromeda, "built" to the computer's instructions, by evoking the nonrational feelings she has inherited with her human component. In Fleming, then, Hoyle and Elliott explore the fluctuating and finally paradoxical relationship between the scientist and his experiment. He begins as director of the project, treating his material, in this case a message, and its resultant products (a computer and subsequently a living creature) as objects; but in the process, he himself becomes the object of the experiment, which is being controlled by an extragalactic intelligence. At best, he and his team are laboratory assistants to the computer; at a more sinister level, they are themselves being studied as part of the computer's data on *Homo sapiens*. Thus Hoyle and Elliot have created a neo-Romantic fable in which the external control of the computer has become a metaphor for the rationalist obsession, and the evil monsters it creates can be exorcised only by irrational means—violence or love.

It is interesting to recall the parallels with the discovery of pulsars at Cambridge five years after the publication of *A for Andromeda*. Jocelyn Bell, a graduate student of Anthony Hewish, was mapping the extraterrestrial signals coming into the radio telescope when she too noticed a regular pattern, which correlated with sidereal time. Like Hoyle's fictional scientists, the Cambridge team used a fast decoder to unscramble the recurring signal, which proved to be a burst of pulses with a period of 1.33 seconds. At that stage there seemed at least a possibility that the pulses were being beamed out by an extraterrestrial civilization; although it proved otherwise,

Hoyle and Elliot's novel implicitly poses the query how far the Cambridge astronomers might have pursued the project out of sheer fascination with the science even if there had been this element of danger in the case.[16]

On 26 April 1986 all hypothetical explorations of runaway technology were instantly eclipsed by the actuality of the world's worst nuclear accident, the explosion of reactor 4 at the nuclear power plant in Chernobyl, a small city in the Ukraine. The explosion started a graphite fire that could have resulted in the disaster that no one wanted to contemplate, core meltdown. Although this was averted, the actual consequences shocked the world into a realization that there could be no protection from the long-term effects of such accidents and that the risk of recurrence could not be eliminated. For weeks the experts gave conflicting advice to terrified populations of neighboring countries as the radioactive material drifted with the winds over much of Europe. One of the first fictional representations of the Chernobyl disaster was *Störfall* (1987), written between June and September 1986 by the (then East) German novelist Christa Wolf.[17] The novel focuses on the widespread confusion and anxiety of ordinary people in the face of the conflicting accounts they were receiving from the "experts." The optimists affirmed that there was no possibility of core meltdown; the pessimists said it could not be ruled out. In the latter case, there would, seemingly, be no end to the destructive potential of the radioactive core of the reactor as it burned its way through the earth to the antipodes, a possibility the scientists, with macabre humor, called the China syndrome.[18] Nor could the experts speaking on television give the assurance their audience wanted for the future. On the contrary, they insisted that absolutely watertight predictions are impossible, that with such a new branch of technology there is always an element of risk.[19]

The magnitude of the Chernobyl disaster was surpassed only by the potential horror of broad-scale nuclear war, with such long-term and all-encompassing consequences as nuclear winter and genetic aberrations, if not the extinction of most life forms from the planet. Yet although this theme has been explored in fiction (one of the earliest treatments being Nevil Shute's prophetic novel of the 1960s, *On the Beach*, which set the worldwide nuclear disaster in 1964), it has rarely been associated with the characterization of particular scientists, except in the stereotypical figure of the mad, evil physicist considered in chapter 12. In the consideration of disaster on such a scale, the involvement of individual scientists is reduced to insignificance, even irrelevance.

Scientists against Authority

Many of the troubled scientists in the literature written during and after the Second World War are struggling for control, not against their inventions, but against powerful authorities who demand their allegiance in projects that their conscience opposes. This was the situation confronting many of the physicists who worked on atomic weapons during the war. They alone realized the full extent of their destructive power and the inhumanity of using such weapons against civilian populations, but their governments, arguing from expedience, demanded that the project be completed, lest the enemy produce the weapons first. Although technically the war ended in August 1945 with the demonstration of superior military power, the race to devise yet more effective methods of mass destruction continued unabated as the cold war, with a large proportion of the superpowers' budgets dedicated to "defense." Scientists in appropriate disciplines, especially physics, were recruited to work in weapons establishments because they could not secure funding in other areas.

The moral struggles of these scientists during and after the war formed the basis for a large number of novels and plays on the theme of the ethically superior scientist anguishing over a moral problem that his government is incapable of understanding. The real enemy is not the bomb itself but the government-military establishment that demands it. The exploration of this situation by writers came first in Germany, where, even in the years preceding the declaration of war, there was a perceived conflict between the scientists' search for truth and the requirements of the Nazi military program.[20] However, parallel restrictions on writers made it impossible for them to explore the issue overtly. Instead, a number of writers resorted to a veiled treatment of the subject through the depiction of historical figures, Kepler and Galileo being obvious parallels. As in the case of Brecht's *Galileo*, the hegemonic authority in these treatments is overtly the medieval church, but the historical situation transparently represents the contemporary one, where the power demanding absolute allegiance is a totalitarian government.

Hans Rehberg's play *Johannes Keppler: Schauspiel in drei Akten* (1933), set during the Thirty Years' War, shows Kepler as a martyr persecuted by Protestants and Catholics alike.[21] Neither of these authorities actually disputes the truth of Kepler's ideas, but they argue that the truth must be subordinate to authority and social order. Kepler the mathematician, on the other hand, admits of no authority higher than the truth, since truth comes from God: "God has directed me and shall I throw away what the everlasting God has given me, for the sake of a duke?"[22]

Max Brod's study of Galileo, *Galilei in Gefangenschaft* (1948), also concentrates on the issue of the allegiance of the scientist to truth as he knows it and the power of authority that tries to suppress that truth. Galileo is less constant than Kepler, whose missionary spirit never succumbed to authority; in the face of the absolute refusal of the church to tolerate any contradiction of Aristotle, Galileo recants his Copernican beliefs, buying time not only to complete his scientific work but also to evolve a philosophical and ethical reassessment of the nature of truth, which has clear implications for Brod's contemporaries. "A truth for which man is not prepared to die is no truth. . . . A man who is not prepared to die for the truth is no man. The truth demands a sacrifice like every great sacrifice demanded of us. Like an epicurean I wanted to avoid the sacrifice, but that does not happen when one takes the thing seriously. Ironically, I was almost a martyr for the Truth against my will though. . . . Now it is different. Now I want it."[23]

Frank Zwillinger's *Galileo Galilei, Schauspiel* (1953) presents a somewhat different portrait of Galileo, as a man who in old age believes he has failed both science (because he did not accept martyrdom for the truth) and the church (because although he recanted outwardly, he has never recanted in his soul) and has therefore lived a double lie. However, Zwillinger's play is more optimistic than most literary treatments of the Galileo story in that his protagonist's dilemma is resolved, albeit somewhat glibly, even dishonestly, when the archbishop of Sienna assures him that, owing to a technicality, the Inquisition's ban on the teaching of the Copernican system is invalid, and hence such teaching is not heretical.[24] Zwillinger thus removes the scientist's anguish by means of an external authority that accepts the responsibility, a resolution that was fundamentally irrelevant to the contemporary situation of the cold war.

In England and the United States it was possible to explore the contemporary political situation more explicitly, although even here an element of self-censorship is suggested by the ambivalence writers display towards their scientist characters involved in a conflict with military and political authorities. These scientists are not portrayed as heroic or even as having any clear idea of alternative courses of action. They are therefore distinct from the idealistic rebel scientists, to be considered in the next chapter.

Two novels of wartime research by the British writer Nigel Balchin depict scientists as neither great discoverers nor potential saviors of their country but merely ordinary men caught up in the protracted state of emergency. Far from defying the establishment, they are manipulated by the politicians, army personnel, or government administrators, whom they despise both intellectually and morally. Their "heroism" can consist of no more than uncompromising honesty in the face of the secrecy, power-seeking, and hypocrisy around them. Sammy Rice, the handicapped physicist of *The Small Back Room* (1943), almost achieves a moment of glory when he is called in to defuse a bomb, but at the crucial moment the physical effort to unscrew it proves too great, and he has to ask for help. His self-disgust, which is apparent in his statement that "the thinking had been all right but when it came to doing anything I'd been gutless as usual,"[25] functions as a signifier of both his heroism and his inadequacy. Thus Balchin implies that although

Rice's determination may be the only possible heroism in the real world, it is clearly a very qualified one.

In *A Sort of Traitors* (1949), its title a quotation from *Richard II*,[26] Balchin explores the various temptations that assail a group of scientists working on antibody control of epidemics during the war. Here the element of external control hinges on government censorship relating to publication of results, lest they be used by the enemy to start an epidemic.[27] Balchin's scientists respond diversely, enacting possible alternative courses of action, none of which is morally satisfactory. One of the young and idealistic members of the team, Marriott, is influenced by a speech given to the Junior Research Association urging the scientists to "throw off the habit of silence and retirement which is the scientist's curse, and insist that the scientist's influence in the world shall be in proportion to his contribution" (39); when he is told of a fictitious international open publishing syndicate, he is almost drawn into a network of scientific espionage. Another scientist, Sewell, decides to subvert the publication ban by openly sending copies of his work to the heads of relevant scientific organizations throughout the world and apprising his government of the fact afterwards. But neither of these desperate expedients comes to fruition. Sewell admits that his allegedly high-minded concern for "the interests of humanity" is basically self-interest. "It isn't really the loss to the human race that hurts us when you come down to it. It's the loss to *us*. . . . I remember old Philip Lowes [saying] . . . 'Science matters, but scientists don't. Sometimes scientists forget that'" (272). Thus Balchin's scientist heroes are essentially ordinary, powerless men, often physically handicapped in some way to emphasize the point. In the world of politics and a military-orientated government, only a limited heroism, involving moral anguish, is permitted to the scientist.

This image of the scientist as essentially honest and well-meaning but unable to counter the machinations of power politics continued into the cold war. Upton Sinclair's topical play *A Giant's Strength* (1948), about an atomic physicist, Barry Harding, who had been involved in the top-secret research on the bomb, was one of the first U.S. works of fiction explicitly to urge upon scientists their particular responsibility for persuading their government of the destructive power of atomic weapons.[28] The play opens just before the dropping of the bomb on Hiroshima and includes documentary clips for immediacy and realism. Thus, immediately after the news flash about the Hiroshima bomb, Sinclair introduces the optimistic voice of General MacArthur on the radio, justifying its use as "a holy mission," thanking "a merciful God that He had given us the faith and the courage and the power from which to mould a victory" (20), while affirming that "the utter destructiveness of war now blots out this alternative. We have had our last chance" (21). This is corroborated by Harding: "There can never be another war like

the old ones. The next war will wipe out every city in the first half hour—
every city on both sides" (21). However, the play goes on to show the
hardening of attitudes through the cold war. Harding, who has joined an
association of nuclear physicists trying to educate the public and guide the
government away from war, is interviewed by the House Un-American
Activities Committee.[29] Eventually an atomic war breaks out, and the characters of the play degenerate to subsistence level in a cave in Dakota. Harding still hopes to exert influence for good through the physicists' organization, but the only real hope expressed at the end of the play is embodied in
his teenage son, who proclaims the need for universal brotherhood and a
world state. Thus, although Harding, the representative scientist, is morally
more aware than most of his society and agonizes over a possible solution,
he is helpless to provide one. The only suggestion Sinclair offers is a moral
one, voiced by a child.

The German playwright Hans Henny Jahnn expresses a similar pessimism in *Trümmer des Gewissens (Der staubige Regenbogen)* (The ruins of conscience: The dusty rainbow) (1959). Here the individual scientist, isolated for
so long from social issues, recognizes too late the evil potential of his research and the military intentions of his government and is thus powerless
to avert the threat of mass destruction. Jahnn's protagonist, Jacob Chervat, is
an atomic physicist living in an Atomstadt, a government-funded science
research ghetto, and refusing to listen to any discussion of the destructive
implications of his research. Even though his son Elia is sexually underdeveloped because of the radiation exposure in the Atomstadt and his wife
suffocates their next baby rather than expose it to a similar fate, he continues
to believe that his discoveries will eventually be for the benefit of humanity
and will never be used for destructive purposes. However, his confrontation
with Dr. Lambacher, a biologist who is determined to extend his genetic
experiments on huge dragonflies and produce a race of supermen, alerts
Chervat to the manipulation of science for destructive purposes by both
individuals and the establishment. He realizes that Lambacher's project is
analogous to the government's plan to subdue the nonwhite races by a
preemptive nuclear strike, using his research. But although Chervat is transformed from a reclusive scientist into a morally committed activist, his
struggle with the authorities is doomed to failure. Acceptance of public
funding has made him government property, and his ethical protests are
regarded as treason. In despair, Chervat attempts to kill an official and then,
appalled at his own violence, commits suicide. Nor does Jahnn leave us with
any hope that his sacrifice will have an effect. The last ironic lines of the play
emphasize this: "You do wrong! You hope!"[30] There is thus no good outcome in a world where science is controlled by a totalitarian military power;
the lone protesting scientist comes too late.

A similar thesis underlies Dino Buzzati's symbolic short story "Appointment with Einstein" (1958). Buzzati's Einstein has just made his break-through in conceiving the idea of curved space. Although usually a humble man, he indulges in a moment of pride, considering himself "above the common herd." Almost immediately he is confronted by a devil (in the person of a Black garage attendant), who announces that his moment of death has arrived. Einstein pleads for time to complete the proofs of his brilliant concept, and after two months, when the work is finished, he re-turns sadly but honorably to the appointed place to die.[31] At this the devil, laughing uproariously, reveals that the threat of impending death was only a satanic device to make him work harder and finish his work: "God knows how long you would've drawn it out if I hadn't scared you two months ago. . . . the big devils run the show. They say your first discoveries have already been extremely useful. . . . Hell has profited greatly from your ideas." Einstein is irritated and incredulous: "Nonsense! Can you find any-thing in the world that's more innocent? They're insignificant formulas, pure abstractions, inoffensive, impartial." But he is assured by the devil, "They're satisfied downstairs. Oh, if only you knew!" (14).

Einstein and Rutherford are the two best-known examples of scientists who, in all innocence, worked on their seemingly obscure theories, believing them to have no sinister significance or destructive potential.[32] In his deeply pessimistic play *Die Physiker* (1961), however, the Swiss playwright Friedrich Dürrenmatt extended the range to include the physicist Johann Wilhelm Möbius as an exemplar of the powerlessness of even the most brilliant scientists to prevent the misuse of their work. In 1956 Dürrenmatt had reviewed Robert Jungk's book *Heller als tausend Sonnen* (Brighter than a thousand suns), which detailed the lives of the physicists who worked on the first atomic bombs and raised the question of the scientists' respon-sibility for the outcome of their research. After the event, it was said by several of the scientists, including Werner Heisenberg, that if even twelve of the physicists had combined to oppose it, they could have prevented the construction of the bomb. Dürrenmatt accepted this but believed that once the first atomic bomb had been constructed, there were no longer any op-tions. The belated warnings of the scientists to society are now useless, merely "demands on an imaginary world, demands not to sin after the Fall."[33]

In the ironic tragicomedy of *Die Physiker*, Möbius is caught in a para-doxical situation where action and nonaction amount to the same thing, just as the Möbius loop has only one surface and one edge.[34] Dürrenmatt re-marked that "the action takes place among madmen and therefore requires a classical framework to keep it in shape." It becomes clear that the concept of a madhouse is relative. A world where atomic physics is directed to-wards weapons research is, by definition, a madhouse; hence the only place in which to preserve one's sanity is an asylum, or so Möbius thinks. "For him, the asylum is no poetic metaphor, but the most normal consequence of an actual development. He examines the broadest form of our as-yet-undiagnosed madness, the hunt for atomic physicists."[35]

Through the three representative "patients" in the Cherry Trees asylum, Dürrenmatt explores three alternative approaches available to scientists confronted by a totalitarian system. Möbius has retreated there to prevent his work from being misused; that is, he has sought moral shelter there from an insane world. Kilton, a British physicist and secret agent who claims to be Newton, and Eisler, an East German secret agent who alleges he is Einstein, have come pretending to be fellow inmates to persuade him to work for their respective governments. Like Newton, Kilton is an optimist about science, believing in freedom of research, universal sharing of discoveries, and freedom from political or social responsibilities. He argues, like the amoral scientists of the previous chapter: "I give my services to any system, providing that system leaves me alone. You talk about our moral responsibility. . . . That is nonsense. We have far-reaching pioneering work to do, and that's all that should concern us."[36] However, Dürrenmatt implies that what was appropriate in Newton's time is no longer valid; it can no longer be assumed that scientific research is beneficial to society. Eisler rejects this view and demands political commitment as the only responsible action, just as Einstein had taken a political position and recommended the construction of the atomic bomb. Eisler argues: "We cannot escape our responsibilities. We are providing humanity with colossal sources of power. That gives us the right to impose conditions. If we are physicists, then we must become power politicians" (42).

Möbius has been in the asylum for fifteen years, during which time he has continued to work on his *Weltformel*, his principle of universal discovery. This work, which combines Newton's gravitational theory and Einstein's uncompleted unified field theory, is the greatest breakthrough in modern physics, but Möbius knows that for the sake of humanity it must be suppressed. "There are certain risks which one may not take: the destruction of humanity is one. . . . Fame beckoned me from the university; industry tempted me with money. Both courses were too dangerous. . . . A sense of responsibility compelled me to choose another course" (44). He has therefore burnt his manuscripts, and the other two "physicists," their plans thereby subverted, reluctantly agree that to remain in the asylum is the only sanity.

> Möbius: Only in the madhouse can we be free. Only in the madhouse can we think our own thoughts. . . . We ought not to be let loose on humanity. . . .
> Newton: Let us be mad, but wise.

EINSTEIN: Prisoners but free.

MÖBIUS: Physicists but innocent.   (45, 47)

Had the play ended here, Möbius would have been cast as a heroic figure, and the tone one of qualified optimism. It would have implied that physicists at least have a choice, albeit limited, concerning the outcome of their research. However, Dürrenmatt denies the possibility of even such a minimal degree of freedom. Ironically, the man whose theory promises such power to its possessor has no choice at all. We learn that the director of the asylum, the mad, resentful Fräulein Doktor Mathilde von Zahnd, knowing Möbius's identity, has secretly copied all his papers as he wrote them and founded a giant cartel ready for world domination through the application of his formulas. Speaking for Dürrenmatt, she says of Möbius: "He tried to keep secret what could not be kept secret. . . . Because it could be thought. Everything that can be thought is thought of some time or other. Now or in the future" (50). Thus the physicists are too late in their decision to assume responsibility. Möbius admits, "What was once thought can never be unthought." Finally the three men do indeed become insane; Kilton now really does believe he is Newton, and Eisler that he is Einstein, while Möbius claims that he is the fallen King Solomon, for whom knowledge has become a curse: "I was a Prince of Peace, a Prince of Justice. But my wisdom destroyed the fear of God, and when I no longer feared God, my wisdom destroyed my wealth. Now the cities over which I ruled are dead. . . . And somewhere, round a small, yellow, nameless star there circles, pointlessly, everlastingly, the radioactive earth" (53).[37]

Among his "Twenty-one Points to *The Physicists*" Dürrenmatt included the following: "The contents of physics are the concern of the physicists, the consequences are the concern of all"; "What concerns everyone, only everyone can resolve"; "Every attempt by an individual to resolve for himself what concerns everyone, must founder." Here Dürrenmatt suggests as a reason for Möbius's helplessness a certain arrogance underlying his assumption that he has the power to resolve the horrifying predicament of the modern world. For the physicists, the Fall has already happened; as Oppenheimer asserted, "The scientists have known sin." There is no return to the Eden of ignorance, and even the extreme sacrifice Möbius is prepared to undertake by way of reparation is not permitted to achieve absolution.

Dürrenmatt's physicists are pitted against a manifest criminal; hence they attain a comparative moral superiority even in their failure. Heiner Kipphardt, on the other hand, is concerned to examine the question of the powerlessness of scientists in a more problematic case: the seemingly highly motivated scientist in conflict with a government that also lays claim to representing the highest values. Kipphardt's play *In der Sache J. Robert Op-*

penheimer (1964) is based on the enquiry in May 1954 into the question of security clearance for Oppenheimer after it had been suspended by the U.S. Atomic Energy Commission (AEC).[38]

During the Second World War Oppenheimer had been one of the leading physicists at Los Alamos working on the development of the atomic bomb (indeed, he was called the "father of the atomic bomb"). Through the questioning of Oppenheimer by State Prosecutor Robb, Kipphardt defines the ironic moral predicament of the physicists: they were trying to develop the bomb because they believed that the German scientists were about to produce it for the Nazis, and they realized too late that the Germans were in fact not anywhere near as far advanced in their development of the bomb. By the time the U.S. atomic bomb was tested it was also too late, in the face of the demand to end the war in the Pacific as quickly as possible, to prevent its use against Japan.[39]

After the deployment of the atomic bomb against Japan, however, Oppenheimer opposed the immediate undertaking of research into a hydrogen bomb. This led to his arraignment for "lack of enthusiasm" for the hydrogen bomb project, as well as to accusations of disloyalty because of his "objectionable associations" with known communists and left-wing sympathizers. In the last analysis, the charge against Oppenheimer was not that he had undertaken any treasonable action (there was no evidence of any complicity with the atom spies) but that he had committed "ideological treason." Robb sums up his case:

> We could already have had the hydrogen bomb four or five years earlier if Dr. Oppenheimer had supported its development. What is the explanation for such a failure in a man so wonderfully gifted? . . . This is the explanation: Dr. Oppenheimer has never entirely abandoned the utopian ideals of an international classless society, he has kept faith with them consciously or unconsciously, and his subconscious loyalty could only in this way be reconciled to his loyalty to the United States. . . . This is a form of treason which is not known in our code of law, it is ideological treason which has its origins in the deepest strata of the personality and renders a man's actions dishonest, against his own will.[40]

Kipphardt's play is structured around Oppenheimer's inability to resolve the conflict between his fascination with the science as an end in itself and his sense of responsibility, which is itself divided between loyalty to a government and loyalty to mankind. Asked about this, Oppenheimer replies: "I would like to put it this way: if governments show themselves unequal to, or not sufficiently equal to, the new scientific discoveries—then the scientist is faced with these conflicting loyalties. . . . In every case I have given undivided loyalty to my government, without losing my uneasiness or losing my scruples, and without wanting to say that this was right" (71).

The feeling of the AEC, however, is that Oppenheimer, because of the influence he exerts through his public position, should not be permitted to voice the ethical struggles he may have as a private individual. The right of the government to dictate the attitudes of scientists is also subtly linked with the fact that it pays for their research. This is expressed baldly by Morgan: "When a state spends enormous sums of money on research, is it to be denied the right to do with that research as it thinks fit?" (73–74).[41]

Of the scientists presented in the play, only Oppenheimer, Evans, Bethe, and Rabi (all physicists except for the chemist Evans) have serious moral scruples about the development of the hydrogen bomb, and all support freedom of conscience.[42] Hans Bethe represents the conscientious scientist plagued by conflicting judgments concerning the correct ethical decision to make in an immoral situation. Like most of the team, he had worked on the project believing it necessary to develop the atomic bomb before the Nazis did so; but after Hiroshima he vigorously campaigned for international control and open information about the weapon, believing this to be the safest course. Totally against the hydrogen bomb project, he was one of the twelve scientists who opposed President Truman's decision to press on with it. With the outbreak of the Korean War, Bethe, believing again that the United States might be in danger, recapitulated and reluctantly returned to work on the project. Yet, as his testimony at the enquiry indicates, he was never comfortable with this decision. "I have helped to create the hydrogen bomb, and I don't know whether it wasn't perhaps quite the wrong thing to do. . . . after a war with hydrogen bombs, even if we were to win it, the world would no longer be the world which we wanted to preserve, and . . . we would lose all the things we were fighting for."[43]

Oppenheimer's final speech is a reaffirmation of his determination to devote himself to pure research in a belated, futile attempt to return to a prewar, prelapsarian state.

> When I think what might have become of the ideas of Copernicus, or the discoveries of Newton under present day conditions, I begin to wonder whether we were not perhaps traitors to the spirit of science when we handed over the results of our research to the military, without considering the consequences. Now we find ourselves living in a world in which people regard the discoveries of scientists with dread and horror, and go in mortal fear of new discoveries. . . . We, the physicists, find that we have never before been of such consequence and that we have never before been so completely helpless. I ask myself whether we, the physicists, have not sometimes given too great, too indiscriminate loyalty to our governments, against our better judgements—in my case, not only in the matter of the hydrogen bomb. . . . We have been doing the work of the Devil, and now we must return to our real tasks. We must devote ourselves entirely to

research again. We cannot do better than keep the world open in the few places which can still be kept open.[44]

By judicious placement and selection of extracts from the hearing, Kipphardt has effectively investigated the investigating committee and transformed the original "trial" of Oppenheimer into a trial of the U.S. government.[45] Evans, the most sympathetic character among the committee members, clearly implies his personal condemnation of Oppenheimer's opponents Robb and Teller and his exoneration of Oppenheimer and Bethe because of their commitment to humanity.[46] Kipphardt's treatment might seem, therefore, to reinstate Oppenheimer as an idealist, but in this play there are no unambiguous heroes, for although the condemnation of the government agencies, and in particular the FBI, is unequivocal, the real charge against the physicists (as opposed to that being investigated by the committee), namely, that they "have known sin," is never revoked. Oppenheimer's final withdrawal to pure research both challenges and is challenged by Brecht's Galileo and Dürrenmatt's Möbius: in an insane and fallen world, can there be a "right" action?

Dürrenmatt's later play Der Mitmacher (The collaborator) (1973) explores the manipulation by industry of a biologist, the significantly nameless Doc, who began his career as an idealist, afire to discover new knowledge: "I wanted to investigate life, to fathom its structure, to pursue its secrets."[47] After a distinguished beginning to his research into the combination of amino acids, the building blocks of proteins, Doc takes a job with a corpse-disposal company and invents a new process, corpse-dialysis, which completely disposes of corpses without pollution. Belatedly he realizes that he has produced not only a fortune for the disposal company but a green light for murderers, who cannot now be traced. Like Frankenstein, he has passed from research into the creation of life to dealing in dead bodies, in the process losing control over the use to which his discovery will be put by others. Despite his economic value to the company, Doc has no ethical influence in or on his society. Symbolically, his "laboratory" (also his "home") is an underground warehouse storage room, where water from the sewer system seeps perpetually—a leaky coffin already buried. At the end of the play Doc, now the head of a chemical empire, epitomizes the scientist as the archconformist, an unintentional collaborator with the criminal elements of society and helpless to "uninvent" his process in the interests of justice. Now that he has given society what it wants—possibilities of limitless crime and wealth—his views are disregarded.

The figure of the brilliant scientist rendered helpless by social forces (as opposed to a runaway invention or government decrees) has become more common in the last two decades. An interesting example is the character of Adam Sedgwick in C. P. Snow's novel In Their Wisdom (1974). Formerly a Nobel prize winner in molecular biology and president of the Royal Society, now a member of the House of Lords, Sedgwick suffers from Parkinson's disease, so that although he retains "one of the most lucid minds extant," his body is racked by uncontrollable spasms and twitches.[48] Sedgwick is therefore Snow's symbol of the scientist in the seventies, a superbly active mind housed in a body that is powerless and uncontrollable, a towering intellect that is ludicrously ineffectual.[49] Thus Snow suggests that importance in the hierarchy of science is of limited worth in the world outside academe. Sedgwick's attempted speech to the House of Lords, for instance, though crisp and coherent in its content and arising from the keenest intellect in the House, nevertheless comes over only as an incoherent sequence of "fumbles of his tongue" (145), an implicit comment on the breakdown of communication between the scientist and the rest of society and in particular between science and government.[50]

A less obviously helpless scientist, but one who is equally powerless in his world, is Professor Victor Jakob, the protagonist of Russell McCormmach's *Night Thoughts of a Classical Physicist* (1982). Jakob was devised as a composite of several actual scientists living in the Germany of September 1918, but he stands equally well as a representative for the would-be pure scientist today whose work is used in unforeseen and abhorrent ways. Although overtly treated by the university authorities with the respect due to his age and scientific reputation, Jakob is out of tune with his contemporaries, for his real allegiance is to a past age, the age of classical physics. Like Einstein, he finds quantum physics, with its emphasis on chance and statistical results, a travesty of the clarity and order that had attracted him to the nineteenth-century discipline. He also believes in pure, rather than "useful," research, an unpopular stance in 1918, when German physicists were expected to adopt a political role and encourage nationalism in their students.[51]

Jakob's long, nocturnal meditation ends on a note of sustained ambivalence about the role and effectiveness of scientists in a world at war. On the one hand, he affirms his belief that "the international spirit of science is evidence that science is stronger than hate, . . . that science is a bearer of understanding between peoples, . . . a reagent of love"; yet he asserts that "it is no betrayal of our scientific creed for us to yield to the dominant pull of patriotism . . . temporarily [to] sacrifice what is essential to our work as scientists" (148). This anachronistic representative of the austere beauty of classical physics (and the analogy with classical Greece, which Jakob also reveres, is explicit) is seen as irrelevant and helpless to influence his society.

More recently, Howard Brenton's provocative play *The Genius* (1983) recapitulates and explores nearly all the sociomoral positions considered here in relation to the cold war. Brenton's protagonist, the brash young American Leo Lehrer, Nobel prize winner in mathematics, has solved the equations necessary to establish Einstein's unified field theory and thereby incidentally has discovered the means of making the ultimate atomic weapon. The play examines his successive attempts to live with the knowledge that "no calculation is pure." Lehrer's first reaction is to feign ignorance in the face of the overtures of the U.S. military, whereupon MIT punishes him by allowing him to be "captured" by an English Midlands university, considered to be the ultimate exile at the end of the world! At this stage he still believes that noninvolvement is possible. "I said—OK, no calculation is pure. Therefore calculate no more. I gave up, . . . I closed down. I exiled me into my own head. If you are shit scared of the damage you can do, do nothing, eh?"[52]

Lehrer soon learns, however, that knowledge cannot be put back in the box. Gilly Brown, a gauche Midlands fresher but also a self-taught mathematical genius, has already solved most of the relevant equations. Seeing her calculations written in the fresh snow of a field, the incredulous Lehrer confronts her with the applications of her work. When Gilly denies any interest in anything but pure maths—"I don't want anything to do with politics; I'm here to study mathematics" (21)—the hitherto irresponsible Lehrer adopts the role of teacher implied by his name and in thirty pages of applied physics equations converts her to activism. Together they explore possible alternatives to helplessness: attempting to shock the "world," represented by the university microcosm, into a realization of the nuclear horror;[53] handing the critical calculations to MI6; trying to equalize the balance of power by giving the same equations to the Soviets.[54] All these courses of action fail to avert a nuclear war. Thus the play examines the basic dilemma of modern science: recognition of the terrible potential consequences of so much research and the simultaneous impossibility of not doing it. As Gilly says, "A thought is a thought. You can't not have it" (21). Brenton presents this view in strongly fatalistic terms, and it is this that gives the play its power and pessimism. Thus Leo asserts:

> The ideas do not love us. I have come to the conclusion that all the investigations into the atom, discoveries, calculations, formulations, nearer and nearer to the description of the force of nature—the scientific quest of the century—is fundamentally malign. . . . If I were the Pope, I think I would announce that we are forbidden to know the true nature of gravity, electromagnetism, and the nuclear forces. But—too late, holiness. We're never going to dig that knowledge out of our lives, out of our thoughts, out of our machines. (22, 37)

All the works discussed so far in this chapter have been to some degree tragic in conception, the helpless scientists for one reason or another losing control over the outcome of their discoveries. All the writers discussed here assume that this must inevitably be disastrous, since no one else, no other power, can be trusted to right the situation. However, there are also optimistic variations on the theme. Some writers suggest that although the scientists may be helpless, there are other survivors.

During the darkest years of the war the American humorist Thornton Wilder wrote his fantasy *The Skin of Our Teeth* (1942), exploring the question of survival in a series of time-telescoping situations in which society, both local and global, seems to have broken down. Science is represented in the person of the whimsical and complacent Mr. Antrobus, an inventor in the Edison tradition, who exemplifies what Wilder saw as the optimistic American trust in the future. Antrobus is busy inventing the wheel, the alphabet, and the answer to ten times ten while the rest of the world struggles to survive an ice age, a war, murders, and general social breakdown.[55] Clearly, Antrobus, closeted with his precursor of the wheel and alphabet-to-be, has nothing to offer in such dire straits; science cannot help us in times of global catastrophe. Wilder's alternative approximates to the instinct for survival by "the skin of our teeth," the innate resourcefulness and endurance of the ordinary individual multiplied over the whole population. This, Wilder believes, is how the human race has hitherto escaped extinction and will perhaps continue to do so. Such optimism, though unrelated to any faith in science, can perhaps still be seen as a characteristically American belief in the power of the ordinary man to create his own future, a faith significantly absent from European writing of the period.[56]

A different outcome again to the problem of the helpless scientist is presented in Arthur C. Clarke's two best-selling space novels, *2001: A Space Odyssey* (1968) and *2010: Odyssey Two* (1982). In the former novel, a highly sophisticated computer named Hal is effectively in charge of the spaceship *Discovery* as it travels towards Saturn on an investigative mission. When two astronauts disappear from the ship, which is left floating in space, it is assumed that Hal has rebelled against his commanders and killed them—an updated version of Capek's *R.U.R.* plot. However, in *2010: Odyssey Two* the team investigating the disappearance finds that Hal has been acting only in obedience to his programmed instructions, which include the obligation of self-preservation. Hal's creator, Chandra, thereupon reprograms Hal so that a command to save human beings overrides the self-preservation order, a procedure consistent with Asimov's Laws of Robotics. However, Clarke's optimism about science, being perhaps less facile than Asimov's, leads to a paradoxical situation in his treatment of scientists. On the one hand, Clarke is one of the most scientifically rigorous science fiction writers, exact in his use of physics and technological details and optimistic about his scientists' ability to achieve seemingly limitless goals through the power of the human mind. Yet there is also a strong metaphysical, even mystical, element in Clarke's best work, as there is in Wells's, and the inescapable suggestion that even the most intelligent scientists are mere children in the scale of the universe. At the end of *2001* mankind itself is symbolized as a fetus; and many of Clarke's stories are shot through with the implication that man is lonely and unfulfilled, crying out, whether consciously or not, for guidance from a lost father figure. Thus in Clarke's stories the scientist is ultimately only a representative human being, having no special insights or status when measured against the scale of the universe.

The helpless scientist, then, may be either reprehensible, insofar as he has failed to foresee the dangerous consequences of his work, or a qualified hero. In this latter case he tries to rescue society from the impending evil but fails because the forces against him are too strong. Given the politically unstable decades of the twentieth century, it is hardly surprising that the enemies most commonly depicted as arraigned against the scientist are social forces inimical to his own moral values. In novels and plays of the cold war period, the most frequent scenario presents the ethically aware scientist (usually a physicist involved in nuclear weapons research) in moral opposition to the militaristic elements of society, which demand the most powerful weapons available. Rarely, if ever, does this figure "win," but like the protagonist of classical tragedy, he assumes a grandeur in defeat.

In recent decades the science-based problems under consideration have multiplied to include genetic engineering, robotics, artificial intelligence, nuclear power, and diverse forms of environmental pollution. In the literary treatments of these issues, especially those written from a feminist perspective, the scientist characters are rarely assigned a leading role. The heroic figure who does battle with the business corporations, the military-industrial complex, or the government administration is more likely to be an otherwise ordinary individual—a mother, a teacher, a farmer, perhaps—with a passion for social justice. This change in focus reflects the general awakening from public apathy that followed the war years. In times of emergency, populations become childlike and accept the view that decisions should be made by "experts" or those in authority. Obedience is perceived to be the only means of survival, and only someone with the status of a scientific expert can challenge the government and military hierarchy. However, during the 1980s, Western nations experienced the effectiveness of "people-power," not only in large-scale operations such as the women's movement, the peace movement, and the environmental movement but in all areas of social activity. The heroic figures of this movement are ordinary citizens rather than scientists or other experts. Interestingly, these new champions of social and environmental justice are more likely to be depicted as successful in their crusade than their predecessors, the noble scientists, were in theirs.

One of the most interesting emphases in recent representations of the helpless scientist has been the implication that the society that promotes science for the sake of the technological advantages it brings must be held jointly responsible with the scientists for the social and environmental consequences. A striking example of this is Christa Wolf's novel *Störfall* (1987), mentioned earlier in this chapter in relation to runaway technology. Wolf's nonindividualized presentation of the scientific "experts" who appear on television to extol the merits and potential of nuclear power suggests that they too are merely faceless units of the establishment, as much victims of the disaster as the rest of the society, who want to receive only reassurance. A clear indication of Wolf's view that the scientists are not really in control is the fact that they are not specifically denounced for their part in the process. On the contrary, she asserts that the society that promotes nuclear power in order to create material products and a utopian life style without working for it must share the moral responsibility for the consequences. "Were we monsters when, for the sake of a utopia—justice, equality and humanity for all—that we didn't want to postpone, we fought those whose interests did not (do not) lie in this utopia and, with our own doubts, fought those who dared to doubt that the end justifies the means? that science, the new God, would deliver to us all the answers we sought from it."[57] However, the society that demands ever-increasing material wealth is presented with only two alternatives: nuclear power, which produces radioactive waste; or power derived from coal or oil by a process that produces acid rain and thus destroys the forests.[58] Ultimately the responsibility for these consequences lies not only, or even mainly, with the immediate perpetrators of these disasters, the nuclear scientists and the chemical engineers, but with the society that demands the products without acknowledging the cost.

The exploration of the responsibility of scientists for the large-scale pollution of the environment has led several contemporary German writers to distribute the blame more widely. In *Flugasche* (1981) Monika Maron wrote about the deluge of 180 tons of ash that falls daily on the city of Bitterfeld as a result of its intense industrialization, while Günter Grass's novel *Die Rättin* (1986) presents a macabre picture of the destruction of the earth by the inundation of garbage produced by society's rampant consumerism.[59] In these most recent treatments of environmental disasters, the blame has shifted from scientists qua individuals to scientists collectively and thence to the society that commissions their research, for society, by its demands for particular products and by the power it exerts through the selective distribution of funding, directs its scientists and technologists into the initiatives it favors.

Just as the image of the powerful scientist holding a society to ransom has been superseded by that of the helpless scientist, so there has been an evolution in the causes of helplessness and, correspondingly, the degree of responsibility ascribed to the scientist characters. In the first decades of the twentieth century scientists were most frequently portrayed as being at the mercy of their own discoveries or inventions. During the decades from 1940 to 1980 writers were more likely to represent scientists as being idealistically engaged in a struggle with the state authorities to ensure that their research was not used for purposes of destruction. The 1980s and early 1990s, however, have been characterized by more generalized depictions of metonymic scientists, who are merely paid and obedient civil servants, and the focus of blame has shifted, correspondingly, to implicate the whole society.

Despite the preponderance of unattractive stereotypes of scientists in this century, it would be misleading to imply that these have been unleavened by any more engaging figures. There have been some important studies of idealistic scientists, and it is significant that these characters are evaluated by their authors almost wholly in relation to their contribution to the community, rather than in terms of their scientific expertise. While never wholly triumphant in their struggles to implement a better society, they do achieve a qualified success and are sometimes held up as offering the only hope we have for a better future.

CHAPTER SIXTEEN

THE SCIENTIST
REHABILITATED

He was a physicist as Pascal had been; but like Pascal he was also
a mystic and a moralist. . . . I saw in him the priest of a new
theology.
—Heinrich Schirmbeck

The preceding four chapters have sufficiently indicated the low esteem in
which scientists have been held by most twentieth-century writers with a
background in the humanities. With the exception of the superficial characters of much science fiction, the dominant picture has been of scientists who
recapitulate the unflattering stereotypes of earlier centuries—the evil scientist, the stupid scientist, the inhuman scientist—or, as a peculiarly twentieth-
century contribution, the scientist who has lost control over his discovery.
This vote of no confidence by writers has not been entirely unanimous,
however. There have been some important studies by eminent writers of
fictional scientists engaged in a struggle to defend their ideals against the
debased values of society or some powerful faction within it. While such
figures are rarely, if ever, wholly successful in their quest (such optimism is
hardly even an option for contemporary writers), in many cases they are
used to explore possibilities of improved communication or to offer an
example to actual scientists, and in some cases they are used to propose to
the wider community an enlarged vision of possibilities beyond the narrow
materialist and economic exploitation of science.

In the 1890s and for the first three decades of the twentieth century, the
principles these idealistic fictional scientists espoused were almost universally associated with the pursuit of the discipline itself. They embodied a
tireless devotion to science, which was usually regarded as being ipso facto
for the benefit of society, and the renunciation of self-interest or personal
gain. Certainly the popular image of real scientists supported this view, for
no one embarked on such a career to obtain wealth or comfort. This idealism
was usually expressed in terms of the lifestyle associated with scientists and
the way it was evaluated. The long hours given over to research were rarely
perceived as infringing on social obligations, unless performed by a woman

(see fig. 25). Rather they were evidence of a noble and ennobling self-sacrifice on the part of scientists. The veneration accorded this lifestyle in the early twentieth century represented a complete reversal of the Romantic evaluation. When the Nobel prize–winning physicist Robert Millikan asserted that scientists offered an example of superhuman moral probity that religion and society would do well to emulate, he was taken seriously.[1]

During the 1930s this view of scientists was epitomized in the popular perception of Marie and Pierre Curie, who were regarded as exemplary in making the supreme sacrifice for science and, by extension, for society, refusing any material profit from their discoveries. Film versions of the Curies' story carefully preempted any possible criticism of Marie Curie as cold and unemotional by emphasizing her romantic relationship with Pierre and by linking her career in some unspecified way with a passionate Polish nationalism.[2]

The novelists and playwrights whose protagonists were idealistic fictional scientists came from a wide variety of backgrounds. Émile Zola had no direct scientific or medical training but was one of the leading figures of French naturalism, a literary movement that attempted to impose scientific procedures and criteria on literature.[3] In his novels *Doctor Pascal* (1893) and *Paris* (1898) the scientifically trained protagonists struggle against the mental oppression, superstition, and violence engendered by the Catholic church, which is presented as the villainous opponent of science, reason, and humanity. Doctor Pascal has spent a lifetime correlating information about insanity and heredity in generations of his own family and produced scientific data of inestimable value to the world; but his mother, aided by a superstitious, priest-ridden servant, is determined to destroy his work lest it disgrace the family. In Zola's view there is no doubt that the idealistic Pascal, representing medical science, stands for life and health, while the way of religion is the way of death. His novel retains a contemporary relevance, for what Pascal envisaged through his observations of heredity is now more immediately accessible through genetic engineering. "If one could only understand it [heredity], master it and make it do one's bidding, one could remake the world at will. . . . no more illness, no more suffering, to be able to limit the causes of death! . . . a new world of perfection and felicity would be hastened by active preventive measures which would make everyone healthy. When all were healthy, strong and intelligent, the human race would be a superior race, infinitely wise and happy."[4] Zola presents this view without perceptible irony, yet it is almost identical with the aims and assumptions of Frankenstein and with the Nazi ideal of a master race.

Zola's other scientist hero, Bertheroy, the celebrated chemist of *Paris,* also affirms his belief that "human happiness can only spring from the furnace of the scientist," that "one single step of science brings humanity nearer to the goal of truth and justice than do a hundred years of politics and

FIGURE 25.
Modern Woman, an
early twentieth-
century photo taken
to deride Marie
Curie.

social revolt."[5] Bertheroy's research has been in the field of explosives, but
his ultimate purpose is not destructive. On the contrary, his dream is to
empower society. This dream is realized when his anarchist colleague
Guillaume, who had planned to blow up the newly completed Sacré Coeur
Basilica at the moment of its consecration, diverts his explosive from purposes of destruction to powering a motor designed by his son.[6] Pointing to
the motor, Bertheroy reaffirms his faith in the constructive impact of science
as opposed to violence and lack of reason: "That is revolution, the true, the
only revolution. . . . It is not by destroying but by creating that you have just
done the work of a revolutionary" (485–86).

Zola's major target as the opponent of science and reason was the
Catholic church, but in twentieth-century literature politicians, the military,
and industry replace religion as the most common opponents of the idealistic scientists, for whom the conflict is frequently internal as well as external.
In his novel *Arrowsmith* (1925) the American writer Sinclair Lewis produced
a full-length portrait of a doctor involved in bacteriological research.[7] Although Lewis himself had no scientific training, he collaborated in the writing of this novel with a young bacteriologist, Paul de Kruif, and it is to his
experience that the novel owes its very accurate portrayal of the laboratory

scenes in research institutes and the professional problems and lifestyle of the medical researchers. The major characters are recognizably based on scientists whom de Kruif had known, but most are simplified in order to emphasize the two principal scientists, both idealists in different ways.[8]

Max Gottlieb is the charismatic German pathologist who inspires Arrowsmith to pursue medical research, and even though Arrowsmith temporarily forsakes this ambition, he retains much of Gottlieb's idealism, experiencing frustration when the clinic where he works proves more interested in exploiting its wealthy patients than in research. Through the figure of Gottlieb, Lewis depicts vividly the stress and loneliness of the researcher, his isolation from family and social contact as he works through the night until his health breaks down and he suffers temporary neurasthenia from the strain of overwork. Lewis also shows how scientific success brings liabilities as acute as failure. The institute, which needs to publicize its new medical discoveries in order to attract research finance, pressures Arrowsmith to publish his startling experimental results on pneumonia staphylococci before rechecking them, a recurrent situation when prestige and subsequent large-scale funding are at stake.[9] Remaining true to Gottlieb's principles, Arrowsmith rejects the demand to publish before he is certain, but while he is painstakingly repeating his experiments, a French scientist publishes a description of just such a bacteriophage, thereby rendering Arrowsmith's work effectively useless.

Although Arrowsmith teeters between devotion to medical research as an end in itself and humanitarian considerations,[10] no such compromises are made by his purist mentor Gottlieb, who enunciates his creed as follows: "To be a scientist—it is not just a different job. . . . It is . . . like mysticism, or wanting to write poetry; it makes its victims all very different from the good normal man. . . . The scientist is intensely religious—he is so religious that he will not accept quarter-truths because they are an insult to his faith. . . . He must be heartless. . . . To be a scientist is like being a Goethe: it is born in you" (267–68).

Gottlieb is a reflection of de Kruif's deep personal admiration for Jacques Loeb, one of the most vigorous twentieth-century proponents of biological mechanism. Loeb, the pure scientist, despised the clinical methodology of medical research in the 1920s, particularly in America, and Gottlieb certainly embodies this attitude. However, he also represents the spirit of German science in general, which at this time held a special allure for Americans. Gottlieb's absolute devotion to an abstract ideal of scientific research rather than its practical potential and his sacrifice of humanitarian interests in the cause of science might well have suggested at least an ambivalent estimate of such idealism as deleterious to human relationships, but one interesting aspect of Lewis's treatment is the transformation of such a character from the villain of Romantic literature to heroic status. In fact, the ebb and flow of Arrowsmith's devotion to Gottlieb is used by the author as a signifier of his moral worth.[11] His successive battles against the cheap popularization of science, the wealth-oriented research of the fashionable clinic, and the temptation to assume administrative power all dignify his cause, but Lewis and de Kruif go further and insist that his final action of leaving his wife and child for a life of monastic research is no less an aspect of his heroism.[12] Gottlieb tells him that "in this vale of tears there is nothing certain but the quantitative method" (36); humanitarian emotions are evidently a luxury with which the scientist has no business. This scientific creed represents a complete reversal of the Romantic evaluation of emotions and personal relationships as a superior guide to truth. If Arrowsmith is heroic at all in the Romantic sense, it is only in his quest for personal integrity, which means, within the ideological framework of the novel, unconditional loyalty to the requirements of pure science. While Gottlieb was the direct inspiration for Arrowsmith, it is likely that the continuing homage popularly accorded to Marie Curie influenced the wider acceptance of the scientist who sacrifices everything and everyone to the cause of research.

The contrary view, namely, that social and moral responsibility should take precedence over the drive for research, is argued by H. G. Wells in a number of novels whose utopian formulations we have considered. His novel *Marriage* (1912) is a fine study of an idealistic scientist who strives to apply the principles of science in the cause of social reform. The protagonist Trafford, a crystallographer, is impatient, as all Wells's ideal scientists are, with accepted formulations and theories; he insists on openness and publication rather than secrecy in research, rejecting the ethics of expediency and secrecy practiced by industry. But although he bargains vigorously with company directors for the right to publish, Trafford is also the first of Wells's "mystical" scientists. Not only does he pursue his research into crystals with a religious devotion and articulate it in the language of mysticism[13] but, like most of Wells's idealists, he rejects the social systems of his time for their failure to engage in "ultimate truth."[14] As he struggles to find a solution to the futility and waste he sees as the major social dilemma of his time, he concedes that his devotion to science for its own sake is a form of selfishness and vanity. He finally leaves his career in colloid chemistry and crystallography to devote his scientific expertise to community reforms, which are presented as the sociological equivalent of his research.

In Holsten, the atomic physicist of *The World Set Free* (1914) who discovers how to construct an atomic bomb, Wells created the prototype of a recurrent figure in twentieth-century fiction, the scientist who has access to the power that could annihilate civilization and agonizes over where his ethical priorities lie. His scientific principles urge him to publish his results openly, while his awareness of the moral inadequacy of society to cope with such knowledge counsels suppression of those results. Unlike most of his successors, Holsten decides to withhold his results, thereby retarding research in the interest of humanity and avoiding the Pandora's-box problem that characterized the scientists of the preceding chapter. *The World Set Free* proved to be prophetic, not just in predicting, some twenty years before it actually happened, the splitting of the atom and the joint possibilities of atomic energy and atomic weapons, but equally in the description of Holsten's inner struggle.[15]

The opposition between scientists intent on using nuclear technology for the good of mankind and politicians who wish to use it for military purposes was further explored in Nichols and Browne's popular interwar play *Wings Over Europe* (1929). Here the dramatic conflict pivots on the question of faith in humanity. The scientist/discoverer Francis Lightfoot remains an optimist because, unlike the politicians, he has not lost his belief in the emotions: "Humanity's not like that. I know it can't be; my heart tells me it can't."[16] This is certainly an unusual case for a scientist to argue; even more unusual is the justification Lightfoot uses, basing it on the comparative smallness of man as revealed by science, in the preceding generation a cause for pessimism and cynicism: "Isolated between the abyss of the unimaginably small, the atom, and the unimaginably great, the night about us, what refuge have we but one another? what future but the future of all? what ethic but the good . . . of Mankind?" (99). Although he is accidentally killed before he can force the British Cabinet to accede to his proposal for a new "Age of Humanity," Lightfoot is ultimately victorious over the forces of reactionary nationalism. The Guild of United Brain-Workers of the World, to whom he has given his secret, takes up his cause and demands a New World for Humanity (160). Despite some skepticism about the misuse of technology, the play ends with the optimistic assumption that free access to information as demanded by the scientists will necessarily prevent the misuse of that knowledge by politicians and military leaders.

Bacon's belief that the open exchange of knowledge would foster a spirit of internationalism and avert war, given new emphasis in the first three decades of the century by Wells's utopian writings, was subscribed to by many real-life scientists of the time. Indeed, the scientific world of the twenties was the closest to an international community of scientists that has yet been seen. Figures such as Ernest Rutherford, James Franck, Niels Bohr, and Peter Kapitza,[17] ignoring the contemporary sociopolitical obsession with the nation-state, remained loyal to an international ideal of science. This brief golden age of internationalism almost certainly contributed to the doubts expressed by many scientists, both real and fictional, about the development of nuclear weapons. It was, however, not strong enough to overcome the nationalist fervor generated by the propaganda machines of the Second World War.

During the 1930s the scientist hero, when depicted at all, was most frequently shown defending ideals that were in conflict with those of his society, and although he might retain a moral advantage, he was rarely seen as achieving a significant victory over the authorities. In the Germany of this period it became increasingly difficult to write anything overtly critical of the government; hence writers commonly resorted to historical parallels to avoid censorship of their contemporary message. Rehberg's play *Johannes Keppler* (1933), Brecht's *Das Experiment* (1939), "Der Mantel des Ketzers" (1939), and *Leben des Galilei* (1938–39), Brod's novel *Galilei in Gefangenschaft* (1948), and Zwillinger's play *Galileo Galilei* (1953) are among the best-known of these overtly historical treatments of a contemporary issue, the role and responsibility of the scientist in an authoritarian state where society is ignorant of the implications of his work. In the majority of these works the scientist protagonist is, as we have seen in the preceding four chapters, too flawed to be regarded as heroic, but in *Das Experiment*, "Der Mantel des Ketzers,"[18] and the first version of his *Galileo*, Brecht portrays his scientists as morally superior to the authorities of their time, whether religious or secular.

Of these examples, *Das Experiment*, based on an idealized reading of Sir Francis Bacon, is the most optimistic, for it looks forward to a new age when science will not have to contend with reactionary authorities, represented here by the church. Brecht makes use of the traditional story about an incident said to have caused Bacon's death from pneumonia, suggesting the added aura of martyrdom in the cause of science. The experiment of the title, which involves freezing a dead hen by filling it with snow to preserve it, symbolizes the practical application of science. Brecht suggests not only a moral but also a temporal victory for these Marxist ideals, for although Bacon dies before the experiment is concluded, his disciple, a simple stable boy, carries on his master's dedication to socially beneficial science.

Brecht's *Leben des Galilei* is the best-known of the works based on a historical scientist. Brecht was writing the first version in 1938–39 in Denmark, where he was exiled from Nazi Germany because of his communist sympathies. When news of the splitting of the uranium atom by Otto Hahn reached him, he was moved to dramatize, in the life of Galileo, the problems confronting German scientists, especially physicists, who were being required to work on the construction of the atomic bomb. This first version of the play is a strongly anti-fascist statement, essentially positive in its attitude towards scientists. Brecht himself said that Galileo is portrayed as "an intellectually heroic figure who fights for progress, who deliberately introduces a new age of scientific truth."[19] The main thrust of the drama is the heroic struggle of scientists against authority, symbolized in the play by the Catholic church. Brecht was concerned, however, to emphasize that the church was only one of many authorities, its twentieth-century counterpart being, obviously, totalitarian governments. He wrote: "The Church functions . . .

simply as Authority. . . . It would be highly dangerous, particularly nowadays, to treat a matter like Galileo's fight for freedom of research as a religious one; for thereby attention would be most unhappily deflected from present-day reactionary authorities of a totally unecclesiastical kind" (342). Thus Galileo becomes Brecht's symbol for the noble scientists trapped in Nazi Germany. Temporarily he succumbs to the threat of torture and publicly denies what he knows to be scientific truth (in the play, the Copernican theory; in Brecht's contemporary terms, knowledge of the consequences of an atomic bomb), but later he returns to his research in secret and smuggles his work across the frontier. Despite his other failings, the Galileo of this version is ultimately heroic in the cause of science. In order to continue his research he cheats, risks any danger, including the plague,[20] and recants solely in order to be able to finish the *Discorsi.* The play was not performed until 1943, in Zurich, and it was not produced in the United States until 1947. Ironically, very soon after this first U.S. performance, Brecht was summoned before the House Un-American Activities Committee to answer charges that this play, like several of his earlier plays, had a Marxist bias. Even more ironically, by that time Brecht had reworked the play, in response to the dropping of the atomic bomb on Hiroshima, to suggest the entirely different and much less heroic estimate of Galileo, as discussed in chapter 14 above.

In Britain it was possible for writers to be more explicit than their German counterparts in their criticism of government policy, and for this purpose several used the theme of an ethical scientist attempting to preserve his discoveries from misuse by an unprincipled government-military complex. One of the earliest in the field was the British dramatist Charles Morgan, who produced two plays exploring this issue of scientists fighting a moral battle against political and military interests. In *The Flashing Stream* (1942) Edward Ferrers, a mathematician working on the development of an aerial torpedo to destroy enemy aircraft, at first cares only about the mathematics of the project. An unusual and interesting aspect of this play is the fact that the only other mathematician capable of working on the problem is a woman, Karen Selby, whose view of science as being only one truth among many is contrasted favorably with Ferrers's conventional objectivity.[21] Here the scientist hero is considerably qualified if not overshadowed by the scientist heroine, whose humanitarian perceptions are clearly nearer to the authorial position. In a later play, *The Burning Glass* (1953), Morgan depicts another scientist, Christopher Terriford, who, having made a discovery of some strategic importance, is harassed by the government to pass over the sole rights to the military. Like Wells's Holsten, Terriford is torn between his duty as a scientist (to publish his results openly), his alleged patriotic duty (to divulge the secret only to his own country, England, and its ally America), and his duty as a moral individual (to save the world from the misuse of such power). "Science has never yet kept back its knowledge. . . . We have always said: 'Ah, but this power has beneficent uses as well. Let's go for it!' We have never yet said: 'We are unfit for it.' The time may have come to say that."[22] As the century progressed, this argument was heard with increasing frequency about other areas of scientific research.

Morgan's protagonists finally achieve their ethical goals, remaining firm in the face of military pressure, and thus both these works end with a certain qualified optimism; but this is an unusual outcome. After the use of the atomic bomb at Hiroshima and Nagasaki, a weapon that so clearly owed its existence to the most brilliant physicists of the day, it became progressively more difficult to believe in the moral superiority of scientists and even more difficult to believe in their ability to initiate a new, peaceful society. After 1945, those scientists who retain their high ethical code are no longer portrayed as figures of power in their society, but as individuals struggling nobly but with limited success against the morality of expedience urged by governments and the military-industrial complex.

The security restrictions of the Second World War and the subsequent cold war, coupled with the growing public perception of scientists as a country's most valuable resource in time of war, challenged as never before the Baconian traditions of open research and international cooperation in science. This situation inspired a spate of novels and plays about the crisis of conscience experienced by the idealistic scientist caught between two opposing ethical codes. In many of these works, the writers clearly use the scientist character as a critical weapon to attack government security policy, which, through increasingly stringent demands for censorship, was affecting their own profession as well. In the United States, the Manhattan Project, begun in 1939, was classified as top-secret, and publication of articles on atomic theory was prohibited. However, before the censorship was imposed, the announcement of the splitting of uranium[235] and the possibility of enormous energy output from the chain reaction so generated inspired a flood of science fiction about the new warfare. The main outlet for this was the U.S. magazine *Astounding Science Fiction,* whose editor, John W. Campbell, Jr., was fascinated by atomic weapons and encouraged stories dealing optimistically with the subject.[23] During the 1930s he himself, under the pseudonym Don A. Stuart, had written a number of stories in which the atomic weapons of the future were shown to be capable of demolishing civilization and even humanity but in the end were used by scientists to impose peace and prosperity.[24] If this seems a curiously naive position to adopt in the decade leading up to the Second World War, it may seem less surprising in the light of Anthony Burgess's statement that "Science Fiction is a child of positivism, which rejects catastrophe."[25] Because of Campbell's editorial policy, the noble and triumphant scientist hero survived in the pages of *Astounding Science Fiction* far longer than in mainstream literature, where the moral victories of scientists had little effect on the world situation.

Mark Bradley, the scientist-protagonist of James Hilton's *Nothing So Strange* (1947), is victimized by military intelligence and forced to rethink the value and role of science. He has believed in "scientific truth as something that transcends frontiers—something that can't be bought or sold or patented or hidden,"[26] but the FBI, suspicious of his former work in Berlin with a Nazi scientist, places him under surveillance, and he is forbidden to continue with his research at Oak Ridge. Hilton's moral, heavily underlined, is that scientists can no longer live in an ivory tower but have social responsibilities, not least of which is to save science from the politicians and administrators who are attempting to manipulate it. However, Hilton himself is unable to imagine how this could be achieved, and his protagonist, although morally cleansed, is politically ineffectual.

One of the first uncompromising treatments of this theme of the heroic scientist catapulted into conflict with the governmental edifice and branded a traitor in consequence was *The Big Secret* (1949), by the American novelist Merle Colby. Her protagonist, Daniel Upstead, is an idealistic, small-town physics teacher who comes to Washington to hear an eminent physicist speak at a meeting of the National Physical Association. He is incensed to learn that the scientist has been prevented from speaking because his research, although theoretical in nature, is considered to be of military importance and thus to come under national security regulations. Like most of the idealistic scientists considered in this chapter, Upstead is an innocent in the evil world of politics, where atomic weapons represent the new power game and each governmental sector hopes to manipulate the situation for its own profit. To the authorities, scientists, with their ethic of universalism, are suspect, and there is a move afoot in Washington to push through a new executive order that would require government clearance on all scientific discoveries before publication. Upstead duly becomes involved in a one-person campaign against a network of political intrigue, maneuvering, and corruption. On behalf of the League for the Advancement of Basic Science, he presents a resolution to the White House affirming that new discoveries in basic theory should be freely and universally available to scientists everywhere and that the principle of secrecy should be confined to technical methods alone. It is not surprising that at a time when Oppenheimer and Einstein were campaigning against the continued development of atomic weapons Colby should choose a physicist as her champion of truth and justice and show him accused of un-American activities for just such a stance as Oppenheimer was to make. But she effectively strengthens her case by identifying her physicist, not with a brilliant researcher, to whom few readers could readily relate, but with the more popular image of the morally upright, American small-town boy. There is never any doubt for the reader that Colby's hero is innocent of the charge of treason, yet his ethic of universalism in science is the same as that which motivated the atom spies during the cold war.

Other writers of the period raised this controversial issue more explicitly and even dared to present as heroic characters who refused to put patriotism before humanity. This is the case in James Aldridge's novel *The Diplomat* (1949), in which Ivre MacGregor, a British micropaleontologist working in Iran, unwillingly becomes involved in the political machinations of governments during the cold war. Caught between two extremes, represented on the one hand by the political expediency of his British superior, who tells him that science must serve political ends (nationalism, political face-saving, economic growth), and on the other by a fellow scientist who advocates complete isolation from public affairs, MacGregor's inclination is to opt out. However, he is recalled to a middle way of political responsibility compatible with his scientific principles by his former Iranian professor, Dr. Aqa (a rare example of a Middle Eastern scientist in literature), and he exposes the true facts about the Iranian situation despite the threat of being prosecuted for treason.

Aldridge's deliberate rejection of the traditional image of the national hero who puts his country before all else represents a politically subversive and peculiarly twentieth-century treatment of the conflict between individual morality and allegiance to the state. It clearly parallels the almost exactly contemporaneous passing of secret information by the atom spies Allan Nunn May and Klaus Fuchs, although Aldridge could not have known about this when he was writing the novel.[27] Science was probably the only possible justification that could be advanced for rejecting the model of patriotism, and the fact that it could be advanced at all is a measure of the degree to which scientific idealism had become accepted in the first decades of the century.

The theme of betrayal of scientific secrets, usually for idealistic reasons, was treated extensively and with many variations in the first two decades after the Second World War in both literature and film. Given the climate of opinion during the McCarthy period, it is the more significant that although in nearly every case the scientist protagonist is presented as idealistic, the authors' interpretation of what constitutes idealism differs considerably. Three examples from this period suggest the variations in political allegiance that the atom spy plot might be made to serve. In Herman Wouk's play *The Traitor* (1949) an American scientist decides to pass atomic secrets to the Russians—again, in order to secure world peace. He later changes his mind and helps to trap the very espionage ring he had been assisting, so that the title comes to have a double meaning. Although this might seem to damn the protagonist twice over, the authorial assumption is that the second betrayal negates the first, leaving the scientist a born-again patriot whose idealism had been temporarily misapplied but who has seen the error of his ways and been redeemed.

A more provocative situation is explored in Ruth Chatterton's complex novel *The Betrayers* (1953), in which a nuclear physicist is charged with having communist sympathies. Although he is successfully defended by a woman lawyer who is convinced of his idealism and innocence, the physicist has in fact been passing atomic secrets to the Soviets for some time, in the belief that peace can come only from an equal distribution of military power. Chatterton is more daring than most of her contemporaries in her choice of a hero who is unambiguously a traitor by the political values of the day, but ultimately she can propose no satisfactory moral solution to the antagonism of nationalism and humanitarianism other than the physicist's suicide. Written a decade later, *Meeting at a Far Meridian* (1961), by Mitchell Wilson, completely reverses the traditional pieties.[28] Although Wilson's nominal protagonist is Nick Rennert, an American cosmic-ray physicist, his moral hero is the Russian atomic physicist Dimitri Gontscharow. Like MacGregor in *The Diplomat*, Rennert is presented with two role models: the cynical, nationalistic Henshel, a character modeled on Edward Teller,[29] and the charismatic Gontscharow, confident, open, and life-affirming. Thus Mitchell's novel effectively vindicates the victims of the McCarthy purges.

Films of this period are more politically conservative in approach. The film industry was quick to capitalize on the paranoia generated by Klaus Fuchs's trial in 1950. *The Atomic City* (1952), *Walk a Crooked Mile* (1952), *Mr. Potts Goes to Moscow* (1953) (also released as *Top Secret*), *Escape Route* (1952), *Kiss Me Deadly* (1955), *The Atomic Man* (1956), and *The Amazing Transparent Man* (1960) all featured nuclear espionage and pandered to the inflamed level of fear elicited by the CIA and the FBI.

A more humorous but nonetheless serious treatment of the scientist's stand against government security policy is C. M. Kornbluth's short story "Gomez" (1955), in which a brilliant seventeen-year-old, self-educated mathematician, Julio Gomez, is compared to Wiener, Urey, Szilard, and Morrison, all scientists who suppressed knowledge that they knew could be used for atomic weapons. Gomez's brilliance in unified field theory is ripe for exploitation in the production of a new weapon that would confer absolute military power on the United States, and he is effectively imprisoned by the military to solve equations. For Gomez, this is at first merely an intellectual game for which he accepts no ethical responsibility, but his innate morality soon surfaces, and wiping his equations off the blackboard, he demands leave to get married. On his return, he claims to have completely forgotten his equations, and no efforts by the government and military agencies can force him to recall them. Kornbluth clearly implies that this is simply his way of defeating the powers that be, but the overtly humorous conclusion, in which Gomez returns to working in his back-street restaurant, suggests a more pessimistic implication, namely, that there is little hope for the physical scientist, even the theoretical mathematician, to preserve his moral integrity while pursuing a career in his discipline.

Although these characters carry authorial approval for their antigovernmental stance and in most cases retain their untarnished moral reputation throughout, they are nevertheless essentially powerless to change the political situation. In contrast to the increasing power ascribed to evil scientists before and during the war, the postwar period in general was marked by the rapid decline of the scientist as an effective, that is, an achieving, hero. Zuckmayer's play *Das Kalte Licht* (1955), which closely follows the case history of Klaus Fuchs, Schirmbeck's *Ärgert dich dein rechtes Auge* (1957), Dürrenmatt's *Die Physiker* (1962), and Kipphardt's semidocumentary play *In der Sache J. Robert Oppenheimer* (1964), discussed in chapters 13–15, chart the increasing helplessness of the scientist who tries to retain humanitarian ideals.

Of all these characters, Prince De Bary in Heinrich Schirmbeck's novel is the most outstanding scientist in moral terms. Yet even he has arrayed against him not only political and economic forces but the combined power of technology and scientific research institutions and, more insidiously, his own unwillingness to leave the artistic perfection of pure mathematics for the turmoil of social involvement. Schirmbeck described De Bary, one of literature's last great scientists, as "in some degree derived from the Duc de Broglie . . . [although he] bears no resemblance of detail or of circumstance to that great man." Internationally famous for his treatise *Light and Matter*, he is also described in charismatic terms by the younger physicist Thomas Grey. "He was a physicist as Pascal had been; but like Pascal he was also a mystic and a moralist. I loved him deeply as I sat there listening entranced. . . . The big, clear eyes, the elegant restraint of his gestures and the controlled fervour of his voice all combined to invest him with an authority which imposed itself upon his audience with an increasingly mystical effect. I saw in him the priest of a new theology."[30]

But De Bary's very perceptiveness makes him deeply critical of what he sees as the spiritual decay underlying the practice of science—"the lust for knowledge, the furious desire to know—to *know* in the Biblical sense— to master and possess. . . . [the scientist] is a hollow man because in place of knowledge he needs wisdom, contemplation as the Eastern philosophers understand it, but as it is not understood in the West" (409–10). Schirmbeck argues through De Bary that there can be no such thing as "pure" science. "Every accession of knowledge so profoundly affects the observed object that in some cases the measurement is rendered invalid. . . . Even the act of pure knowledge has a moral significance, and the recognition of this must in its turn influence the ethical personality of the scientist. It must serve to break down the idealised picture of science, the Pontius Pilate attitude, and set in its place a new conception of science having a definite place in the moral world" (412–13). Yet De Bary himself is guilty of the same sin, of withdrawal from the world through his obsession with the abstract perfection of pure mathematics. Despite his nobility, he remains essentially a figure of a past age, a critic rather than a reformer, fleeing from the arena of military-oriented science to his country estate to meditate upon the "original sin of physics," the separation of primary and secondary characteristics, objective and subjective. It is left to Schirmbeck's narrator Grey to carry this awareness into the future. In the last sentences of the postscript Grey concludes, "I have learnt one thing at least, that for me there is no escape. I may not pluck out the eye that offends me; for to be worthy of the world is to look it in the face."

This muted heroism and ambivalence about the role of the scientist can also be glimpsed in Douglas Stewart's poem "Rutherford" (1962). Stewart presents his distinguished fellow New Zealander in his study in Cambridge at the height of his scientific career. Well aware of the seductive pleasures of the scientific fraternity,

> he had grown to like
> This life of power where scientists met together
> And felt they were priests and rulers. He liked to talk
> With his great peers that language wrapped in mystery.[31]

Rutherford meditates on the nature of power and responsibility and the possible destructive applications of his research into nuclear fission:

> You could pay dearly
> For probing too deeply into that dark resistance
> Where light lay coiled in stone. He had seen clearly
> In flashes of the mind each atom exploding the next
> To the end of the world, and the light came out of the distance
> Like a wave upon him, towering.   (176)

Although he tries optimistically to concentrate on the benefits of his work to humanity in medicine and engineering, Stewart's Rutherford cannot quite shut out the awareness of the danger and is on the point of dismantling his apparatus and retiring to a farm in New Zealand. However, in an abrupt volte-face he decides that there is an inner power that, "like force in the atom . . . filled him with its radiance," directing him. Whether or not this is merely self-justification is never finally clear, but Rutherford's decision to continue the work is presented as an act of quiet nobility, the scientist's equivalent of the white man's burden, which the man of integrity must undertake for the greater glory of mankind.

More recent depictions of heroic scientists, and they are few, do not look back to the modes of idealization found in pre-1945 literature, but explore new paths whereby scientists might make a contribution to the well-being of society at a level beyond the merely technical. A new wave of optimism concerning the ability of science to save mankind from the ills of the past is to be found in much science fiction, notably that of Arthur Clarke, Isaac Asimov, and James Hogan, all of whom are fascinated by the theoretical possibilities of science and predict a fruitful symbiosis between people and technology. But in contrast to the unalloyed enthusiasm of these science fiction writers, mainstream novelists have been more cautious in their reinstatement of the scientist. Rarely is any attempt made to idealize scientists or even to portray them as successful in their aims; at best they point a way forward and make a plea for communication and tolerance. Unlike the macho scientists of science fiction tradition, they are unashamed of expressing their emotions and hence can be seen as contributing to the emergence of a new kind of hero. It is perhaps significant that a number of such studies have been produced by women writers.

In her complex novel *The Transit of Venus* (1980) Shirley Hazzard creates a young astronomer Ted Tice as her exemplar of humane emotions, independent thought, and social responsibility.[32] Tice, as idealist, is contrasted with an older, established astronomer, Sefton Thrale, who has falsified his "scientific" findings concerning the optimal siting for a telescope in order to accommodate what is politically expedient and who judges other scientists only by their professional success. Thrale's arrogant pragmatism and Tice's concern with intention rather than overt results are contrasted through their respective responses to the story of the eighteenth-century French astronomer Guillaume Legentil. Traveling to India to observe the transit of Venus in 1761, Legentil had arrived in good time at Pondicherry only to find the town fallen to the British. He therefore set out to Mauritius as an alternative observation site but arrived too late for the transit. He waited eight years in the East for the next transit in 1769 and on this occasion reached Pondicherry in good time. However, when the day of the transit dawned, visibility was so poor that he could make no observations. For Thrale this story is one of humiliating defeat, best forgotten. For Tice, honoring the faith not the failure, "his story has such nobility that you can scarcely call it unsuccessful" (16).

Whereas Thrale represents the scientific obsession with definition and clarity, an urge Hazzard identifies with the stereotyped male need to conquer and possess, Tice values complexity, seeing art, emotions, and imagination as integral parts of the truth. In this sense he is an idealized figure, the ethical hero of the novel; but the price he pays for rejecting the conventional standards is relative failure in both science and emotional relationships. Tice thus represents an inversion and reworking not only of the conventional image of the unemotional scientist but also of the hero stereotype, since his heroism consists essentially in a kind of defeat.

Perhaps the most interesting and innovative depictions of the scientist in recent literature are those that reject the Cartesian dualism that has characterized Western science since the seventeenth century and urge instead a return to a holistic understanding of the world. One of the first literary presentations of this philosophy was Aldous Huxley's last novel *Island* (1962), the counterproposal to his earlier, intensely pessimistic *Brave New World* (1932). In *Island* Huxley explored the possibility of uniting the insights of modern science and ancient yoga, the cognitive and spiritual perceptions, to foster a more creative life. His chief exponent is Dr. Robert MacPhail, surgeon and naturalist (these two pursuits signify his integrated approach), who promotes the holistic view of science as a relationship. Thus on Pala, the island of the title, the children learn science from an ecological rather than an analytical base. The school creed is that nothing exists in isolation: "All living is relationship."[33] Bacon's premise of a socially useful science is interpreted as a natural extension of Palan ecological principles rather than as an exploitation of nature. Unfortunately, despite the philosophical interest of *Island*, Huxley's characters never come to life in a convincing manner;[34] indeed, it is possible that the failure in credibility results from Huxley's own problems in imagining how Brave New World could be transformed into Pala. Significantly, the ending of *Island* is far from optimistic about the survival of the utopian community, as its powerful neighbors stand poised to take over and exploit Pala's oil deposits. In the global village of the twentieth century islands, however ideologically sound, cannot exist in isolation.

A more optimistic and innovative portrait of the idealistic scientist searching for wholeness is that of Shevek, protagonist of Ursula LeGuin's acclaimed futuristic novel *The Dispossessed* (1974). LeGuin draws parallels between the Taoist search for the unity of opposites as the path to understanding and the fundamental paradoxes inherent in quantum physics. Like Hazzard's character Ted Tice, Shevek, a brilliant physicist of a future world, Anarres, rejects easy simplification and false clarity. Instead, he applies to his study of time the unifying concepts that Einstein tried to establish in his search for a theory of general relativity. Shevek's General Theory of Temporality thus involves a synthesis of his earlier Principles of Simultaneity with the Principles of Sequency, concepts as difficult to hold together as the wave and particle theories of light. This search for a theoretical unity in temporal physics parallels Shevek's struggle to bring unity and tolerance to the polarized political systems of Urras and its moon, Anarres. Each society has its strengths but is so obsessed with its difference from the other that Shevek finds his idealist principles threatened by both. His defense of openness in scientific research and the free sharing of ideas with the inhabitants of other planets—his attempts to "unbuild walls" between individuals and societies—are regarded by both societies as treachery.

The Dispossessed thus becomes an exploration of the position of the scientist in a socialist society such as obtains on Anarres, where Shevek's theoretical physics is regarded as a self-indulgent substitute for real work. After a sojourn on the rich, capitalist Urras, where, by contrast, scientists are highly respected but a social conscience is taboo, Shevek returns to Anarres as an ambassador for an idea, knowing that he is regarded as a traitor but still determined to continue "unbuilding" walls. He represents, therefore, a new and much modified hero, in LeGuin's terms the only hope for the future. In his qualified success in immediate terms and his rejection of violence to achieve his purposes, he stands at the opposite pole from the scientist heroes popular at the beginning of the century, who never doubted the ethics of nationalism and aggression, whether against the alien creatures of another planet or against the equally abhorrent fellow humans of a different political stripe. Shevek's goal (symbolized by his unifying theory of temporality and by the ansible, the communication device he has constructed as a practical spin-off from his research) is to facilitate interplanetary communication and overcome the many different kinds of walls that divide people. "For seven generations there had been nothing in the world more important than that wall. Like all walls it was ambiguous, two-faced. What was inside it and what was outside it depended upon which side of it you were on."[35]

LeGuin has effectively replaced the traditional image of the physicist associated with disintegration (the atomic bomb or some destructive precursor) with a character involved in a project that is essentially creative in both the scientific and the social sense. Through the figure of Shevek she explores new interpretations of utopia, rejecting both the nineteenth-century ideal of material well-being and the self-negating ideal of uniformity, suggesting instead the possibility of creative freedom, where the individual is voluntarily in harmony with the society. Thus her idealization of the scientist and his newly defined heroic role represents both a radical rethinking in terms of ideology and also, arguably, the most realistic hope for the future of humanity.

From our discussion of the characters in this chapter it is apparent that the concept of idealism in relation to the figure of the scientist has been a rapidly changing one. Early in the century it signified the single-minded pursuit of the discipline as an end in itself. Nothing else was required of the ideal scientist but objectivity, rationality, and devotion to science to the exclusion of all else. Towards the end of the Second World War some writers put forward the radical proposal that integrity might be identified with noncompliance with the orders of the government-military establishment, that is, with treason as conventionally defined. This interpretation of idealism in science implicitly shifted the criterion from within the discipline itself to an external principle, in this case the rejection of militarism in some form or other. Later it acquired an increasingly sociological component. By the 1970s the idealism of the scientist (in literature as opposed to life) was assessed very little in relation to devotion to research and very considerably in relation to social conscience. More recently, the noble scientist was characterized by an ability to conceive of a viable future society in which the claims of the individual and the state were integrated, allowing scope for emotional and spiritual wholeness without sacrificing rationality and efficiency. That is, the ideal scientist of the seventies and early eighties was required to be a philosopher and an effective communicator with nonscientists. This latter role in particular was emphasized in literature long before the need was acknowledged by real-life scientists, most of whom considered the popularization of science an activity to be denigrated, if not actively opposed.

During the eighties and early nineties, as the perception of the world's problems changed from a political to an environmental focus, the ideal scientist in literature was required to be committed to restoring the ecological health of the planet, channeling technological expertise into the solving of environmental problems and combating the economic imperative of multinational corporations. This particular emphasis in literature predated, in many cases, the parallel integrated vision of much contemporary science, where the holistic perspective, rather than the reductionist premise, is only now coming to be regarded as important.

A number of the writers considered in this chapter have contributed to evolving a new role model for the scientist hero, replacing the crude, macho Martian-basher with a multidisciplinary and socially aware communicator who, like Shevek, "unbuilds walls." Whether that ideal is consonant with the pressures currently placed on scientists in the competitive world of peer review and limited research funding is arguable, and most writers suggest that it may be necessary for society to change those requirements before these exemplars from literature can eventuate in real life. The message from literature would seem to be that a society produces the scientists it deserves.

IMPLICATIONS

May you now guard Science's light
Kindle it and use it right
Lest it be a flame to fall
Downward to consume us all.
—Brecht, *Life of Galileo*

Literature does not merely hold up the mirror to nature and express more eloquently what is already well known and understood; it allows the exploration of what is perceived only dimly, if at all, the subversive anxieties that cannot be directly stated, because they challenge too vigorously the mores and taboos of society.

Many of the fictional scientists examined in this book have an importance far beyond their already considerable literary interest. Some of the writers referred to have provided Western culture with enduring myths to explore and express the deep-rooted but often irrational fears their society has held, usually inarticulately and perhaps even subconsciously, with respect to science and technology. The archetypal figures of Faustus, Frankenstein, Jekyll/Hyde, the Time Traveller, Doctor Moreau, and the Invisible Man have allowed the construction of cultural myths that each successive generation deconstructs for its own situation. It is not accidental that the Gothic novel *Frankenstein*, which has continued to acquire new relevance throughout the twentieth century, emanated from Mary Shelley's own subconscious fears concerning the future progress of science, the effect that the obsessive pursuit of power and knowledge might have on the emotional state of the scientist, and the eventual consequences for his family and community.[1] While the details have changed, the essential fears remain: deep-rooted fears of the new, of a loss of emotional roots and even of extinction of the entire human race; fears concerning loss of individuality and the stability engendered by accepted values; fears of the cargo cult of technology, bringing with it immense power and unanswered questions about its control.

Theodore Roszak spoke for the majority of the writers whose scientist characters are discussed in this book:

Dr. Faustus, Dr. Frankenstein, Dr. Jekyll, Dr. Cyclops, Dr. Caligari, Dr. Strangelove. The scientist who does not face up to the warning in this persistent folklore is himself the worst enemy of science. In these images of our popular culture resides a legitimate public fear of the scientist's stripped-down depersonalised conception of knowledge—a fear that our scientists, well-intentioned and decent men and women all, will go on being titans who create monsters. What is a monster? the child of knowledge without gnosis, of power without spiritual intelligence.[2]

Yet, as we begin to contemplate the twenty-first century as a reality rather than as a term from science fiction, there are signs of a new attitude on the part of nonscientists towards two very different aspects of science, computers and environmental science. Already there are clear indications that computers, used by many of the writers discussed in chapter 13 as a signifier for the impersonal, noncommunicative scientist characters who use them, are becoming a potent agent of change for "unbuilding walls" in the way suggested by LeGuin's Shevek. Already some four million scientists around the world, and some thirty million people overall, are linked to interconnected computer networks, called collectively Internet. By offering virtually instantaneous communication between users at minimal cost, Internet eliminates the former tyranny of time, distance, and economic inequalities in the field of communication. Productivity and interaction between scientists who may never otherwise meet have increased dramatically as the latest research results explode across the networks instantaneously. What is perhaps more important for the acceptance of science, these benefits are not confined to scientists. The proliferation of personal computers has revolutionized the science-society interface, converting nonscientists of all professions into computer users. The very term *personal computer*, which less than two decades ago would certainly have been considered a self-contradiction, suggests the domestication of what was widely regarded as the most dehumanizing of scientific tools. The challenge for writers now may be to explore how these user-friendly tools affect the growth of the imagination and relate to creative endeavor.

During the 1980s a new and powerful motivation for communication between scientists and others emerged: the need to collaborate on environmental issues. The image of the scientist as the perpetrator of environmental damage through development of dangerous substances and destructive technology has shifted to one of the scientist as an ally in finding the solution to the crisis of the earth's fragile ecosystem. In both these areas science has gone some way towards implementing the ideal enunciated with considerable literary power by the most famous scientist of our century, Albert Einstein:

A human being is a part of the whole, called by us the "universe," a part limited in time and space. He experiences himself, his thoughts and feelings, as something separated from the rest—a kind of optical delusion of his consciousness. This delusion is a kind of prison for us, restricting us to our personal desires and affection for the few persons nearest to us. Our task must be to free ourselves from this prison by widening our circle of compassion to embrace all living creatures and the whole world of nature in its beauty.

NOTES

Introduction

1. Margaret Mead and Rhoda Métraux, "Image of the Scientist among High-School Students," *Science* 126 (30 Aug. 1957), 384–90.

2. In a sample of 4,807 schoolchildren, aged five to eleven years, who were asked to draw a scientist, only 28 drew female scientists. These female scientists were all drawn by girls, who constituted 49 percent of the sample. See D. W. Chambers, "Stereotypic Images of the Scientist: The Draw-A-Scientist Test," *Science Education* 67, no. 2 (1983), 255–65.

3. Philip Hills and Michael Shallis, "Scientists and Their Images," *New Scientist*, no. 964 (28 Aug. 1975), 471–74. Their survey results showed an overwhelming bias towards the male figure in sections where respondents were asked to describe their "image of a scientist." Of the twenty most frequently named scientists, only one, Marie Curie, was a woman. Perhaps the most interesting aspect of this analysis is that the authors failed to comment on this sexist stereotyping.

4. See W. J. Megaw, "Gender Distribution in the World's Physics Departments," *Australian and New Zealand Physicist* 29, nos. 1 and 2 (1992), 15–18. Megaw's survey, which obtained responses from physics departments in 403 universities, indicates that only 22 percent of honors graduates are women and that only 13 percent of faculty members are women.

5. A consequence of quantum physics discovered by Werner Heisenberg in 1927, the uncertainty principle states that it is impossible to measure both the position and the momentum of a particle simultaneously with more than strictly limited precision. (The uncertainty in position multiplied by the uncertainty in momentum must always exceed Planck's constant.) This is a special case of Nils Bohr's complementarity principle, which states that experimenting on one aspect of a system of atomic dimensions destroys the possibility of learning about a "complementarity" aspect of the same system.

6. For a full discussion of the words *science* and *scientist* see Sydney Ross, "Scientist: The Story of a Word," *Annals of Science* 18 (1962), 65–85.

7. Francis Bacon, "The New Organon; or, True Directions Concerning the Interpretation of Nature" [1620], in *The Works of Francis Bacon*, ed. James Spedding, Robert

Leslie Ellis, and Douglas Denon Heath, 14 vols. (London: Longman & Co., 1857–74), 4:47–248.

8. Ross, "Scientist," 82.

9. [W. Whewell,] review of *On the Connexion of the Physical Sciences*, by Mrs. Somerville, *Quarterly Review* 51 (1834), 58–61. In this article, Whewell made the analogy with *sciolist, economist*, and *atheist*, thereby subverting the suggestion from the start, but seven years later he made the suggestion again, less frivolously and *in propria persona*, comparing the term with *artist* and *physicist* (*The Philosophy of the Inductive Sciences*, vol. 1 [London: Parker, 1847], cxiii).

10. Ross, "Scientist," 66.

11. See below, chapter 8.

Chapter 1. Evil Alchemists and Doctor Faustus

1. *Chem*, or *kmt*, was the ancient Egyptian name for the black land of the Nile delta and, by extension, for Egypt itself. *Chyma* was a Greek word for fusing or casting metals. It therefore seems likely that the Arabic name *alkimia* was a combination of both these derivations, as both concepts are implied.

2. See, e.g., R. Cummings, *The Alchemists: Fathers of Practical Chemistry* (New York: David McKay, 1966), 17–27.

3. Cummings, *Alchemists*, 42.

4. See, e.g., Lyndy Abraham, *Marvell and Alchemy* (Aldershot, Hampshire: Scolar, 1990), chap. 1.

5. These acts culminated in the formal edict *Spondent*, of Pope John XXII, denouncing all alchemists as tricksters and *falsarii* (counterfeiters). The edict is quoted in translation by E. H. Duncan in "The Literature of Alchemy and Chaucer's *Canon's Yeoman's Tale*: Framework, Theme, and Characters," *Speculum* 43 (1968), 636–37. Duncan sums up the stereotype of alchemy and alchemists as advertised by religious organizations in the fourteenth century as follows: "Alchemy in practice always fails and may be impossible. It both deludes its practitioners, whose motivating force is avarice, by wasting their goods and making them deceivers, and is also the cause of scandalous dangers to others who are drawn into its sphere of activity. Its secrecy and addiction to enigmatic language are dangerous because based on ignorance, or vicious because intended to deceive" (638).

6. See Abraham, *Marvell*, 4–6.

7. For a comprehensive discussion of this see John Cohen, *Human Robots in Myth and Science* (London: George Allen & Unwin, 1966), 43–48.

8. Goethe retained this aspect of the legend in his *Faust*, pt. 2, act 2.

9. The notion of the golem is first found in the story of Rabbah in *Sanhedrin* 65b of the *Talmud*. In Talmudic tradition Adam was designated as a golem, made from dust gathered from all parts of the earth. See Cohen, *Human Robots*, 41–43, 48.

10. See Betty Jo Teeter Dobbs, *The Foundations of Newton's Alchemy* (Cambridge: Cambridge University Press, 1975), 47.

11. There is a brief, allegedly autobiographical account of the adventures of Salomon Trismosin and his search for the philosophers' stone in a collection of alchemical tracts entitled *Aureum Vellus oder Guldin Schatz und Kunstkammer*, printed at Rohrschach in 1598.

12. Chaucer's unfinished *Treatise on the Astrolabe* (1391) indicates that he had a comprehensive knowledge of the science of his day.

13. For the involvement of religious orders in alchemy see Duncan, "Literature of Alchemy," 633–38; and Marie P. Hamilton, "The Clerical Status of Chaucer's Alchemist," *Speculum* 16 (1941), 103–8.

14. Geoffrey Chaucer, "The Canon's Yeoman's Prologue," in *The Canterbury Tales*, trans. Neville Coghill (Harmondsworth: Penguin, 1957). Quotations in the text are from the "Prologue" or the "Tale" and are identified accordingly.

15. S. Foster Damon has argued that not only was Chaucer defending the true alchemists in the last fifty-four lines of the tale but he must himself have been an initiate in alchemy and was so regarded by later alchemists ("Chaucer and Alchemy," *PMLA* 39 [1924], 782–88). This view is supported by Duncan, "Literature of Alchemy," 656.

16. See, e.g., Erich Kahler, "Doctor Faustus from Adam to Sartre," *Comparative Drama* 1, no. 1 (1967), 75–83; and J. W. Smeed, *Faust in Literature* (New York: Oxford University Press, 1975), 13.

17. Christopher Marlowe, *The Tragical History of Doctor Faustus*, 1.1.47, 51–60, in *Marlowe's Plays and Poems*, ed. M. R. Ridley (London: Dent, 1963), 122. Subsequent quotations from the play are from this edition.

18. Gy. E. Szőnyi points out that in this sense Marlowe goes against the neo-Platonic trend of Renaissance thinking, which treated "white" magic as a religious adjunct ("The Quest for Omniscience: The Intellectual Background of Marlowe's *Doctor Faustus*," *Papers in English and American Studies* 1 [1980], 147).

19. The device also occurs in H. G. Wells's treatment of Griffin, the Invisible Man, which owes much to Faustus. See below, chapter 10.

20. The passage in which Faustus is shown longing to be less than a man ("This soul should fly from me, and I be chang'd / Unto some brutish beast! all beasts are happy" [157]) is carefully constructed to recall the opening monologue, where he twice desired to transcend human limitations ("A sound magician is a mighty god: / Here, Faustus, try thy brains to gain a deity" [122]).

21. Jonson's command of alchemical terms and technicalities is considered one of the best. *The Alchemist* gives an accurate representation of the ideas and vocabulary of seventeenth-century alchemy.

22. Dee is alluded to in act 2, sc. 6. For a study of Dee's activities and influence see Peter French, *John Dee: The World of an Elizabethan Magus* (London: Routledge & Kegan Paul, 1972). Edward Kelley's association with Rudolf II of Bohemia is mentioned in act 4, sc. 1, 89–91. Simon Forman (1552–1611), self-styled physician, seer, necromancer, astrologer, and alchemist, was still alive when the play was written. He is also mentioned by name in Jonson's earlier play *Epicoene* (1609).

23. See John S. Mebane, "Renaissance Magic and the Return of the Golden Age: Utopianism and Religious Enthusiasm in *The Alchemist*," *Renaissance Drama* 10 (1979), 121; and Katharine Maynard, "Science in Early English Literature, 1550 to 1650," *Isis* 17 (1932), 113.

24. Ben Jonson, *The Alchemist* 2.1.81–90, in *The Complete Plays of Ben Jonson*, ed. G. A. Wilkes, 4 vols. (Oxford: Clarendon, 1981–82), 3:256. Moses's sister, Maria prophetissa, was believed to be a preeminent female alchemist. The "book of alchemy" referred to was Trismosin's *La Toyson d'Or*.

25. *The Alchemist,* act 3. See Mebane, "Renaissance Magic," 123.

26. Nebogipfel is the time-traveling protagonist of "The Chronic Argonauts," the story Wells later recast as "The Time Machine."

Chapter 2. Bacon's New Scientists

1. Francis Bacon, *De Interpretatione Naturae Proaemium,* in *The Works of Francis Bacon,* ed. James Spedding, Robert Leslie Ellis, and Douglas Denon Heath, 14 vols. (London: Longman & Co., 1857–74), 3:505–20.

2. Sir Thomas More had argued this in similar terms in *Utopia.* See, e.g., Robert P. Adams, "The Social Responsibilities of Science in *Utopia, New Atlantis,* and After," *Journal of the History of Ideas* 10 (1949), 374–98.

3. Francis Bacon, *Of the Interpretation of Nature, Works,* 3:222. Bacon was familiar with the work of Paracelsus, who believed that in the biblical millennium to come, Mankind would regain the domination over nature that Adam had enjoyed before the Fall. While Bacon rejected many of Paracelsus's more mystical ideas, he clearly adopted this view as the basis for his "Great Instauration" (see Peter M. Heimann, "The Scientific Revolution," in *The New Cambridge Modern History,* vol. 13 [Cambridge: Cambridge University Press, 1979], 252).

4. Francis Bacon, *The Advancement of Learning,* bk. 1, *Works,* 3:274.

5. Ibid., 294.

6. Francis Bacon, *New Atlantis, Works,* 3:145.

7. This college for research was based on ideas Bacon had evolved some years earlier when he still cherished hopes of being appointed master of some existing foundation in Oxford or Cambridge, but it was also possible that part of his inspiration came from reports of the Academy of Lincei, in Rome, founded in 1603 for research into physics and mathematics.

8. Francis Bacon, *De Augmentis Scientiarum,* bk. 7, chap. 1, *Works,* 5:8.

9. Francis Bacon, *Magna Instauratio,* preface, *Works,* 4:20–21 (my italics).

10. Francis Bacon, *Novum Organum,* Aphorism 63, *Works,* 4:64–65.

11. Ibid., Aphorism 13, *Works,* 4:49 (my italics).

12. Although Bacon's theory of experimental method is still closest to the popular conception of what scientists do, it has since fallen into disrepute amongst philosophers of science. In his Jayne Lectures for 1968 Sir Peter Medawar analyzed the relative importance of induction and intuition in scientific thought and showed the deficiencies of the Baconian method of observation, experimentation, and induction in both the theory and the practice of science (P. B. Medawar, *Induction and Intuition in Scientific Thought* [London: Methuen, 1969]; see also Max Charlesworth, *Science, Non-Science, and Pseudo-Science* [Geelong, Victoria: Deakin University Press, 1982]).

13. Bacon, *New Atlantis, Works,* 3:156–66.

14. Every twelve years, six of the Brethren of Salomon's House are sent on a fact-finding mission "whose errand was only to give us knowledge . . . especially of the sciences, arts, manufactures, and inventions of all the World; And withal to bring unto us, books, instruments, and patterns" (ibid., 164).

15. Thus Salomon's House includes the "Merchants of Light," who travel abroad to collect information; the "Depredators," who collect the experiments already published; the "Mystery-Men," who collect the experiments of all mechanical arts and

liberal sciences; the "Pioneers" or "Miners," who try out new experiments; the "Compilers," who collate the results from all these sources; the "Dowry Men" or "Benefactors," who seek good applications for these experiments; the "Lamps," who direct new experiments "of a higher light more penetrating into Nature than the former"; the "Inoculators," who execute these experiments and report on them; and finally the "Interpreters of Nature," who raise the former discoveries into greater observations, axioms, and aphorisms.

16. Bacon described his own philosophy as mediating "between the presumption of pronouncing on everything and the despair of comprehending anything."

17. Bacon, *New Atlantis, Works,* 3:156.

18. See, e.g., Brian Easlea, *Science and Sexual Oppression: Patriarchy's Confrontation with Women and Nature* (London: Weidenfeld & Nicolson, 1981).

19. "He held up his bare hand, as he went, as blessing the people," and two attendants walk before his litter bearing "the one a crosier, the other a pastoral staff like a sheep-hook" (Bacon, *New Atlantis, Works,* 3:155).

20. See, e.g., G. C. Moore Smith, Introduction to *New Atlantis.*

21. "We have certain hymns and services, which we say daily, of laud and thanks to God, for his marvellous works: and forms of prayers, imploring his aid and blessing, for the illumination of our labours, and the turning of them into good and holy uses" (Bacon, *New Atlantis, Works,* 3:166).

22. Ibid., 165.

23. Bacon, *Novum Organum,* Aphorism 129, *Works,* 4:115.

24. Bacon, *The Advancement of Learning,* bk. 3, chap. 2, *Works,* 4:341.

25. Ibid.,2:95, 97. Despite these pronouncements, Bacon's immediate successors persisted in regarding his long-term program as further reason for praise of the Creator. Thus Joseph Glanvill wrote of the members of the Royal Society, who were regarded as the heirs of New Atlantis: "They believe, there is an inexhaustible variety of treasure which Providence hath lodged in things, that to the world's end will afford fresh discoveries, and suffice to reward the ingenious industry and researches of those that look into the works of God, and go down to see his wonders in the deep" (*Plus ultra,* ed. J. I. Cope [Gainesville: Scholars' Facsimiles and Reprints, 1958], 7).

26. Rudolph Metz writes: "In the great process of setting learning free from shackles of faith and secularising it, which begins with the Renaissance and continues its advance with humanism, Bacon's teaching plays an important part. . . . Bacon wished to strengthen science by confining religion within its proper boundaries, and his efforts led to an essential strengthening of the secular at the expense of the spiritual. . . . the course which he adopted in religious matters led directly to the deism of the Enlightenment, and further to the scepticism of Hume. . . . More than any of his contemporaries, he succeeded in freeing philosophy from the clerical tutelage of the Middle Ages, and establishing the one-time *ancilla theologiae* in a new position as her own mistress" ("Bacon's Part in the Intellectual Movement of His Time," trans. Joan Drever, in *Seventeenth-Century Studies Presented to Sir Herbert Grierson,* ed. J. Dover Wilson [New York: Octagon, 1967], 32).

27. In Metz's opinion, "He vindicated it to the philosophic consciousness of his age. . . . He lent this new science the power of his word and the authority of his personality" (ibid., 25).

28. Bacon argued, "It is very probable that the *Motion of Gravity* worketh weakly,

both farre from the Earth and also within the Earth; The former, because of the Appetite of Union of Dense Bodies with the Earth, in respect of the distance is more dull; . . . For as for the moving to a *Point* or place (which was the Opinion of the Ancients) it is a meere Vanity" (Francis Bacon, *The History of Dense and Rare, Works,* 5:340).

29. "The ayre in that place I found quiet without any motion of wind, and exceeding temperate, neither hot nor cold. . . . As for that imagination of the Philosophers, attributing heat together with moystnesse unto the ayre, I never esteemed it otherwise than a fancy. . . . As for that Region of Fire our *Philosophers* talke of, I heard no news of it, mine eyes having sufficiently informed me there can be no such thing" (Francis Godwin, *The Man in the Moone: or a Discourse of the Voyage thither by Domingo Gonsales, the Speedy Messenger* [1638; reprint, Menston, Yorkshire: Scolar, 1971], 54, 66).

30. This, of course, is what Copernicus had predicted, but what is interesting here is the means whereby Gonsales/Godwin professes to refute the Ptolemaic system, namely, observation: "*Philosophers* and *Mathematicians* I would should now confesse the wilfulnesse of their owne blindnesse. They have made the world beleeve hitherto, that the Earth hath no motion. And to make that good they are fain to attribute unto all and every of the celestial bodies, two motions quite contrary to each other. . . . I say, allow the Earth his motion (which these eyes of mine can testifie to be his due) and these absurdities are quite taken away, every one having his single and proper Motion onely" (ibid., 58, 60). Amusingly, Gonsales realizes at this juncture that he might be accused of indulging in the very polemics for which Bacon had condemned the Aristotelians, and he hastily withdraws: "But where am I? At the first I promised an History, and I fall into disputes before I am aware."

31. Ibid., 101.

32. See Betty Jo Teeter Dobbs, *The Foundations of Newton's Alchemy* (Cambridge: Cambridge University Press, 1975), 62–63.

33. Abraham Cowley, *A Proposition for the Advancement of Learning* (London, 1661). In this work Cowley attempted to translate Bacon's House of Salomon into practical contemporary terms. He envisaged a college in which twenty philosophers or teachers and sixteen young scholars would be furnished with living quarters, servants, and equipment to carry out research and to "give the best education in the world gratis" to some two hundred pupils. Following Bacon's plan for his "Merchants of Light," four of the college's professors would be always traveling, collecting, and dispersing knowledge (see Dorothy Stimson, *Scientists and Amateurs* [New York: Greenwood, 1968], 21–22).

34. Thomas Sprat, *History of the Royal Society of London for the Improving of Natural Knowledge* [1667], ed. J. I. Cope and H. W. Jones (London: Routledge, 1959), 35.

35. "From these and all long Errors of the way, / In which our wandering Predecessors went, / And like the old Hebrews many years did stray / In Desarts but of final extent, / Bacon, like Moses, led us forth at last, / The barren Wilderness he passt" (Abraham Cowley, "To the Royal Society," stanza 5, in ibid.).

36. "While the Old could only bestow on us some barren Terms and Notions, the New shall impart to us the uses of all the *Creatures,* and shall inrich us with all the Benefits of *Fruitfulness* and *Plenty*" (Sprat, *History of the Royal Society,* 438).

37. See Teeter Dobbs, *Foundations of Newton's Alchemy,* 194–95.

38. Denis Diderot, *Oeuvres complètes de Diderot*, 20 vols. (Paris: Garnier, 1875–77), 13:133.

39. See Teeter Dobbs, *Foundations of Newton's Alchemy*, 59.

40. Bacon, *Novum Organum*, Aphorism 129, *Works*, 4:114.

41. See, e.g., Easlea, *Science and Sexual Oppression;* Genevieve Lloyd, *The Man of Reason: "Male" and "Female" in Western Philosophy* (Minneapolis: University of Minnesota Press, 1984); Evelyn Fox Keller, "Baconian Science: A Hermaphroditic Birth," *Philosophical Forum* 11, no. 3 (1980), 299–308; and idem, "Feminism and Science," *Signs* 7, no. 3 (1982), 589–602.

Chapter 3. Foolish Virtuosi

1. See R. H. Syfret, "Some Early Critics of the Royal Society," *Notes and Records of the Royal Society of London* 8 (1950), 45.

2. See, e.g., R. F. Jones, "The Background of the Attack on Science in the Age of Pope," in *Pope and His Contemporaries*, ed. J. L. Clifford and L. A. Landa (Oxford: Oxford University Press, 1949), 96–113; see also idem, *Ancients and Moderns: A Study of the Background of the Battle of the Books* (St. Louis: Washington University Press, 1961).

3. Galen had taught that the blood originates in the heart and ebbs and flows through the body like a tide.

4. William Harvey, *Movement of the Heart and Blood in Animals*, trans. K. J. Franklin (Oxford: Oxford University Press, 1957), 7.

5. For a more detailed account of the factors leading to the amalgamation of these groups and the nature of the early Royal Society see, e.g., Michael Hunter, *Science and Society in Restoration England* (Cambridge: Cambridge University Press, 1981), 32–58. Charles II was something of a virtuoso himself, having his own chemistry laboratory, for which he outlaid considerable sums. He seems, however, not to have contributed anything more substantial than moral support to the rather impecunious Royal Society. In this it is distinguished from the French Academy of Sciences, which, established as an agency of the royal government of Louis XIV, enjoyed a lavish income (see H. Brown, *Scientific Organizations in Seventeenth-Century France, 1620–1680* [Baltimore: Williams & Wilkins, 1934]; Joseph Ben-David, *The Scientist's Role in Society: A Comparative Study* [Englewood Cliffs, N.J.: Prentice-Hall, 1971], chaps. 5, 6; J. T. Merz, *A History of European Thought in the Nineteenth Century*, 4 vols. [New York: Dover, 1965], 1:41, 89–109; and Maurice Crosland, *Science under Control: The French Academy of Sciences, 1795–1914* [Cambridge: Cambridge University Press, 1992]).

6. Robert Hooke, Manuscript Papers, quoted in C. R. Weld, *History of the Royal Society*, 2 vols. (London: John W. Parker, 1848), 1:146.

7. Thomas Sprat, *History of the Royal Society of London for the Improving of Natural Knowledge* [1667], ed. J. I. Cope and H. W. Jones (London: Routledge, 1959), 99.

8. William Wotton, *Reflections Upon Ancient and Modern Learning* [1697] (London: Tim Goodwin, 1705), 30.

9. These factors included the emphasis on classical writers and Greek philosophy, the influence of Descartes, and, perhaps most immediately, the reaction against Puritan fanaticism.

10. Robert Merton suggests several reasons why the new scientific values and

concepts should have been congenial to the Puritans (see Robert K. Merton, "Science, Technology, and Society in Seventeenth-Century England," *Osiris* 4 [1938], 360–632). Michael Hunter, on the other hand, argues that Puritanism "was not specially favourable to natural philosophy. One of the few specifically Puritan societies in the seventeenth century—colonial New England—had no extraordinary commitment to science and this is equally true of the most characteristic, clerical wing of English Puritanism" (*Science and Society in Restoration England*, 114).

11. Jones, "Background of the Attack on Science in the Age of Pope," 104.

12. Sprat, *History of the Royal Society*, 417.

13. See Hunter, *Science and Society in Restoration England*, 174.

14. Both Robert Boyle and John Wallis, also a fellow of the Royal Society, wrote books attacking Hobbes's views. Boyle's *Examen of Mr T. Hobbes his Dialogus Physicus De Natura Aeris* (1662) refuted Hobbes's theory of the nature of air, and Wallis's *Hobbius Heauton-timorumenos* (1669) was intended to show that Hobbes's ideas were speculative, dogmatic, and hence unscientific in Bacon's sense.

15. The desire to dissociate themselves from Cartesian dualism led several Royal Society members to espouse the ideas of the Cambridge Platonists. Among the most prominent of these were Ralph Cudworth (*The True Intellectual System of the Universe* [1678]), Joseph Glanvill (*Saducismus Triumphatus: A full and plain Evidence Concerning Witches and Apparitions* [1681]), and John Ray (*The Wisdom of God Manifested in the Works of the Creation* [1691]).

16. Quoted in Hunter, *Science and Society in Restoration England*, 175.

17. Richard S. Westfall, *Science and Religion in Seventeenth-Century England* (New Haven: Yale University Press, 1958), 145.

18. It has even been suggested that it was the overlap of Charles II's alchemical-chemical interests with those of the Royal Society that led to his patronage of the society. The king had a private laboratory in which he conducted alchemical experiments, and there have been speculations that he died from mercury poisoning induced through a desire to "fix" mercury (see Betty Jo Teeter Dobbs, *The Foundations of Newton's Alchemy* [Cambridge: Cambridge University Press, 1975], 78).

19. The term *virtuoso* was often applied more widely to any scientist (see, e.g., Westfall, *Science and Religion in Seventeenth-Century England*, 13–14; and Walter E. Houghton, Jr., "The English Virtuoso in the Seventeenth Century," *Journal of the History of Ideas* 3 [1942], 52–58).

20. Even Henry Oldenburg, the first secretary of the Royal Society, believed that these private museums "will at length make up such a Store-house, as our Society designeth for an Universal History of Nature" (Oldenburg to Lister, Oct. 1671, quoted in Hunter, *Science and Society in Restoration England*, 67; for a comparison between the virtuoso "cabinets" and the "repository" of the Royal Society see Hunter, 66–67).

21. Michael Hunter quotes references to the orders placed by wealthy virtuosi (see ibid., 68–69). By the eighteenth century the market for scientific instruments had expanded as their usefulness in trade, industrialization, navigation, and surveying was realized. Many of the most famous makers of mathematical instruments were based in London, having the Board of Longitude, the Royal Observatory, and the Royal Society as patrons (see, e.g., J. A. Bennett, *The Divided Circle: A History of Instruments for Astronomy, Navigation, and Surveying* [Oxford: Phaidon. Christie's, 1987], 88–92).

22. Wotton, *Reflections Upon Ancient and Modern Learning*, 356. "Though the Royal Society has weathered the rude Attacks of such sort of Adversaries as Stubbe, who endeavoured to have it thought, That Studying of Natural Philosophy and Mathematicks was a ready Method to introduce Scepticism at least, if not Atheism into the World: Yet That no great things have ever, or are ever like to be performed by the *Men of Gresham*, and have so far taken the Edge of those who have opulent Fortunes, and a Love of Learning, that Physiological Studies begin to be contracted amongst Physicians and Mechanicks" (356–57).

23. See Bennett, *Divided Circle*, 65–68.

24. *The Blazing World* was originally appended as a fictional illustration to Cavendish's *Observations upon Experimental Philosophy*.

25. Margaret Cavendish, duchess of Newcastle, *The Description of a New World called the Blazing World* (London: A. Maxwell, 1666), 27–28.

26. The Danish ambassador visited the Royal Society and "was entertained with experiments on Mr. BOYLE's air-pump" (*Journal of the Royal Society*, 13 Feb. 1660–1661).

27. "Ballad of Gresham Colledge," stanzas 2, 6, 8, 9, 11, quoted in Dorothy Stimson, "Ballad of Gresham Colledge," *Isis* 18 (1932), 108–17.

28. Factors affecting the production of woolen cloth, a cheaper method of tanning leather, a diving bell, the deleterious effects of coal fires on health and environment, aids to navigation, etc.

29. The clue to the author's intention is in the lines "These be the things with many more / Which miraculous appere to men / The Colledge intended: The like before / Were never donne, nor wil be agen" (ibid., stanza 28).

30. See, S. Bruun, "Who's Who in Samuel Butler's 'The Elephant in the Moon,'" *English Studies* 50 (1969), 381–89.

31. Samuel Butler, "A Satire on the Royal Society," lines 505–20, in *The Poetical Works of Samuel Butler*, ed. J. Mitford (London: Bell & Dalby, 1866), 138.

32. Ibid., lines 85–104, pp. 159–60. Butler's continuing hostility to science can be seen also in "An Occasional Reflection upon Dr. Charleton's feeling a Dog's Pulse at Gresham College" and "Character of a Virtuoso."

33. Henry Fielding, *The History of Tom Jones*, 2 vols. (1749; reprint, London: Dent, 1957), vol. 2, bk. 12, chap. 5, 171.

34. See, e.g., A. S. Borgman, *Thomas Shadwell: His Life and Comedies* (1928; reprint, New York: B. Blom, 1928); Claude Lloyd, "Shadwell and the Virtuosi," *PMLA* 44 (1929), 472–94; Marjorie H. Nicolson, Introduction to *The Virtuoso*, by Thomas Shadwell, ed. M. H. Nicolson and D. S. Rodes (London: Edward Arnold, 1966; quotations from *The Virtuoso* are from this edition); Joseph M. Gilde, "Shadwell and the Royal Society: Satire in *The Virtuoso*," *Studies in English Literature, 1500–1900* 10 (1970), 469–90.

35. *The Virtuoso*, 2.1.304. "The College" is Gresham College, the precursor of the Royal Society, and the name continued to be applied to the latter institution.

36. The blood transfusion mentioned between a spaniel and a bulldog was described by Thomas Cox at a meeting of the Royal Society in May 1667 and published in the *Philosophical Transactions*, 2:451 (see Jessie Dobson, "Doctors in Literature," *Library Association Record* 71 [1969], 270).

37. Shadwell's mockery of the abstruse language cultivated by the virtuosi was another aspect taken up and amplified by William King, lord bishop of Derry, in his

satires *The Transactioneer* (1700) and *Useful Transactions in Philosophy* (1709). King, who professed to be "mov'd by the Respect I have for Natural Studies," specifically attacked Sir Hans Sloane, then secretary of the Royal Society, for his less than impeccable English, and in his dialogues between a Gentleman and a Virtuoso and between a Gentleman and a Transactioneer he satirized the inflated and obscure language of the *Philosophical Transactions of the Royal Society*. In his *History of the Royal Society* (1667) Thomas Sprat had affirmed its "constant Resolution, to reject all the amplifications, digressions, and swellings of style; to return back to the primitive purity, and shortness, when men deliver'd so many *things*, almost in an equal number of *words*" (113). Ironically, according to C. J. Horne, "the style of the *Philosophical Transactions* of the Royal Society in the eighteenth century was a poor return for the hopes Sprat had entertained when he wrote his manifesto" ("Literature and Science," in *The Pelican Guide to English Literature*, vol. 4, ed. Boris Ford [Harmondsworth: Penguin, 1973], 196). For an extended discussion of the effect of science on eighteenth-century style see ibid., 188–202. Walter Houghton, Jr., makes the point that "if the roots of metaphysical poetry lie in the disintegration of the scholastic mind, may this not also, in part, explain the passion of the virtuosi for mechanical subtleties?" ("English Virtuoso in the Seventeenth Century," 199).

38. Sir Arthur Oldlove in Thomas D'Urfey's *Madam Fickle* (1677), Periwinkle, "a kind of silly Virtuoso," in Susannah Centlivre's *A Bold Stroke for a Wife* (1718), and Doctor Boliardo in Aphra Behn's *The Emperor in the Moon* (1687).

39. Susannah Centlivre, *The Basset-Table* [1705], in *Plays*, ed. Richard Frushell (New York: Garland, 1982), 97–98.

40. James Miller, *The Humours of Oxford: a Comedy* [1726], 2d ed. (London: J. Watts, 1730).

41. Other famous scientists have been incorporated into fiction, notably the Curies, Thomas Edison, and Albert Einstein, but they were treated as fictional characters rather than addressed in their own person.

Chapter 4. Newton: A Scientist for God

1. John Maynard Keynes, first baron Keynes, "Newton, the Man," in *The Royal Society Newton Tercentenary Celebrations, 15–19 July 1946* (Cambridge: Cambridge University Press, 1947), 27.

2. Betty Jo Teeter Dobbs, *The Foundations of Alchemy, or "The Hunting of the Greene Lyon"* (Cambridge: Cambridge University Press, 1975); idem, *The Janus Faces of Genius: The Role of Alchemy in Newton's Thought* (Cambridge: Cambridge University Press, 1991).

3. See, e.g., Henry Guerlac, "Where the Statue Stood: Divergent Loyalties to Newton in the Eighteenth Century," in *Aspects of the Eighteenth Century*, ed. E. R. Wasserman (Baltimore: Johns Hopkins University Press, 1965), 325–28.

4. Jean Lerond D'Alembert, *Eléments de philosophie* [1759], in *Oeuvres philosophiques, historiques, et littéraires de d'Alembert*, ed. J. F. Bastien, 8 vols., 2d ed. (Paris: A. Belin, Readex Microprints, 1821), 1:121; Pierre Simon, Marquis de Laplace, *The System of the World* [1796], trans. J. Pond (London: R. Phillips, Readex Microprints, 1830), vol. 2, bk. 5, 324.

5. Isaac Newton, *Opticks: or, a Treatise of the Reflexions, Refractions, Inflexions and*

Colours of Light (London: Printed for Sam. Smith & Benjamin Walford, Printers to the Royal Society, 1704).

6. Richard G. Olson, ed., *Science as Metaphor: The Historical Role of Scientific Theories in Forming Western Culture* (Belmont, Calif.: Wadsworth, 1971), 71.

7. Voltaire to Cideville, in *Voltaire's Correspondence*, ed. Theodore Besterman, 10 vols. (Paris: Gallimard, 1977–88), vol. 4, no. 838, 48–49.

8. The idea that Newton personally forced Nature to obey his mathematical laws becomes explicit in the poems of James Thomson and William Tasker quoted below.

9. "For if Nature be simple and pretty conformable to herself, causes will operate in the same kind of way in all phenomena, so that the motions of smaller bodies depend upon certain smaller forces just as the motions of larger bodies are ruled by the greater force of gravity. It remains therefore that we inquire by means of fitting experiments whether there are forces of this kind in nature, then what are their properties, quantities and effects" (Isaac Newton, draft "Praefatio," in *Unpublished Scientific Papers of Isaac Newton: A Selection from the Portsmouth Collection in the University Library, Cambridge*, ed. and trans. A. Rupert Hall and Marie Boas Hall [Cambridge: Cambridge University Press, 1962], 304, 307).

10. Betty Jo Teeter Dobbs, *The Foundations of Newton's Alchemy* (Cambridge: Cambridge University Press, 1975), 230.

11. Newton, *Opticks*, bk. 1, 374 n. 9.

12. See Joseph Ben-David, *The Scientist's Role in Society: A Comparative Study* (Englewood Cliffs, N.J.: Prentice-Hall, 1971), 69–74.

13. See Richard S. Westfall, *Force in Newton's Physics: The Science of Dynamics in the Seventeenth Century* (London: Macdonald, 1971), 375–77. The same ideas, expressed as elective affinities, remained current in chemistry into the nineteenth century as an explanation of reactivity. Goethe's popular novel *Die Walverwandschaften*, or *Elective Affinities* (1809), is based on this concept.

14. Olson, *Science as Metaphor*, 72.

15. Samuel Johnson, 30 July 1763, quoted in James Boswell, *The Life of Samuel Johnson, L.L.D.* (New York: Oxford University Press, 1953), 276.

16. E. N. da C. Andrade, "Newton and the Science of his Age," *Proceedings of the Royal Society, Series A* 181, no. 986 (May 1943), 241–42.

17. Wright emphasizes the logic of the mechanism (an eidouranion, or transparent orrery), with its elliptical metal bands encircling space and catching the light from the hidden candle, which represents the sun. The figure of the natural philosopher (or scientist), standing at the center and dominating the group, fashionably dressed in a green gown with red sleeves, has an air of remoteness, amplified in the figure of the alchemistlike demonstrator in *An Experiment on a Bird in the Air-Pump* (see fig. 11 below).

18. James Thomson, "Poem Sacred to the Memory of Sir Isaac Newton," lines 82–84, in James Thomson, *The Castle of Indolence and Other Poems*, ed. A. D. McKillop (Lawrence: University of Kansas Press, 1961).

19. William Tasker, "An Ode to Curiosity," *Poems* (London, 1779), quoted in W. P. Jones, "Newton Further Demands the Muse," *Studies in English Literature* 3 (1963), 304.

20. *A Philosophic Ode on the Sun and the Universe* (London, 1750), quoted in Jones, "Newton Further Demands the Muse," 297. Even at the end of the century, the same

promise of a universal order and harmony inspired Henry Moore of Liskeard: "Now boldly soars among the stars to stray, / While Newton's mighty genius points the way: / Thro' Nature's dread immense he darts his eyes, / And new unnumber'd wonders round him rise; / What well-proportion'd pow'rs the planets roll, / How various parts compose one beauteous whole; / While in her centre thron'd, blest harmony / Tunes her immortal strings and charms them to agree" ("Private Life: A Moral Rhapsody," in *Poems, Lyrical and Miscellaneous* [London, 1803], 144–53).

21. Isaac Newton, *Mathematical Principles of Natural Philosophy and System of the World*, ed. R.T. Crawford (Berkeley: University of California Press, 1934), 398.

22. Thomson, "Poem Sacred to the Memory of Sir Isaac Newton," lines 64–67, 94–98, 36–38.

23. Newton himself had not aspired to such certainty as was accorded him; he frequently qualified his conclusions, admitting that the propositions arrived at by induction were "very nearly true" and that "although arguing from Experiments and Observations by Induction be no Demonstration of general conclusions; yet it is the best way of arguing which the Nature of Things admits of."

24. Micromégas, an eight-league-tall exile from Sirius, and his companion from Saturn encounter a group of philosophers and scientists in a ship in the Baltic. They question them at first about metaphysics and are overcome with amazement and mirth at the ignorance and dissension these tiny "insects" reveal about such matters. They then question them about scientific matters and are just as astounded to discover that the philosophers are suddenly in agreement and give extraordinarily accurate answers (see *Micromégas*, in *Works of Voltaire* [New York: St. Hubert Guild, 1901], vol. 2, pt. 1, 46–47).

25. Edmund Halley, "Ode to the Illustrious Man, Isaac Newton," translated from the Latin by Leon J. Richardson in Newton, *Mathematical Principles*, xv.

26. Basil Willey, *The Eighteenth Century Background* (London: Chatto & Windus, 1974), 4.

27. Olson, *Science as Metaphor*, 73.

28. Sir Thomas Browne, *Religio Medici* (1642; reprint, London: Dent, 1962), 17.

29. John Dryden, *Annus Mirabilis*, lines 657–60, in *Poems and Fables of John Dryden*, ed. James Kinsley (London: Oxford University Press, 1962), 81.

30. See, e.g., E. W. Strong, "Newton and God," *Journal of the History of Ideas* 13 (1952), 146–67. Cotes's own preface to the second edition of the *Principia* also stresses the implications for religious faith: "Therefore we may now more nearly behold the beauties of Nature, and entertain ourselves with the delightful contemplation; and, which is the best and most valuable fruit of philosophy, be thence incited the more profoundly to reverence and adore the great MAKER and LORD of all."

31. Isaac Newton, *Opticks, or, a Treatise of the Reflections, Refractions, Inflections and Colours of Light*, 3d ed. (1721; London and New York: Innys Readex Microprint), Query 28, bk. 3, pt. 1, 345.

32. John Ray, *The Wisdom of God Manifested in the Works of the Creation* [1691] (London: W. Innys, 1743).

33. Richard Blackmore, *Creation. A philosophical Poem in Seven Books*, bk. 7, reprinted in *A Book of Science Verse—The Poetic Relations of Science and Technology*, ed. W. Eastwood (London: Macmillan, 1961), 62–64.

34. See Michael Hunter, *Science and Society in Restoration England* (Cambridge:

Cambridge University Press, 1981), 184–87. William Whiston, Newton's successor at Cambridge, produced several editions of *A New Theory of the Earth,* which supplied poets with a simplified version of the solar system and the comforting reflection: "Which things, considering that the number of them is continually found to be greater, according as the telescopes we use are longer and more perfect, do vastly aggrandise the idea of the visible universe; and ought proportionably to raise our admiration of the Great Author of the whole to the highest degree imaginable" (4th ed., rev. [London: B. Tooke, 1696], 33). George Cheyne, Richard Bentley, Roger Cotes, William Wollaston, and William Derham also helped to publicize Newton's work during his lifetime, but the spate of commentary, both scientific and theological, increased markedly after his death in 1727. Henry Pemberton's *A View of Sir Isaac Newton's Philosophy* (1728) contained Richard Glover's eulogy, "A Poem on Newton."

35. See Isaac Newton, "Thirteen Letters from Sir Isaac Newton, Representative of the University of Cambridge to John Covel, D.D., Vice-Chancellor, Master of Christ's College," in *Isaac Newton's Papers and Letters on Natural Philosophy and Related Documents,* ed. I. Bernard Cohen (Cambridge: Cambridge University Press, 1958), 303; cf. *Four Letters from Sir Isaac Newton to Doctor Bentley, Containing Some Arguments in Proof of the Deity* (London: R. & J. Dodsley, 1756).

36. Allan Ramsay, "Ode to the Memory of Sir Isaac Newton: Inscribed to the Royal Society," in *Poems by Allan Ramsay and Robert Fergusson,* ed. Alexander Manson Kinghorn and Alexander Law (Edinburgh: Scottish Academic Press, 1974).

37. "See where he mounts the high, diurnal sphere, / And leaves a trail of light above the air; / The stars accost him, as he soars along, / And souls of wand'ring sages round him throng. / Mark where he halts on Saturn, tipt with snow, / And pleas'd surveys his theory below; / Sees the five moons alternate round him shine, / Rise by his laws and by his laws decline. / Amidst the vast infinity of space" (Samuel Bowden, "A Poem Sacred to the Memory of Sir Isaac Newton," *Poetical Essays on Several Occasions,* vol. 2 [London: J. Pemberton, 1735], 1–16).

38. "With faculties enlarg'd he's gone to prove / The laws and motions of yon worlds above; / And the vast circuits of th'expanse survey, / View solar systems in the Milky Way" (Jane Brereton, "Merlin: a Poem . . . by a Lady" [London, 1735], quoted in W. P. Jones, *The Rhetoric of Science* [London: Routledge & Kegan Paul, 1966], 104).

39. This reversal of the consequences of the Fall is analogous to that proposed by Bacon's restoration to mankind of power over nature, which had been lost by the Fall of Adam.

40. "Have ye not listen'd while he bound the SUNS, / And PLANETS to their Spheres! th'unequal Task / Of human-kind till then. Oft had they roll'd / O'er erring Man the Year, and oft disgrac'd / The Pride of Schools before their Course was known / Full in its Causes and Effects to him, / All-piercing Sage! Who . . . / . . . bidding his amazing Mind attend, / And with heroic Patience Years on Years / Deep-searching, saw at last the SYSTEM dawn, / And shine, of all his Race, on him alone" (Thomson, "Poem Sacred to the Memory of Sir Isaac Newton," lines 17–29).

41. Henry Grove, *Spectator,* no. 635 (20 Dec. 1714).

42. The abbé Jacques Delille wrote: "O pouvoir d'un grand homme, et d'une âme divine! / Ce que Dieu seul a fait, Newton seul l'imagine; / Et chaque astre répète en

proclamant leur nom: / Gloire à Dieu qui créa les mondes et Newton!" (*Oeuvres complètes*, vol. 9 [Paris, 1824], 7).

43. David Mallett, *The Excursion: a Poem in two Books*, bk. 2 (London: J. Walthoe, 1728), reprinted with changes in Alexander Chalmers, ed., *The Works of the English Poets*, 21 vols. (London, 1810), 14:16–24.

44. Thomson, "Poem Sacred to the Memory of Sir Isaac Newton," lines 190–206.

45. Abraham Cowley, "To the Royal Society," in Thomas Sprat, *History of the Royal Society of London for the Improving of Natural Knowledge* [1667], ed. J. I. Cope and H. W. Jones (London: Routledge, 1959).

46. James Thomson, *The Seasons*, "Summer," lines 1531–63, in *The Poetical Works of James Thomson*, 2 vols. (London: Bell & Daldy, 1860), 1:92. Ironically, William Blake was later to link Bacon, Locke, and Newton as the infernal trinity who had reduced a spiritual universe to a mechanical one (see chapter 6 below).

47. Halley's successful prediction that the Great Comet of 1682 would return in 1758 was important in vindicating Newton's theories of motion.

48. Richard Savage, *The Wanderer*, in *Works of the English Poets*, ed. Samuel Johnson, 68 vols. (London, 1779–81), ed. G. B. Hill, rev. ed. (Oxford: Oxford University Press, 1905), 45:12.

Chapter 5. Arrogant and Godless: Scientists in Eighteenth-Century Satire

1. The famous passage is in *Essai philosophique sur les probabilités:* "A mind that in a given instance knew all the forces by which nature is animated and the position of all the bodies of which it is composed, . . . could embrace in one single formula the movements of the largest bodies of the Universe and of the smallest atoms; nothing would be uncertain for him; the future and the past would be equally before his eyes" (see F. A. von Hayek, "The Rise of Scientism," in *Science as Metaphor: The Historical Role of Scientific Theories in Forming Western Culture*, ed. Richard G. Olson [Belmont, Calif.: Wadsworth, 1971], 105).

2. Alexander Pope, *Dunciad*, bk. 4, lines 469–76, in *Selected Poetry and Prose*, ed. W. K. Wimsatt (New York: Holt, Rinehart & Winston, 1962), 443.

3. Alexander Pope, *Epistle I*, ibid., 33–34.

4. Pope was himself a student of Newtonian astronomy and Cartesian cosmogony (see Marjorie Hope Nicolson and George S. Rousseau, *This Long Disease, My Life: Alexander Pope and the Sciences* [Princeton: Princeton University Press, 1968]).

5. Jonathan Swift, *The Mechanical Operation of the Spirit*, in Swift, *Collected Works*, 16 vols. (Oxford: Basil Blackwell, 1957), 1:174. "Squaring the circle," or constructing a square equal in area to a given circle, was a mathematical preoccupation of the period, but Swift's mention of it involves a specific reference to Newton, whose invention of the calculus provided the closest approximate answer to this longstanding riddle.

6. Jonathan Swift, *The Tale of a Tub* (1704; reprint, London: Dent, 1939), 108.

7. The Royal Society was by that time a highly respected body, represented by such distinguished scientists as Boyle, Hooke, Halley, and, preeminently, its president, Sir Isaac Newton. Not only was Newton effectively a national monument in his lifetime but, as Master of the Mint and a visitor at court, he also had considerable political power. The manifest success of England's productivity as a result of her

scientific inventions made it difficult any longer to satirize scientists as foolish, bumbling virtuosi.

8. Swift himself wrote of *Gulliver's Travels*, "It is a book in which man is the measure of all things" (*The Correspondence of Jonathan Swift*, ed. H. Williams [Oxford: Clarendon, 1963], 128).

9. Marjorie Nicolson and Nora Mohler have explored in detail the parallels between the Academy of Projectors in Lagado and the Royal Society in London. The professors of the academy "contrive new rules and methods of agriculture and building, and new instruments and tools for all trades and manufactures," however useless these prove to be (Jonathan Swift, *Gulliver's Travels* [London: Dent, 1956], bk. 3, chap. 4, 189). Nicolson draws the comparison with Sprat's *History of the Royal Society of London for the Improving of Natural Knowledge*: "They have propounded the composing a Catalogue of all Trades, Works, and Manufactures, wherein Men are employ'd . . . by taking notice of all the physical Receipts or Secrets, the Instruments, Tools, and Engines . . . and whatever else belongs to the Operations of all Trades" (3d ed. [London, 1722], quoted in Marjorie Nicolson, "The Scientific Background of Swift's Voyage to Laputa," in *Science and Imagination* [New York: Cornell University Press, 1956], 134–35). Like the Baconian optimists of the Royal Society, they promise a utopian future if their methods are followed, and their theories and experiments, ridiculous as they may seem, are closely modeled on contemporary preoccupations of the Royal Society, as recorded in the *Philosophical Transactions* of the Society (see ibid., 135–54). The overall framework of Gulliver's voyages and his descriptions of his travels reflects not only the general contemporary interest in travel but, more pointedly, the disproportionate space devoted to travel accounts in the Royal Society *Transactions* and the dangerous speculations contingent upon them.

10. See Nicolson, "Scientific Background of Swift's Voyage to Laputa," 134–35.

11. For the former see M. Nicolson, *Voyages to the Moon* (New York: Macmillan, 1960), 193; and R. Philmus, *Into the Unknown* (Berkeley: University of California Press, 1970), 10–12. For the latter see M. Nicolson and N. Mohler, "Swift's 'Flying Island' in the *Voyage to Laputa*," *Annals of Science* 2 (1937), 299–335.

12. The fact that Gulliver is "quite stunned with the noise" as the Laputans all play to the music of the spheres is of course an indication that they have not attained to spiritual perfection, or indeed to any felicity in musical practice, whatever their proficiency in theory. This is also an implied reference to the Tower of Babel.

13. Samuel Johnson, *The History of Rasselas, Prince of Abissinia*, in Johnson, *Rasselas, Poems and Selected Prose*, ed. B. H. Bronson (New York: Holt, Rinehart & Winston, 1958), chap. 40, 590.

Chapter 6. Inhuman Scientists: The Romantic Perception

1. Descartes ascribed his belief in the ultimate triumph of mathematics over all human and social problems to a dream he experienced on 10 November 1619. He was, however, almost certainly influenced as well by the English philosopher Thomas Hobbes, who regarded the process of reasoning as nothing more than a complex form of arithmetic.

2. The first calculating machine was created by the mathematician-philosopher

Pascal. Leibniz, influenced by his reading of Lull and Descartes, also dreamed of a reasoning machine that would be able to derive a complete mathematical system of the universe, perhaps as an answer to Newton.

3. "On voit qu'il n'y en a qu'une dans l'Univers, & que l'Homme est la plus parfaite. Il est au Singe, aux Animaux les plus spirituels, ce que la Pendule Planétaire de Huygens, est à une Montre de Julien le Roi" (Julien Offray de La Mettrie, *L'Homme machine* [1747], in *La Mettrie's L'Homme machine: A Study in the Origins of an Idea*, ed. Aram Vartanian [Princeton: Princeton University Press, 1960], 190). Julien Leroy (1686–1759), a French watchmaker and mechanist, improved the construction and precision of repeater watches and applied his methods to the improvement of pendulum clocks. From 1739 he held the title *horloger du roi*. Christian Huygens's "pendule planétaire" was a mechanical model of the solar system that faithfully duplicated the movements of the planets and their satellites (ibid., 246). La Mettrie was forced to leave Holland, where he had been living when his book was published, and after being refused permission to return to France, he found refuge at the court of Frederick the Great of Prussia. Condillac (1715–80) developed La Mettrie's ideas in his *Traité de sensations* (1754).

4. Joseph Ben-David, *The Scientist's Role in Society: A Comparative Study* (Englewood Cliffs, N.J.: Prentice-Hall, 1971). See also Donald Davie, *The Language of Science and the Language of Literature, 1700–1740* (London: Sheed & Ward, 1963); and Bernard Smith, *European Vision and the South Pacific* (Sydney: Harper & Row, 1988).

5. As noted in chapter 4, one of the early criticisms of Newton's *Principia* was its failure to account for the force of gravity except in terms that could be construed as differing little from the occult ideas of medieval alchemy.

6. Goethe's novel *Die Wahlverwandschaften* (Elective affinities) (1809) is based on the explicit application of the theory of chemical affinities to personal relationships.

7. Friedrich Schiller, Johann Gottlieb Fichte, Georg Wilhelm Hegel, the two Schlegel bothers, Clemens Brentano, Ludwig Tieck, Wilhelm von Humboldt, and Friedrich Schelling were all part of the Jena group, while Goethe, Christoph Martin Wieland, and Johann Gottfried Herder were in nearby Weimar. See also E.T.A. Hoffmann's descriptions of the automaton Olimpia in "Der Sandmann" below. Derek de Solla Price, a historian of technology, argues that the actual production of mechanical devices such as complex clockworks and automatons preceded, and almost certainly prompted, philosophical speculation about their implications for the nature of the animate world. "It seems as if mechanistic philosophy . . . led to mechanism rather than the other way about. We suggest that some strong innate urge toward mechanistic explanation led to the making of automata, and that from automata has evolved much of our technology, particularly the part embracing fine mechanism and scientific instrumentation. When the old interpretation has been thus reversed, the history of automata assumes an importance even greater than before" ("Automata and the Origins of Mechanism and Mechanistic Philosophy," in *Science since Babylon* [New Haven: Yale University Press, 1975], 50).

8. This was not unusual in the German education system of the time. Schiller had studied medicine; Heinreich von Kleist, mathematics; Achim von Arnim, mathematics, physics, and chemistry.

9. Friedrich Schlegel wrote a laudatory poem to Ritter (see Friedrich Schlegel,

Kritische Friedrich Schlegel Ausgabe, ed. E. Behler and H. Eichner, 35 vols. [Munich: F. Schoningh, 1958–91], 2:485).

10. Goethe observed this phenomenon, which was later to fascinate the Impressionist painters, during his ascent of Brocken in the winter of 1777 and described it in the diary of his journey to Italy. It was to form the basis of his most important scientific work, *Die Farbenlehre*, and subsequently of his historical supplement to this, *Materialen zur Geschichte der Farbenlehre* (1810) (see Karl J. Fink, *Goethe's History of Science* [Cambridge: Cambridge University Press, 1992]).

11. In the preceding generation Gotthold Ephraim Lessing had written a *Faust* of which only a fragment has survived; but in its time it had enormous influence on the Romantic writers. Lessing's Faust is essentially a pure and noble character (indeed that is why Satan regards him as a challenge), and not the least of his virtues is his single-minded devotion to wisdom. Moreover, although in Lessing's plot Faust fell prey to Satan because of his lust for knowledge, he was eventually to be saved because "die Gottheit hat dem Menschen nicht den edelsten der Triebe gegeben, um ihn ewig unglücklich zu machen" (the Deity has not given Man the most noble of instincts in order to make him eternally wretched) (Lessing, fragment of the otherwise lost tragedy *Faust*, *Gotthold Ephraim Lessings Sämtliche Schriften*, ed. Karl Lachmann, 23 vols. [Stuttgart: G. J. Göschensche Verlagshandlung, 1886–1924], 3:236). This is the first statement in a Faust play of the Enlightenment belief that the quest for knowledge cannot be intrinsically evil or lead to damnation, since it is the noblest attribute of man. The theme is developed further in Paul Weidmann's play *Johann Faust* (1775), an allegorical treatment of the cosmic struggle between good and evil powers for the soul of man. Here again Faust, the exemplar of intellectual curiosity, is saved. These depictions of Faust as essentially noble in his quest for knowledge paved the way for the more extreme treatments of the *Stürmer und Dränger*, who transformed him into a Promethean hero.

12. See J. W. Smeed, *Faust in Literature* (New York: Oxford University Press, 1975), 8.

13. "Daß ich erkenne was die Welt / Im Innersten zusammenhält" (pt. 1, lines 382–83).

14. Faust's attempt to create a homunculus, referred to in chapter 1 above, was a satirical attack by Goethe on the reductionist view that life could be resolved into specifiable components and reproduced in the laboratory.

15. "Zwei Seelen wohnen, ach! in meiner Brust" (pt. 1, line 1112).

16. Goethe retained some of the traditional elements in the person of Faust's famulus Wagner, the experimentalist preoccupied with his own small field in the realms of theory. But Wagner remains a comic, minor character, in the tradition of Swift's Projectors. Even his apparent success in finally creating a homunculus (significantly only a *little* man) is a delusion, insofar as it is actually Mephistopheles who has engineered its appearance. For a detailed account of the alchemical references in *Faust* see, e.g., R. D. Gray, *Goethe the Alchemist* (Cambridge: Cambridge University Press, 1952).

17. Smeed, *Faust in Literature*, 19. They thus come to embody the more general Romantic longing for transcendence, not merely in the intellectual realm but in all areas of experience. In the subsequently added *Prolog im Himmel*, so reminiscent of the Book of Job, Goethe not only stresses the universality of Faust as a dramatization

of an inner struggle within man between God and Satan but has God affirm from the outset that Faust, despite his errors and sins, will never finally fall victim to Mephistopheles but will eventually return to the right road.

18. Smeed characterizes the most important as being those by Chamisso (1804), Platen (1820), Lenau (1836), and Woldemar Nürnburger (1842). However, they contribute little to the portrait of the scientist and are therefore not discussed here.

19. This change is evident in Adalbert Stifter's *Der Nachsommer* (The Indian summer) (1857). The protagonist Heinrich Drendorf begins as a Goethe-like figure intending to study all the sciences and arrive at an integrated world-view, to become a "Wissenschafter im Allgemeinen," an all-round scientist. Like Goethe, he believes that a proper understanding of nature will establish an irrefutable basis for ethics and aesthetics. But Stifter, although an admirer of Goethe, is too honest to espouse an unquestioning Goethean idealism in the next century. In emphasizing the technical achievements of the age, he foresees, without fear or regret, the inevitability of a new fragmentation and complexity in science (Adalbert Stifter, *Der Nachsommer* [1857; reprint, Leipzig: Insel Verlag, 1910]).

20. The major exception was Percy Bysshe Shelley, whose training in science and particular fascination with chemistry have been well documented in Carl Grabo, *A Newton among Poets: Shelley's Use of Science in Prometheus Unbound* (New York: Cooper Square, 1968).

21. Byron's *Manfred* is frequently cited as being closely parallel to, if not actually based on, Goethe's *Faust* (but see Smeed, *Faust in Literature*, 226–27). Manfred, like Faust, is dissatisfied with the current offerings of learning and the disparity between knowledge and life, but he does not desire more knowledge, or even more experience, only oblivion, escape from his tormenting guilt, a characteristically Byronic affliction. However, the aspiration after limitless knowledge seems to have been almost an obligatory sin for Gothic heroes. The protagonist of William Beckford's fantasy *Vathek* is said to have "studied so much . . . as to acquire a great deal of knowledge; though not a sufficiency to satisfy himself: for he wished to know everything; even sciences that did not exist." After his suitably terrible death, we are told: "Such is . . . the chastisement of blind ambition, that would transgress those bounds which the Creator hath prescribed to human knowledge" (William Beckford, *Vathek* [1786], in *Shorter Novels of the Eighteenth Century*, ed. P. Henderson [London: Dent, 1963], 196, 278). But this theme is hardly present in the main part of the novel, being overshadowed by Vathek's obsession with luxury, adultery, cruelty, blasphemy, and most other sins. Similarly, the title character of Charles Maturin's Gothic novel *Melmoth the Wanderer* (1820) proclaims in his deathbed confession, "Mine was the great angelic sin—pride and intellectual glorying! It was the first mortal sin—a boundless aspiration after forbidden knowledge!" (chap. 32), but again this aspect is not developed in the novel.

22. For a more detailed discussion of these premises of the English Romantic poets see, e.g., Alfred North Whitehead, *Science and the Modern World* (Cambridge: Cambridge University Press, 1953), chap. 5; and M. H. Abrams, *The Mirror and the Lamp* (New York: W. W. Norton, 1958), 184.

23. It was this latter aspect that prevented English Romantic thought from becoming wholly subjective and embracing Berkeley's philosophical idealism. Instead, it

remained essentially naturalistic, grounded in a response to, and respect for, particular experience.

24. See, e.g., Donald Ault, *Visionary Physics: Blake's Response to Newton* (Chicago: University of Chicago Press, 1974).

25. Paley's "evidence" of God as creator of the world mechanism by analogy with a watch having required a watchmaker exemplified the deists' attitude. The watchmaker image, however, was to prove less reassuring to the next generation. Arthur Hugh Clough's poem "The New Sinai" casts science as the antitype of Moses, destroying religious faith by its doctrine that "the heart and mind of human kind / [is] A watch-work as the rest!" (*A Choice of Clough's Verse*, ed. Michael Thorpe [London: Faber & Faber, 1969], 34).

26. William Blake to Thomas Butts, 22 Nov. 1802, 2d letter, lines 83–88 (*The Poetical Works of William Blake*, ed. G. Keynes [London: Oxford University Press, 1966], 818–19).

27. William Blake, *Jerusalem*, pl. 54, lines 15–18, ibid., 685.

28. Ibid., pl. 15, lines 11–16, 636.

29. William Blake, "Mock on, mock on, Voltaire, Rousseau," ibid., 418.

30. Ault, *Visionary Physics*, 2–3.

31. S. Foster Damon, *A Blake Dictionary* (London: Thames & Hudson, 1979), 299.

32. Blake, *Jerusalem*, pl. 98, line 9, p. 745.

33. Apart from his specific attacks on Newton and theoretical physics, Blake, like Swift, feared that scientists would indulge in irresponsible and destructive experiments. In the unfinished prose satire *An Island in the Moon* (1784–85) several characters can be identified with different aspects of science. Suction the Epicurean represents the philosophy of the senses, and Sipsop the Pythagorean, mathematical science. Inflammable Gass, "the wind-finder," obviously represents Joseph Priestley, and his "dangerous experiments" refer to Priestley's search for new gases, in particular oxygen, which Priestley had isolated some thirteen years earlier. Inflammable Gass boasts, "I have got a bottle of air that would spread a Plague"; and when, subsequently, the "happy Islanders" meet at his house, he produces his apparatus as a toy for the diversion of his guests, an exercise that ends in disaster when the pestilence, Flogiston, escapes from the bottle. Inflammable Gass is merely a dealer in tricks and gadgets, unable to control his experiments; when it appears that disaster may eventuate, he is the first to flee the scene. Underlying the comedy is the serious assertion that science is destructive and that scientists, treating it as a game, are irresponsible about the consequences of their work.

34. William Wordsworth, *Prelude*, bk. 3, lines 60–63, in Wordsworth, *Poetical Works*, ed. T. Hutchinson (Oxford: Oxford University Press, 1978), 509.

35. William Wordsworth, Preface to the second edition of *Lyrical Ballads*, ibid., 738.

36. William Wordsworth, "The Tables Turned," ibid., 377.

37. Cf. Wordsworth, *Prelude*, bk. 13, line 198, where he speaks with approbation of the "feeling intellect," which enters into a sympathetic relationship with nature.

38. The air pump had been invented in 1654 by Otto von Guericke, of Magdeburg, and was soon recognized as a device for demonstrating the effect of a near-vacuum on living organisms. It was further developed in England by Boyle and Hooke (see above, chap. 3) and became an element of popular entertainment just after the mid-eighteenth century. In 1760 James Ferguson gave a series of lectures entitled *Lectures*

on Select Subjects in hydrostatics, optics, and mechanics, including one on the air pump. The instrument Ferguson describes in the published lecture is very similar to that in Wright's painting. Ferguson goes on to describe a series of experiments that can be performed with the instrument. Number 18 reads: "If a fowl, a cat, rat or bird, be put under the receiver, and the air be exhausted, the animal is at first oppressed as with a great weight, then grows convulsed, and at last expires in all the agonies of a most bitter and cruel death. But as this Experiment is too shocking to every spectator who has the least degree of humanity, we substitute a machine called the lungs-glass in place of the animal; which, by a bladder within it shows how the lungs of animals are contracted into a small compass when the air is taken out of them" (quoted in Benedict Nicolson, *Joseph Wright of Derby: Painter of Light* [London: Routledge & Kegan Paul, 1968], 114).

39. Percy Bysshe Shelley, *A Defence of Poetry*, ed. Roland A. Duerksen (New York, Appleton-Century-Crofts, 1970), 190.

40. Walt Whitman, "When I Heard the Learn'd Astronomer" [1865], *Leaves of Grass* (New York: Viking, 1958), 266. W. B. Yeats, who regarded himself as one of the "last Romantics," expresses a similar view in his "Song of the Happy Shepherd" [1889]: " . . . Seek, then, / No learning from the starry men, / Who follow with the optic glass / The whirling ways of stars that pass— / Seek, then, for this is also sooth, / No word of theirs—The cold star-bane / Has cloven and rent their hearts in twain, / And dead is all their human truth" (Yeats, *Collected Poems* [London: Macmillan, 1965], 8).

41. Thomas Carlyle, "Signs of the Times" [1829], in *English and Other Critical Essays* (London: Dent, 1915), 228.

42. "Boz" [Charles Dickens], "Full Report of the First Meeting of the Mudfog Association for the Advancement of Everything," *Bentley's Miscellany* 2 (1837), 397–413.

43. Pierre Moreau de Maupertuis, a friend of La Mettrie's, carried Locke's idea of a mathematical basis for all human reasoning into the realm of the emotions; in his *Essai de philosophie morale* he devised a calculus of pleasure and pain. The idea that pleasure and pain were quantifiable was to form the basis for the nineteenth-century Utilitarian philosophy of Jeremy Bentham and James and John Stuart Mill.

44. W.J.M. Rankine, "The Mathematician in Love," in Rankine, *Songs and Fables*, ed. W. J. Millar (London: Griffin, 1881). The English mathematician and engineer Charles Babbage, whose Difference Engine, built around 1829, is regarded as the prototype of the modern computer, provides a real-life example of insensitivity to nonmathematical considerations. Having read Tennyson's lines "Every minute dies a man / Every minute one is born," he wrote to the poet pointing out the mathematical inaccuracy of the statement and suggesting an alternative. "I need hardly tell you that this calculation would tend to keep the sum total of the world's population in a state of perpetual equipoise, whereas it is a well-known fact that the said sum total is constantly on the increase. I would therefore take the liberty of suggesting that in the next edition of your excellent poem the erroneous calculation to which I refer should be corrected as follows: 'Every moment dies a man, / And one and a sixteenth is born.' I may add that the exact figures are 1.167, but something must, of course, be conceded to the laws of metre" (Joel Shurkin, *Engines of the Mind: A History of the Computer* [New York: W. W. Norton, 1984]).

45. Honoré de Balzac, *The Quest of the Absolute* [1834], trans. Ellen Marriage (London: Newnes, n.d.). Balzac several times extends the frame of reference to the general case of genius. Thus Cläes is compared explicitly to "a struggling artist" (92), and Balzac remarks, "What is genius but a form of excess which consumes time and money and health and strength?" (21).

46.   "Unluckily, such affinities as these are too rare, and the indications are too slight to be submitted to analysis and observation."

"And this," she said, coming closer for a kiss, to put chemistry, which had returned so inopportunely at her question, to flight again, "is this to be an affinity?"

"No, a combination; two substances which have the same sign produce no chemical action."

Ibid., 85. This is almost certainly a reference to Goethe's *Die Wahlverwandschaften* (Elective affinities) (1809).

47. Charlotte Dacré, *Zofloya: the Moor* (London: Longman, Hurst, Rees & Orne, 1806), 260.

48. Ignaz Denner is a physics professor whose father, a *Wunderdoktor*, has entered into a pact with the Devil to obtain knowledge about nature as a means of acquiring wealth and doing evil. He is unambiguously a demonic figure whose deeds bring only suffering and destruction. "Der Sandmann" also links the quest of the father figure for knowledge with a mysterious collaborator. See E.T.A. Hoffmann, "Der Sandmann," in *Fantasie und Nachtstücke*, ed. Walter Müller-Seidel (Munich: Winkler-Verlag, 1960). An English translation by J. T. Bealby appears in *Isaac Asimov Presents: The Best Science Fiction of the Nineteenth Century*, ed. I. Asimov, M. Greenberg, and C. G. Waugh (New York: Beaufort Books, 1981).

49. Hoffmann is believed to have been inspired to write his stories of mechanical inventions after seeing an exhibition of automatons in Dresden in 1813 (see Patricia S. Warrick, *The Cybernetic Imagination in Science Fiction* [Cambridge, Mass.: MIT Press, 1980], 34). He may also have been inspired by Jean-Paul Richter's novel *The Death of an Angel*. Hoffmann's "Sandmann" in turn inspired Adolphe Adam's *La Poupée de Nuremberg* and the ballet *Coppelia, or The Girl with the Enamel Eyes* (1810).

50. Quoted in T. Martin, *Nathaniel Hawthorne* (Boston: Twayne, 1983), 99.

51. One of the few exceptions is the character Ethan Brand, in the short story of that name, who embarks on a quest for the "Unforgivable Sin." This sin turns out to be "the sin of an intellect that triumphed over the sense of brotherhood with man and reverence for God and sacrificed everything to its own mighty claims." Once a simple lime-burner, Brand had initially revered "the heart of man as a temple originally divine, and, however desecrated, still to be held sacred by a brother; . . . Then ensued that vast intellectual development, which, in its progress, disturbed the counterpoise between his mind and heart. . . . So much for the intellect! But where was the heart?" When, as the climax of his life, Brand throws himself into a lime kiln, his heart symbolically remains as pure white lime, an indication that it was made of marble ("Ethan Brand," in *Complete Short Stories of Nathaniel Hawthorne* [New York: Hanover House, 1959], 478).

52. Nathaniel Hawthorne, "Rappaccini's Daughter" (ibid., 271).

53. Ibid., 25. Less overtly malevolent but nonetheless manipulative of the personalities of others is the protagonist of "Dr. Heidegger's Experiment" (1837), who gives

his elderly guests a concoction allegedly from the Fountain of Youth, a typically alchemical potion, and then cynically observes their ludicrous attempts to emulate their younger selves.

54. His laboratory is complete with "distilling apparatus and the means of compounding drugs and chemicals which the practised alchemist knew well how to turn to purpose" (Nathaniel Hawthorne, *The Scarlet Letter* [1850], ed. Brian Harding [Oxford: Oxford University Press, 1990], 126).

55. Nathaniel Hawthorne, *Septimius Felton, or the Elixir of Life*, in *The Dolliver Romance, Fanshawe and Septimius Felton* (Boston: Houghton Mifflin, 1886), 296–97.

56. Edward Bulwer, Lord Lytton, *Zanoni* (London: Routledge, 1853), chap. 4.

57. Similarly, in *What will he do with It?* (1858) Bulwer Lytton satirized Charles Darwin in the character of Professor Long, who, like Darwin, had written "two huge volumes on limpets" (Darwin had spent eight years on a study of Cirripedia). Lytton believed that the inductive processes of scientific method were as fictitious as any novel. In *Kenelm Chillingly* (1873) he asserted that Darwin was a greater romance writer than Scott or Cervantes.

58. Quoted in Allan Conrad Christensen, *Edward Bulwer-Lytton: The Fiction of New Regions* (Athens: University of Georgia Press, 1976), 177.

59. The popularity of this novel and its topical issues is indicated by the fact that an enterprising British manufacturer devised a beef extract that he marketed under the name Bo-vril to suggest its powerful, energy-promoting properties. The product is still available.

Chapter 7. Frankenstein and the Monster

1. Ken Russell's film *Gothic* (1986) and the opera *Mer de Glace* (1991), libretto by David Malouf.

2. Mary Shelley, Introduction to the 1831 edition, *Frankenstein, or The Modern Prometheus*, ed. J. Kinsley and M. K. Joseph (Oxford: Oxford University Press, 1980), 9–10. Subsequent quotations from *Frankenstein* are cited in the text.

In her introduction to a recent edition of *Frankenstein*, however, Marilyn Butler has pointed out that the original (1818) edition of the novel carried no such moral implications (Introduction to *Frankenstein, or The Modern Prometheus* [London: Pickering & Chatto, 1993]). The scientific references were to the celebrated public debate of 1814–19 carried on between John Abernethy and William Lawrence, two professors at London's Royal College of Surgeons, on the origins and nature of life. Abernethy rejected materialist explanations and opted for an added force, "some subtile, mobile, invisible substance" analogous equally to the soul and to electricity. Lawrence, who was Percy Shelley's physician, put forward the materialist position as being the only intellectually respectable one. His views had considerable influence on both Mary and Percy Shelley, and his aggressive materialism was strongly represented in the first edition of *Frankenstein*. It seems certain that the discussion between Percy Shelley and Byron described by Mary in her 1831 introduction was concerned with the vitalist debate, and Butler further suggests that the Frankenstein of the first edition, the blundering scientist attempting to infuse life by means of an electric spark, is a contemptuous portrait of Abernethy, while the unhealthy relationships of the aristocratic Frankenstein family recall Lawrence's research on heredity and sexual selection. When Lawrence's *Lectures on Physiology, Zoology and the Natural History of Man* elicited a virulent review in the influential *Quarterly Review* of November 1819 and Lawrence himself was suspended from the Royal College of Surgeons until he agreed to withdraw his book, Mary Shelley feared the same fate would befall *Frankenstein*. She therefore revised it extensively, removing all controversial references, adding suitably remorseful statements by Frankenstein and, of course, the introduction, with its indication that we should read the novel as a frightful "human endeavour to mock the stupendous mechanism of the Creator of the world." This 1831 edition has been the one most commonly reproduced and consequently the one that has colored the various successive interpretations of the novel.

3. Shelley would certainly have known part 1 of Goethe's *Faust*, published in 1808.

4. Benjamin Franklin's most famous discovery—that lightning is caused by the discharge of static electricity in the atmosphere—was made as a result of the highly dangerous experiment whereby he tied a metal key to a silk conducting thread attached to a kite which he flew in a lightning storm. By touching the key, Franklin showed that electrical sparks would leap from the metal. Several scientists in Europe who tried to repeat his experiment were electrocuted; however, the experiment led to the use of the lightning rod.

5. See T. J. Hogg, *Life of Shelley*, quoted in Richard Holmes, *Shelley: The Pursuit* (London: Quartet Books, 1976), 44–45. Galvani, whose work is referred to explicitly, mistakenly believed that the effect he observed when animal muscle was brought into contact with two different metals was a kind of "animal electricity" inherent in the tissue and released by the metals. The voltaic pile, precursor of the modern battery, was produced by Volta in 1800. Ohm's experiments on the flow of current through a wire were conducted between 1825 and 1840, and Faraday's experiments on electromagnetic induction were not performed until 1831.

6. See also Francis Bacon, *New Atlantis*, in *The Works of Francis Bacon*, ed. James Spedding, Robert Leslie Ellis, and Douglas Denon Heath, 14 vols. (London: Longman & Co., 1857–74), vol. 3, and chapter 2 above.

7. The Monster responds to the beauties of nature, to the joys of domesticity, and the ideas of great books, all things that Frankenstein has put aside for his research. But it also kills Frankenstein's younger brother William, his fiancée Elizabeth, and his friend Clerval, people whom Frankenstein is duty-bound to love but whom he subconsciously wishes to be rid of because they attempt to distract him from his obsession.

8. Many critics have discussed the Doppelgänger relationship between Frankenstein and his creation. See, e.g., Harold Bloom, "Frankenstein, or the New Prometheus," *Partisan Review* 32, no. 4 (1965), 611–18; George Levine and U. C. Knoepflmacher, eds., *The Endurance of Frankenstein* (Berkeley and Los Angeles: University of California Press, 1979); and Masao Miyoshi, *The Divided Self: A Perspective on the Literature of the Victorians* (New York: New York University Press, 1969), 79–89.

9. Note also the play on the word *confinement*, since Frankenstein is abrogating the female as well as the male role in presuming to create a man by himself.

10. It is surely no mere coincidence that in 1816, the year in which Mary Shelley began *Frankenstein*, and in the same venue as that in which Frankenstein first confronts his Doppelgänger-Monster, Percy Shelley wrote "Mont Blanc," "composed under the immediate impression of the deep and powerful feelings excited by the

objects which it attempts to describe" (Percy Bysshe Shelley and Mary Shelley, *History of A Six-Weeks' Tour* [1817], in *Complete Works of Shelley*, ed. Roger Ingpen and Walter E. Peck, 10 vols. [London: Benn, 1965], 6:87–143). Chief among such impressions is the poet's fascination with the sense of immeasurable power embodied in Mont Blanc: "awful scene / Where Power in likeness of the Arve comes down"; "Mont Blanc yet gleams on high:—the power is there / The still and solemn power of many sights / And many sounds, and much of life and death" (Percy Bysshe Shelley, "Mont Blanc," *Poetical Works* [London: Macmillan, 1890], 493–95).

11. This is a clear echo of Goethe's Faust ("Daß ich erkenne was die Welt / Im Innersten zusammenhält") and, more generally, an expression of the Romantic quest to transcend the limitations of the natural world and discover an underlying unity.

12. This is another Romantic attribute of Frankenstein's. The English Romantic poets had generally affirmed the poetic experience as one of creation, and Coleridge had explicitly compared the creative faculty of the imagination to God's creation of the universe (see *Biographia Literaria* [1817], chap. 13).

13. For a detailed study of this political reference to France see Lee Sterrenburg, "Mary Shelley's Monster: Politics and Psyche in Frankenstein," in Levine and Knoepflmacher, *Endurance of Frankenstein*, 143–71.

14. William Godwin, *Enquiry concerning Political Justice and its Influence on Morals and Happiness* (1793).

15. "Remember that I am thy creature; I ought to be thy Adam; but I am rather the fallen angel whom thou drivest from joy for no misdeed. Everywhere I see bliss, from which I alone am irrevocably excluded. I was benevolent and good; misery made me a fiend" (100). Cf. Milton's Adam: "Did I request thee, Maker, from my clay / To mould me Man? did I solicit thee / From darkness to promote me?" (*Paradise Lost*).

16. In response to the Monster's entreaties, Frankenstein only replies, "There can be no community between you and me; we are enemies" (100).

17. It is significant that Frankenstein apparently ignores the warning of the Monster that he will be with him on his wedding night. The neglect of this warning is so incredible, given the previous murder, that we are surely intended to see it as a suppressed wish by Frankenstein.

18. See, e.g., Anthony Frewin, *One Hundred Years of Science Fiction Illustration, 1840–1940* (London: Bloomsbury Books, 1988). The menacing monsters range from robots to Martians, but the point is the same. The prevalence of the image suggests that the threatening Monsters represent what the scientist-heroes, rendered temporarily helpless and thus reduced to voyeurs, would like to do—attack the helpless female—but are prevented from doing by the restrictions of their decent, rational civilization.

19. George Levine, "The Ambiguous Heritage of *Frankenstein*," in Levine and Knoepflmacher, *Endurance of Frankenstein*, 15.

20. Carlos Clerens, *An Illustrated History of the Horror Film* (New York: Capricorn, 1967), 64.

21. *Son of Frankenstein* (1938), *The Ghost of Frankenstein* (1942), *Frankenstein Meets the Wolf Man* (1943), *House of Frankenstein* (1944), *Abbott and Costello Meet Frankenstein* (1948), *I was a Teenage Frankenstein* (1957), *The Curse of Frankenstein* (1957), *The Revenge of Frankenstein* (1958), *Frankenstein's Daughter* (1958), *Frankenstein 1970* (1958), *El Testamento del Frankenstein* (1964), *The Evil of Frankenstein* (1964), *Jesse James Meets Franken-*

stein's Daughter (1965), *Frankenstein Conquers the World* (1965), *Frankenstein Meets the Space Monster* (1965), *Frankenstein Created Woman* (1967), *Frankenstein Must be Destroyed* (1969), *Gothic* (1986), *Frankenstein, the Real Story* (1993). See Dennis Gifford, *Movie Monsters* (London: Studio Vista, 1974).

Chapter 8. Victorian Scientists: Doubt and Struggle

1. For a comprehensive discussion of these issues see, e.g., J. Hillis Miller, *The Disappearance of God* (Cambridge, Mass.: Harvard University Press, 1963); Walter E. Houghton, *The Victorian Frame of Mind, 1830–1870* (New Haven: Yale University Press, 1957); and David R. Oldroyd, *Darwinian Impacts: An Introduction to the Darwinian Revolution* (Kensington: New South Wales University Press, 1980).

2. See Leonard G. Wilson, "Science by Candlelight," in *The Mind and Art of Victorian England*, ed. J. L. Altholz (Minneapolis: University of Minnesota Press, 1976), 94–106. The founding of University College, London (1826), and King's College (1828) provided a number of paid positions for scientists, although without significant research facilities.

3. The mathematician Charles Babbage, the designer of the first digital computer, was a leader in the movement to reform science in Britain. He and like-minded young scientists criticized the government and the antiquated educational system, which still adhered rigidly to the doctrines of Newton, ignoring the more recent developments in mathematics in Europe. In 1830 he published the provocative *Reflections on the Decline of Science in England*, which attacked the complacency and restricted membership of the Royal Society. He was a founding member of the British Association.

4. For example, Arthur Schopenhauer (see J. W. Smeed, *Faust in Literature* [New York: Oxford University Press, 1975], 211–12).

5. The novels of Flaubert and Balzac in France, Thackeray, George Eliot, Thomas Hardy, and George Moore in England, and Henry Handel Richardson and Joseph Furphy in Australia can be seen as expressing the literary counterpart of biological and social determinism.

6. "So ist der Mensch die Summe von Eltern und Amme, von Ort und Zeit, von Luft und Wetter, von Schall und Licht, von Kost und Kleidung" (Jacob Moleschott, *Kreislauf*, letter 19, quoted in Smeed, *Faust in Literature*, 201 [Thus Man is the sum of heredity and nurture, of time and place, of air and weather, of sound and light, of food and clothing]).

7. Many novels dealt with the dilemma facing clergymen who felt they could no longer honestly subscribe to a fundamentalist understanding of the Creation as described in the book of Genesis or of the miracles. For a discussion of these see, e.g., Margaret M. Maison, *Search Your Soul, Eustace* (London: Sheed & Ward, 1961).

8. As Zola argued in his *Roman expérimental* (Paris, 1880), realism is essentially concerned with the study of man's relation to the material world, that is, with his physical, social, and psychological interactions rather than with an idealist, transcendental, or allegorical picture.

9. Dr. George Henry Kingsley studied medicine at Edinburgh and Paris and was wounded on the Paris barricades in the 1848 revolution before setting out to explore the South Seas and the Rocky Mountains as doctor-companion to various young

noblemen (see Susan Chitty, *The Beast and the Monk* [London: Hodder & Stoughton, 1974]).

10. Charles Kingsley, *Two Years Ago* (1857; London: Macmillan, 1889), 74, 150.

11. Kingsley encouraged the Ladies' Sanitary Association and wrote and lectured continually on what he called the science of health (see, e.g., Charles Kingsley, *Sanitary and Social Lectures and Essays*, in *Collected Works*, vol. 18 [London: Macmillan, 1880]).

12. Tertius Lydgate, in George Eliot's *Middlemarch* (1871), is usually cited as the first such character but in fact is a later creation (*Two Years Ago* having been written in 1857), although the later novel is set twenty years earlier. There are, as Harold Gillespie points out, many parallels between the two characters and their situations (see his "George Eliot's Tertius Lydgate and Charles Kingsley's Tom Thurnall," *Notes and Queries* 2 [1964], 226–27). It is also interesting that although Thurnall has lived for some years in America, he (and therefore presumably Kingsley) is apparently ignorant of the new treatment for *delirium tremens* proposed in the United States by Dr. John Ware in 1831, which Lydgate knows and follows. Thurnall prescribes morphine and laudanum for Squire Trebooze's *delirium tremens*, a treatment Ware criticized.

13. The title character of Trollope's novel *Doctor Thorne* (1858) and the heroine's doctor-father in Mrs. Oliphant's *Miss Marjoribanks* (1865) are also said to be endowed with a scientific spirit rare in medical practice at the time, and they too are presented as driven by noble and altruistic motives; but they are not shown at work on research, their interest for their respective authors lying in their approach to their patients, so they will not be discussed here. An interesting later vignette of a woman doctor is that of Dr. Mary Prance in Henry James's novel *The Bostonians* (1866). Although Dr. Prance is only a minor character, she is morally above reproach and appears to have been based on Dr. Mary Walker, an eminent if eccentric doctor distinguished for her work during the American Civil War (see R. E. Long, "A Source for Dr Mary Prance," *Nineteenth-Century Fiction* 19 [1964], 87–88).

14. Kingsley to F. D. Maurice, 29 Mar. 1863, in *Charles Kingsley: His Letters and Memories of his Life*, edited by his wife, vol. 2 (London: Henry S. King & Co., 1877), 181.

15. In *The Water Babies* Kingsley refers respectfully to Owen, Huxley, Murchison, Darwin, Sedgwick, and Faraday, although he pokes gentle fun at a Professor Ptthmllspths, who is so busy classifying species that he cannot see them for what they are. The professor classifies Tom the chimney sweep as a cephalopod hunting for the Hippopotamus Major, a comic reference to the contemporary *hippocampus minor* controversy.

16. The wise Dean Winnstay, of *Alton Locke* (1850), and at least one of Kingsley's heroic soldiers, Major Campbell of *Two Years Ago*, are such morally elevated amateur naturalists.

17. Charles Kingsley, *Glaucus; or, the Wonders of the Shore* (London: Dent, 1949), 232–35.

18. Elizabeth Gaskell, *Wives and Daughters* (Harmondsworth: Penguin, 1969), 75. His mother, with her literary leanings, had regretted that Roger was "not a reader," ignoring the fact that he read scientific books with great interest and perseverance.

19. We know from one of Elizabeth Gaskell's letters that the parallel with Charles Darwin, although anachronistic in terms of the time the novel is set, was intentional (Elizabeth Gaskell to George Smith, 3 May 1864, in *Letters of Mrs Gaskell*, ed. J.A.V.

Chapple and Arthur Pollard [Manchester: Manchester University Press, 1966], 732).

20. See, e.g., Matthew Arnold, "Literature and Science," in *Discourses in America* (London: Macmillan, 1885), 82.

21. Gustave Flaubert's *Madame Bovary* (1856) is interesting in its depiction of the whole spectrum of medical practitioners in early nineteenth-century France, from Homais, the local pharmacist, to Larivière, the Paris surgeon with an international reputation; but the qualities necessary for keeping up with medical research are most notable by their absence in the character of Charles Bovary, who is complacent, unambitious, and stolid, failing to read his medical journals or even his former textbooks.

22. Anna T. Kitchel, ed., *Quarry for Middlemarch* (Berkeley and Los Angeles: University of California Press, 1950).

23. "This morning I finished the first chapter of *Middlemarch*. . . . I am reading Renouard's *History of Medicine*" (Journal entry, Aug. 1869, in *George Eliot's Life as Related in her Letters and Journals*, ed. J. W. Cross, 3 vols. [Edinburgh and London: Blackwood, 1885], 3:97). A month later she commented, "I have achieved little during the last week except reading on medical subjects—the Encyclopedia about the Medical Colleges, Cullen's *Life*, Russell's *Heroes of Medicine*, etc." (99).

24. This is made explicit in the novel when Lydgate's reactionary colleagues discuss the program of reform publicized by Wakley, the proprietor of the *Lancet*. Dr. Sprague says: "I disapprove of Wakley, no man more; he is an ill-intentioned fellow, who would sacrifice the respectability of the profession, which everyone knows depends on the London colleges, for the sake of getting some notoriety for himself" (*Middlemarch: A Study of Provincial Life* [1871–72; reprint, 2 vols., London: Dent, 1959], 1:136; parenthetical references in the text are to this edition).

25. Even as late as 1865, Claude Bernard felt the need to argue against those who claimed that "aucun fait ne démontre qu'on puisse atteindre en médicine la précision scientifique des sciences expérimentales" (No one has proved that one can extend into medicine the scientific precision of experimental science) (*Introduction à l'étude de la médicine expérimentale* [Paris: Bordas, 1966], 286–87).

26. See, e.g., Charles Newman, *The Evolution of Medical Education in the Nineteenth Century* (London: Oxford University Press, 1957).

27. In Paris, Lydgate has met Broussais and Pierre-Charles Louis, whose "new book on Fever" he studies so avidly far into the night. See, e.g., C. L. Cline, "Qualifications of the Medical Practitioners of *Middlemarch*," in *Nineteenth-Century Perspectives*, ed. C. de L. Ryals (Durham, N.C.: Duke University Press, 1974), 271–82.

28. Lydgate diagnoses Casaubon's heart condition with the help of a stethoscope. The details concerning the discovery of this instrument had been published by René Laënnec only ten years earlier, in 1819.

29. This was another reform due largely to Laënnec. See *Middlemarch*, 1:123.

30. John Ware, "Remarks on the History and Treatment of Delirium Tremens" (1831) (see Kitchel, *Quarry for Middlemarch*, 35–36).

31. See *Middlemarch*, 1:128; cf. Cline, "Qualifications of the Medical Practitioners of *Middlemarch*," 275, 280.

32. Evolutionary theory, with its emphasis on change and process rather than on fixed entities, had its counterpart in the philosophy of William Whewell, who stressed the "subjective" elements of scientific discovery in opposition to the strictly

"objective" view of Mill. For a more detailed discussion of these opposing philosophies see, e.g., M. Y. Mason, "*Middlemarch* and Science: Problems of Life and Mind," *Review of English Studies*, n.s. 22, no. 116 (1971), 151–69. George Lewes, Eliot's de facto husband, had also worked through this new concept of the necessarily subjective element of scientific procedure and had concluded that "perception is the assimilation of the Object by the Subject, in the same way that Nutrition is the assimilation of the Medium by the Organism. . . . Hence the search after the thing in itself is chimerical: the thing being a group of relations, it is what these are" (George Henry Lewes, *Problems of Life and Mind*, 5 vols. [London: Trübner, 1874–83], 1:189, 2:26).

33. Bichat can also be seen as an inheritor of the "vitalistic materialism" of Harvey's disciple Francis Glisson, modified through the eighteenth century by von Haller, La Mettrie, and Diderot.

34. Eliot, in fact, deliberately makes this analogy with a woven fabric the central image of her novel, using the recurrent and complex metaphor of the web at many levels to describe and link the structures of the body, the interrelations of social intercourse, and the process of writing the novel itself ("I at least have so much to do in unravelling certain human lots, and seeing how they were woven and interwoven, that all the light I can command must be concentrated on this particular web" [1:122]).

35. Chapter 45 gives a vivid description of the insidious marshaling of medical and lay forces against Lydgate. The network of gossip spreads inaccurate rumors and superstition at all levels, from the mayoral dinner table to the local public house.

36. Lydgate, who had originally intended to exploit the wealthy Bulstrode's "notions," is soon popularly regarded as Bulstrode's man, so that the latter's subsequent disgrace destroys Lydgate also.

37. There is a further medical irony concerning the reason for Lydgate's final condemnation by Middlemarch. He treated a case of *delirium tremens* according to Ware's new theory (see above, n. 30), supported by his own observations; that is, he forbade the patient, Raffles, alcohol and all but the minimum dosage of opium. Had this prescription been carried out, Raffles would probably have recovered as Lydgate expected. However, Bulstrode, who wanted Raffles dead, secretly countermanded these instructions and, to Lydgate's surprise, Raffles died. Suspecting that Bulstrode may have disobeyed his orders, Lydgate admitted that in other circumstances he would have made a rigorous enquiry into the matter; but because Bulstrode had just lent him one thousand pounds, Lydgate, to his own self-disgust, kept silent. The irony resides in the fact that the other Middlemarch doctors, having discovered that Raffles was in fact offered brandy and repeated doses of opium, find no fault with such treatment, since it is their own current practice. Instead, they accuse Lydgate of keeping secret Raffles's assumed confession, which would have implicated Bulstrode. That is, Middlemarch condemns Lydgate for a crime of which he is innocent, while absolving him of the more serious crime of which he bitterly accuses himself.

38. In 1862 G. H. Lewes had written the article "Physicians and Quacks" advocating the institution of facilities and grants for scientist-doctors to pursue research full-time, unhampered by the need to support themselves with a time-consuming practice. In 1879, after his death, George Eliot endowed just such a scheme in his memory, instituting a studentship in physiology to be awarded to a young man qualified and

eager to carry out research, but without the financial means to do so (see Eliot to Mme Bodichon, 8 Apr. 1879, quoted in Cross, *George Eliot's Life*, 3:355).

39. Robert A. Greenberg points out that having read Raspail's book *Nouveau système de chimie organique fondé sur des nouvelles méthodes d'observation*, Eliot would have known that Raspail himself suffered hardships worse than Lydgate's. He participated in the 1830 revolution in the cause of humanity and was forced to continue his work in prison. Raspail, however, unlike Lydgate, did not lose hope in the future. The dedication of the book carried the address "Maison d'arrêt de Versailles," but Raspail nevertheless wrote, "l'avenir me console" (Greenberg, "Plexuses and Ganglia: Scientific Allusion in *Middlemarch*," *Nineteenth-Century Fiction* 30 [1976], 46–47).

40. W. J. Harvey, "The Intellectual Background of the Novel: Casaubon and Lydgate," in *Middlemarch—Critical Approaches to the Novel*, ed. B. Hardy (London: Athlone, 1967), 25–37.

41. Harvey points to the further irony that Eliot herself was limited in her biological knowledge by her time. Biochemistry has, after all, brought us back from cell theory as she knew it to something not unlike Lydgate's notion of primitive tissue.

42. Alfred Tennyson, *In Memoriam A.H.H.*, stanzas 3, 15, 16.

43. See, e.g., Leo J. Henkin, *Darwinism in the English Novel, 1860–1910: The Impact of Evolution on Victorian Fiction* (New York: Russell & Russell, 1963).

44. Chambers had tried to convey the ideas of the embryologist Karl Ernst von Baer, who showed the parallelism between stages of embryonic development, the Linnaean categories of classification, and the sequence of fossils. Although Chambers himself probably understood von Baer's theory, it was incorrectly interpreted by many of his contemporaries to mean a rapid recapitulation of evolution during gestation.

45. Benjamin Disraeli, *Tancred or the New Crusade* (London: Peter Davis, 1927), 112–13.

46. Edmund Gosse, *Father and Son* [1907], ed. Peter Abbs (Harmondsworth: Penguin, 1986), 103.

47. *Omphalos* is Greek for "navel." The reason for this apparently esoteric title relates to the necessary conclusion of Gosse's theory, namely, that Adam too, though created *de novo*, would have been endowed with a navel as though he had undergone a foetal existence.

48. Charles Kingsley wrote *The Water Babies* specifically to introduce children to evolutionary theory, which he believed to be entirely compatible with Christianity, without recourse to such sophistry as Gosse's.

49. Knight's profession is, appropriately, that of literary critic, and this has become his habitual attitude to the world. He observes social gatherings "with quiet and critical interest." Hardy thus implies, although he does not actually state the connection, a parallel between the scientific and the critical attitude.

50. Thomas Hardy, *A Pair of Blue Eyes* [1873], ed. Roger Ebbatson (Harmondsworth: Penguin, 1986), 271.

51. Hardy applied to the Astronomer Royal on 26 November 1881 for permission to visit the Greenwich Observatory (F. E. Hardy, *The Early Life of Thomas Hardy* [London: Macmillan, 1928], 195).

52. In his preface to the 1895 edition, Hardy explained that his intention was "to set the emotional history of two infinitesimal lives against the stupendous back-

ground of the stellar universe, and to impart to readers the sentiment that of these contrasting magnitudes the smaller might be the greater to them as men" (Thomas Hardy, *Two on a Tower* [London: Macmillan, 1922], i).

53. Cf. "If you could but share my fame,—supposing I get any, which I may die before doing" (ibid., 68).

54. It was not accepted until the 1920s that these "dark spaces" were actually light-absorbing dust clouds rather than empty space.

55. This passage could almost be read as a gloss on Tennyson's stanzas in *In Memoriam A.H.H.* quoted above. An interesting contrast to this view is provided in *Miriam's Schooling* (1890), a novel written by William Hale White under the pseud-onym Mark Rutherford. White's novels are as pessimistic, in their way, as Hardy's, but in *Miriam's Schooling* a realization of the regularity and beauty of the solar system and the "fixed stars" beyond is presented as being as much a cause for celebration and optimism as it was in Newton's time. The amateur astronomer-cleric Mr. Arm-strong explains to Miriam, "Suppose we were in the centre of the planetary system, all these irregularities would disappear; but we are outside, and therefore it looks so complicated" (*Miriam's Schooling and Other Papers* [London: T. Fisher Unwin, 1890], 139). Indeed, Hale White may be directly contesting Hardy's view in *Two on a Tower* when Armstrong continues, "If you can once from your own observation *realise* the way the stars revolve—why some near the pole never set—why some never rise, and why Venus is seen both before the sun and after it—you will have done yourselves more real good than if you were to dream for years of immeasurable distances, and what is beyond and beyond and beyond, and all that nonsense. The great beauty of astronomy is not what is incomprehensible in it, but its comprehensibility—its geo-metrical exactitude" (139–40). Yet another interpretation of celestial order by a con-temporary writer is that of Arthur Hugh Clough, already referred to briefly in chap-ter 6. Clough derives from such a premise the conclusion that there is no God, only one vast mechanism: "And as of old from Sinai's top / God said that God is One, / By Science strict so speaks He now / To tell us, There is None! / Earth goes by chemic forces; Heaven's / A Mécanique Céleste! / And heart and mind of human kind / A watch-work as the rest!" ("The New Sinai," *A Choice of Clough's Verse*, ed. Michael Thorpe [London: Faber & Faber, 1969], 34).

56. Eclipsing variables, known to be eclipsing binary stars, were still being cata-loged at the time the novel was written. The discussion of variable stars in Camille Flammarion, *Astronomie populaire* (Paris: Flammarion & Cie, 1880), may well have precipitated Hardy's intention to write a novel with an astronomer hero, since the concept of an unstable universe chimes in well with his pessimistic view of other aspects of nature.

57. T and W Gemini are Cepheids and R and X Gemini are Mira-type variables.

58. "And thus they talked about Sirius, and then about other stars ' . . . in the scrowl / Of all those beasts, and fish, and fowl, / With which, like Indian planta-tions, / The learned stock the constellations'" (Hardy, *Two on a Tower*, 32; the quota-tion is from Samuel Butler, *Hudibras*, 431–32).

59. Holdsworth finds it merely amusing that Phillis reads Italian and knows Latin and Greek. Unable to realize that she reads Dante because she delights in the poetry, he assumes that she does so merely for the pragmatic reason of learning Italian.

60. It is interesting that, although reprehensible in many of his actions, not least in

his marriage to the innocent Grace Melbury, Fitzpiers resembles Hardy in several of his views—in his scientific determinism and in his belief that marriage is merely a civil contract.

61. George Gissing, *Born in Exile* (1892; reprint, London: Gollancz, 1970). Geologists, with their insistence on the vast age of the earth, the disappearance of species through past catastrophes, and the development of species from simple to more complex through time, were implicated in the popular mind with biologists as being responsible for the overthrow of religious faith.

Chapter 9. The Scientist as Adventurer

1. Alfred North Whitehead, *Science and the Modern World* (Cambridge: Cambridge University Press, 1953), 120.

2. Many of the wonderful-journey stories of the nineteenth century employed scientific or pseudo-scientific details as a means of taking the adventurers to the unknown place, without necessarily involving characters who could reasonably be called scientists. George Tucker's *A Voyage to the Moon* (1827), Edgar Allan Poe's "Unparalleled Adventure of One, Hans Pfaal" (1835) and *The Narrative of Arthur Gordon Pym* (1838), and Matthew Phipps Shiel's *The Lord of the Sea* and *The Purple Cloud* (both 1901) are of this kind. They were to be followed in the early twentieth century by numerous stories of space travel led by heroes exemplified by the Buck Rogers and Flash Gordon stereotypes. But again, although there is frequently a pretense of scientific authenticity, there is little attempt to characterize the alleged scientists, and so these figures will not be discussed in detail here.

3. Lewis Mumford, *Technics and Civilization* (London: Routledge, 1934), 184.

4. Verne's publisher, Pierre-Jules Hetzel, rigorously censored the texts to ensure that they conformed to the morally didactic intention of the journal to "augment both the knowledge and the wholesome ideas, the logic and the taste, the compassion and the wit which constitute the moral and intellectual fabric of French youth" (quoted in Arthur B. Evans, *Jules Verne Rediscovered: Didacticism and the Scientific Novel* [New York: Greenwood, 1988], 34).

5. Roland Barthes, "The *Nautilus* and the Drunken Boat," in *Mythologies*, trans. Annette Lavers (London: Jonathan Cape, 1972), 65–67.

6. Ibid., 65.

7. The best-known example is Captain Nemo's *Nautilus*, furnished with a library, an organ, and the best culinary offerings, but few of Verne's explorers venture far without corresponding symbols of cultural superiority. In *The Steel House*, a similar function is served by the *Steel Giant*. This locomotive, hidden within the exterior façade of an elephant, appears less threatening through the seeming association with a living creature.

8. *Cinq semaines en ballon* (1863), translated as *Five Weeks in a Balloon* (1870).

9. So fascinated was Verne with this ingenious invention that he failed to see the extreme danger involved in a gas leak.

10. *Les Aventures du Capitaine Hatteras* (1864), translated as *The Adventures of Captain Hatteras* (London: Ward Lock, 1894), containing "The English at the North Pole" and "The Ice Desert."

11. "Humphry Davy came to see me on his way through Hamburg. Among the

questions, we spent a long time discussing the hypothesis of the liquid nature of the terrestrial nucleus. We were agreed that this liquidity could not exist, for a reason which science has never been able to refute. . . . this liquid mass would be subject, like the sea, to the attraction of the moon, and consequently, twice a day, there would be internal tides which, pushing up the earth's crust, would cause periodical earthquakes" (Jules Verne, *Journey to the Centre of the Earth*, trans. Robert Baldick [Harmondsworth: Penguin, 1965], 39–40).

12. Roland Barthes, "The *Nautilus* and the Drunken Boat," 65.

13. Similar clarification is offered by Doctor Clawbonny concerning the mysterious red snow in *The Adventures of Captain Hatteras*, by Arronax in explanation of the phosphorescent water in *Twenty Thousand Leagues under the Sea*, by Harbert (medicinal plants) and Cyrus Smith (the measurement of the cliff) in *The Mysterious Island*.

14. *L'Ile mysterieuse* (1875), translated by Lowell Blair as *The Mysterious Island* (London: Corgi, 1976), 292. Almost the same words are used by the geographer Paganel in *The Children of Captain Grant* (1868).

15. An interesting and perhaps unexpected aspect of the characterization of both Lidenbrock and Axel is their evolutionary stance. Verne gratuitously included mention of the recent discovery of prehistoric fossils then believed to be 100,000 years old and, in the 1867 edition, daringly added details of the latest discoveries of fossil men. Lidenbrock, to his intense joy, stumbles upon a specimen of Quartenary man and delivers an extempore lecture on the current ideas of geology, supporting this estimate of the age of the fossils—and this despite the Catholic church's insistence at that time that such findings were incompatible with the Genesis account of the Creation. Axel too is associated with evolutionary thinking, through his hallucinatory dream in which he travels in reverse through the past history of the earth, back through the evolution of life to the birth of the planet as a gaseous nebula, a dream that both recalls Alton Locke's feverish nightmare (Charles Kingsley, *Alton Locke: Tailor and Poet* [1850], chap. 36, "Dreamland") and looks forward to Wells's story "Time Machine" (1895).

16. See, e.g., Peter Costello, *Jules Verne: Inventor of Science Fiction* (London: Hodder & Stoughton, 1978), 108.

17. The Belgian zoologist Bernard Heuvelmans believes that "Fulton was an idealist with utopian ideas: he thought it was possible to put an end to all wars by making them insupportable by means of his redoubtable weapon of the deeps" (ibid.). In the sequel to *Twenty Thousand Leagues Under the Sea*, *The Mysterious Island* (1875), we discover that Captain Nemo was formerly an Indian prince, Dakkar, which sufficiently "explains" his hatred of British imperialism without explicit recourse to French nationalism.

18. Percy Greg's *Across the Zodiac* (1880) employs the interplanetary voyage to considerable effect in the manner of Verne, paying scrupulous attention to science in hypothesizing about the gravitational effects and the topography of Mars. Verne's *Mysterious Island* may also have been the inspiration for Edward Bellamy's *Looking Backward* (1888), which is often cited as the first example of the modern technological Utopia.

19. H. G. Wells, "The Chronic Argonauts," reprinted in B. Bergonzi, *The Early H. G. Wells* (Manchester: Manchester University Press, 1961), 190.

20. See, e.g., Charles Higham, *The Adventures of Conan Doyle: The Life of the Creator of Sherlock Holmes* (London: Hamish Hamilton, 1976), 47.

21. Arthur Conan Doyle, *The Poison Belt* (London: Hodder & Stoughton, 1913), 90.

22. "His colleagues, and those about him observed the change without clearly perceiving the cause. He was a gentler, humbler, and more spiritual man. Deep in his soul was the conviction that he, the champion of scientific method and of truth, had, in fact, for many years been unscientific in his methods, and a formidable obstruction to the advance of the human soul through the jungle of the unknown" (ibid., 516).

23. See chapter 6 above.

24. See chapter 11 below.

25. Doyle made a further assertion of the emotional character of the apparently impassive scientist in the short story "The Physiologist's Wife" (1929). The protagonist, Professor Grey, is introduced as the Romantic stereotype of the scientist. His particular area of study is "the mesoblastic origin of excito-motor nerve roots," that is, the physical basis of feelings, whereby he intends to discount the validity of the emotions. Predictably, he regards women as lesser beings, because of their excessive cultivation of sensibility, and he denies the possibility of a spiritual dimension, his only faith being in evolution. "I believe in the differentiation of protoplasm," he affirms guardedly. On hearing that his wife has fallen in love with his best student, he departs punctually for his scheduled classes and proceeds to make clinical observations of his own physical state as his wife leaves, in order to disprove the importance of the emotions. Outwardly, therefore, he confirms his theories, but his sudden death suggests that this cold, "unfeeling" scientist has in fact suffered intensely through his emotional attachment to his wife and thereby finally disproves his theory.

26. The latter half of the nineteenth century saw a surge in popularity of various forms of spiritualism, often presented as having scientific credibility. Nathaniel Hawthorne was greatly attracted to the fringe sciences of mesmerism, phrenology, parapsychology, and physiognomy; and W. B. Yeats's belief in the physical existence of fairies is equally well known. In England, several reputable scientists, including Alfred Russel Wallace, found no incompatibility between science and spiritualism.

Chapter 10. Efficiency and Power: The Scientist under Scrutiny

1. See chapter 6 above.

2. John Tyndall, "The Belfast Address," *Fragments of Science*, 2 vols. (New York: Appleton, 1892), 2:201.

3. See, e.g., Charles Darwin to Asa Gray, 22 May 1860, in Francis Darwin, *Life and Letters of Charles Darwin*, vol. 2 (London: Murray, 1887), 312.

4. H. L. Sussmann, *Victorians and the Machine* (Cambridge, Mass.: Harvard University Press, 1968), 155.

5. E. M. Forster, "The Machine Stops," *Collected Short Stories* (Harmondsworth: Penguin, 1977), 131.

6. Poe was almost certainly drawing on the multitude of rumors that had grown up around the bogus "automatic" chess player devised by von Kempelen (1734–1804). This device was in the form of a life-size Turk in which was concealed a human player sitting in front of a chess board. The Turk won all the games he played in Europe and America (see John Cohen, *Human Robots in Myth and Science* [London:

George Allen & Unwin, 1966], 90–91). Poe also wrote a story, "Von Kempelen and his Discovery" (1849), about a chemist who has purportedly turned lead into gold.

7. Ambrose Bierce, "Moxon's Master," in *Science Fiction—Thinking Machines*, ed. Groff Conklin (New York: Vanguard, 1954), 8.

8. A similar assumption—and a similar, moralistic outcome—underlies Tickner-Edwardes's story "The Man who Meddled with Eternity" (in Hilary Evans and Dik Evans, *Beyond the Gaslight: Science in Popular Fiction, 1895–1905* [London: Frederick Muller, 1976], 110–17) about a scientist, Professor Starkie, who attempts to study and meddle with the soul. His research is exploited by the evil Stephen Gallier, who offers himself as a guinea pig for an experiment of soul transfer and willingly enters the body of his rival, while retaining his own evil soul.

9. G. K. Chesterton, *Robert Louis Stevenson* (London: Hodder & Stoughton, 1927), 72–73.

10. In popular imagination, the double, being unnatural, was believed to represent the evil side of the personality. The relationship between Mephistopheles and Faust is of the same kind.

11. For probable causes both personal and social see Arthur B. Evans, *Jules Verne Rediscovered: Didacticism and the Scientific Novel* (New York: Greenwood, 1988), 79–82.

12. *L'Ile à hélice* (1895), translated as *The Floating Island* (1896).

13. Physical defects symbolize their mental and moral failings. The members of the Gun Club have cracked skulls and rely on artificial limbs, while Ofranik has no eyes. It would seem, however, that Verne ultimately passed beyond this bleak vision. The scientists of his last work, "The Eternal Adam," are wise and benevolent, intent only on guiding mankind into a peaceful future.

14. See chapter 9 above.

15. One of the most powerful images of these conditions is Charles Dickens's description of the machines in the factories of Coketown as "melancholy-mad elephants" that work "monotonously up and down like the head of an elephant in a state of melancholy madness"; as their heads "went up and down at the same rate, in hot weather and cold, wet weather and dry" (*Hard Times*, 1.5.65, 1.11.107).

16. H. G. Wells, "The Time Machine," *Selected Short Stories* (Harmondsworth: Penguin, 1970), 63.

17. B. Bergonzi, *The Early H. G. Wells* (Manchester: Manchester University Press, 1961), chap. 1.

18. See R. D. Haynes, *H. G. Wells: Discoverer of the Future* (London: Macmillan, 1980), 25.

19. That this antipathy emanated chiefly from Hapley's compulsive aggression is implied by his attempts to forget Pawkins. He tries to begin on a new branch of science, thinking immediately that "if he could get up a vigorous quarrel with Halibut, he might be able to begin life afresh and forget Pawkins" (H. G. Wells, "The Moth," *Short Stories* [London: Ernest Benn, 1949], 306).

20. Bergonzi, for example, has seen the story as a meditation on religion and as "an image of the encounter between east and west" (Bergonzi, *Early H. G. Wells*, 70).

21. H. G. Wells, "The Lord of the Dynamos," *Selected Short Stories*, 184.

22. Vivisection was a highly topical issue of the time; see, e.g., R. D. French, *Antivivisection and Medical Science in Victorian Society* (Princeton: Princeton University Press, 1975), 71.

23. H. G. Wells, *The Island of Doctor Moreau* (Harmondsworth: Penguin, 1967), 114.

24. "I'm itching to get to work again—with this new stuff," Moreau says, aggrieved at the interruption caused by Prendick's arrival. "And here I have wasted a day saving your life" (ibid., 44, 105).

25. This is emphasized, if emphasis were necessary, when Prendick hears Montgomery calling out the name as two separate syllables, "Mor-eau."

26. T. H. Huxley, "Evolution and Ethics," in *Evolution and Ethics*, by J. S. Huxley and T. H. Huxley (London: Pilot, 1947), 80, 82.

27. See R. D. Haynes, "Wells's Debt to Huxley and The Myth of Dr. Moreau," *Cahiers Victoriens et Edouardiens* 13 (Apr. 1981), 31–41.

28. It is no accidental irony that Moreau, who seems to have transcended human limitations, should uncover the innate bestiality of man. Further, although Moreau has been concerned with demonstrating this continuity in only one direction, namely, the "elevating" of beasts to human form, he is forced to realize that the continuity is equally valid in the contrary direction: the "human" beasts revert to their previous, bestial state. Nor, the novel implies, is such reversion confined to the Beast Folk. We have had forewarning of this in the lifeboat of the *Lady Vain*, when the three castaways agree to draw lots for the sacrifice of one of their number. The impending cannibalism is prevented only by two of the men falling overboard. When Prendick is rescued by Montgomery, he revives by drinking "a dose of some scarlet stuff, iced" (13) and reacts with carnivorous enthusiasm to the boiled mutton offered him: "I was so excited by the appetising smell of it, that I forgot the noise of the beast forthwith" (16). We see the significance of this later when Prendick hears the Beast Folk reciting the law that is designed to keep them "human": "Not to eat Flesh nor Fish; that is the Law. Are we not Men?" (85). (Wells himself was a vegetarian and believed that all society would soon follow suit, because it would be impossible to persuade anyone to be a butcher.) Moreover, as the end of the novel demonstrates, Prendick comes to believe that this devolution is not confined to the Beast Folk. The respectable members of civilized British society seem to be reverting to the beast before his horrified gaze. In effect, the whole novel involves an extended deconstruction of the title of Darwin's controversial book *The Descent of Man* (1871).

29. "Each time I dip a living creature into the bath of pain, I say, 'This time I will burn out all the animal, this time I will make a rational creature of my own'" (Wells, *The Island of Doctor Moreau*, 112).

30. "Poor brutes! I began to see the viler aspect of Moreau's cruelty. I had not thought before of the pain and trouble that has come to these poor victims after they had passed through Moreau's hands" (ibid., 138).

31. Gulliver remarks: "I began last week to permit my wife to sit at dinner with me, at the farthest end of a long table; and to answer (but with the utmost brevity) the few questions I asked her. Yet the smell of a Yahoo continuing very offensive, I always keep my nose well stopped with rue, lavender, or tobacco-leaves" (Jonathan Swift, *Gulliver's Travels* [1726], rev. ed. [1735; reprint, London: Dent, 1956], 317).

32. H. G. Wells, *The Invisible Man* [1897] (London: Ernest Benn, 1926), 110.

33. Griffin is proud of having robbed his father, since he despises affection as sentimental illusion compared with the world of his laboratory. "Re-entering my room was like the recovery of reality. There were the things I knew and loved" (ibid., 153).

34. Wells was aware of the endless potential for irony in those Greek myths that hinge on the belief that even the gods cannot take back their gifts when these are found to be abhorrent. Prayers and wishes, once granted, cannot be revoked. Similarly, Wells shows that knowledge cannot be unlearned.

35. The almost illiterate tramp Marvel, who, having inherited the money that Griffin has stolen, puzzles in secret over Griffin's notes about the process for inducing invisibility, desiring the wealth and power he believes invisibility will confer, represents, in turn, a parody of the intellectually brilliant Griffin.

36. See chapter 11.

37. Lewisham, the science student in *Love and Mr. Lewisham* (1900), is a third problematic case from this period, but the scientist aspect of his character is less prominent, and for reasons of space I have omitted him.

38. Bergonzi describes Cavor as "a wholly benign figure of the traditional 'comic scientist' type: by 1901 Wells seems to have finally exorcised the 'Frankenstein' archetype that has appeared in Moreau and Griffin" (Bergonzi, *Early H. G. Wells*, 159).

39. H. G. Wells, *The First Men in the Moon* [1901] (London: Nelson, 1924), 142.

40. This incident is closely paralleled by an incident at the end of the novel when another victim of cavorite, the inquisitive boy who climbs into the sphere and closes the blinds, thereby flying off into space, is dismissed by Bedford with similar callousness. Again, Bedford's halfhearted search for Cavor after he himself has attained the safety of the sphere is reflected in Cavor's minimal concern over, as he believes, Bedford's certain death.

41. Given Cavor's description to the incredulous Grand Lunar of mankind's delight in war and other forms of destruction (the scene is by design reminiscent of that in which Gulliver describes such practices to the King of Brobdingnag) and the further premise that he alone possesses the secret whereby man could arrive on the moon, Cavor's death becomes, to the Selenites, as logically necessary as the conclusion of a syllogism. The proponent of pure rationalism becomes its victim. The Selenites have, after all, been conditioned to a flawless consistency.

42. H. G. Wells, *Tono-Bungay* [1909] (London: Unwin, 1964), 329.

43. In *First and Last Things* Wells sees a nightmare vision of misrule spreading over the whole earth, which he describes in terms that are equally appropriate for *Tono-Bungay:* "I see humanity scattered over the world, dispersed, conflicting, unawakened. . . . I see human life as avoidable waste and curable confusion. . . . Their disorder of effort, the spectacle of futility, fills me with a passionate desire to end waste, to create order, to develop understanding" (H. G. Wells, *First and Last Things* [London: Unwin, 1926], 277).

44. H. G. Wells, "Religion and Science," *Guide to the New World* [London: Gollancz, 1941], 103).

Chapter 11. The Scientist as Hero

1. Carl Becker, *Progress and Power* (New York: Vintage Books, 1965), 2.

2. Herbert Spencer had regarded progress as "not an accident, but a necessity" (see Sidney B. Fay, "The Idea of Progress," *American Historical Review* 42, no. 2 [1947], 237).

3. See, e.g., the ongoing debate between Matthew Arnold and T. H. Huxley over

priorities in education. A description of the stages of this debate may be found in William Irvine, *Apes, Angels, and Victorians: Darwin, Huxley, and Evolution* (New York: Meridian, 1959), 283–86.

4. Rush Welter, "The Idea of Progress in America," *Journal of the History of Ideas* 16 (1955), 401–5.

5. Francis Galton, *English Men of Science: Their Nature and Nurture* [1872] (London: Cass, 1970), 260.

6. See Frank M. Turner, "Public Science in Britain, 1880–1919," *Isis* 71 (1980), 589–608.

7. The English poet W. H. Auden admitted, "When I find myself in the company of scientists, I feel like a shabby curate who has strayed by mistake into a drawing-room full of dukes" ("The Poet and the City," in *The Dyer's Hand and Other Essays* [London: Faber & Faber, 1962], 81).

8. Edison was dismissed from school for being retarded, and his mother proceeded to teach him herself, apparently with some considerable effect. It was probably as a result of this background that Edison despised theorists of science; certainly this chimed in with the American popular image of the theorist as useless, if not sinister, usually European in origin, and vastly inferior to the practical inventor. The Edison myth passed through several phases, from the wizard image (arising out of his invention and exploitation of the phonograph), through the "innocence and power" stage, associated with the experiments with electricity, to the stage of the "practical, democratic individualist," identified with the great American (see, e.g., Wyn Wachhorst, *Thomas Alva Edison: American Myth* [Cambridge, Mass.: MIT Press, 1981]).

9. The only other scientist of this century whose name has become a comparable household word is Einstein, whose popular reputation was from the start more problematic because of the doubtful benefit of the atomic bomb, which was associated with his famous equation.

10. Robert Cromie, *A Plunge into Space* (1890; reprint, Westport, Conn.: Hyperion, 1976), 70.

11. A similar moral comparison is implicit in one of the earliest Martian invasion stories, *Auf zwei Planeten* (Two planets) [1897], by the German writer Kurd Lasswitz. Lasswitz's Martians are immeasurably superior to humanity not only in technological skills but in social development, and they establish a protectorate over Earth in order to teach their social and cultural skills to their warring, underdeveloped neighbors. Lasswitz's scientists can do little more than learn from the Martians, for like Cromie, he ascribes to the Martians many of the social reforms he himself advocated; thus his novel is essentially propagandist in intention. A similar moral disparity underlies Mark Wicks's *To Mars via the Moon* [1911], which was dedicated to the astronomer Percival Lowell and professed to extrapolate from Lowell's work. Wicks's Martians, though technologically inferior to human beings, are ethically superior. They tell their visitors from Earth that Earth is merely passing through an unfortunate intermediate stage in its evolution to a higher spiritual plane.

12. Garrett P. Serviss, *Edison's Conquest of Mars* (Los Angeles: Carcosa House, 1947), 179–80.

13. Tesla's prolific inventions included the induction motor, based on his discovery that a rotating magnetic field could be produced by using an alternating current. In his promotion of the AC system, he ran into headlong conflict with Edison, who

championed the direct-current system. Tesla went on to invent a tele-automatic boat, arc lighting, and the Tesla coil, still used for long-distance radio and TV transmission.

14. Weldon J. Cobb, *A Trip to Mars* (New York: Street & Smith, 1901), 107.

15. See, e.g., Anthony Frewin, *One Hundred Years of Science Fiction Illustration, 1840–1940* (London: Bloomsbury Books, 1988), chaps. 4, 5. The covers of these pulp magazines characteristically display a mixture of horror, violence, and helpless Anglo-Saxon women about to be attacked by aliens.

16. The most important promoter of the genre was John W. Campbell, Jr., a student at M.I.T. who in 1937 became editor of *Astounding Science Fiction* and was himself also one of the most prolific and successful contributors.

17. Ursula K. LeGuin, "American SF and the Other," *Science Fiction Studies* 2, no. 3 (1975), 210, 209.

18. Bram Stoker, *Dracula* [1897] (New York: Airmont, 1965), 99.

19. Stanley Waterloo, *Armageddon* (Chicago and New York: Rand McNally, 1898), 258–59.

20. See, e.g., the discussions reported in William Broad, *Star Warriors* (New York: Simon & Schuster, 1985).

21. Simon Newcombe, *His Wisdom, the Defender* (New York: Harper, 1900), 110–11.

22. Garrett P. Serviss, "The Second Deluge" [1912], *Fantastic Novels Magazine* 2 (July 1948), 8, 115.

23. Bernhard Kellermann, *Der Tunnel* (Berlin: S. Fischer Verlag, 1913), 398.

24. Wells's novel proved to be prophetic in several ways. Wells was originally inspired by Frederick Soddy's *Interpretation of Radium*, but in 1913, when *The World Set Free* was written, physicists had no proposals for splitting the atom; yet 1933, the year in which the scientists of the novel first succeed in constructing fission bombs to be used in the 1956 holocaust, was actually the year in which Frédéric and Irène Joliot-Curie first produced radioactive phosphorus by bombarding aluminum with beta particles, the first step in tapping atomic energy. In the novel, the scientists discover a substance called carolinium, which has the same properties and uses as plutonium, an element that was isolated only much later and that, with uranium, is the most important source of nuclear fuel and component of nuclear weapons. The physicist Leo Szilard acknowledged a debt to Wells's novel both for the idea of how to liberate atomic energy on a large scale and for the realization of the probable consequences. He therefore did what Holsten in the novel had done: he attempted to stop the spread of the knowledge by applying for a patent to cover his invention. "Knowing what this would mean—and I knew it because I had read H. G. Wells—I did not want this patent to become public" (Leo Szilard, *Perspectives in American History*, vol. 2 [Cambridge Mass.: Harvard University Press, 1968], 102).

25. Pax's first message makes this clear: "To all mankind. I am the dictator of human destiny. . . . I command the cessation of hostilities and the abolition of war upon the globe. I appoint the United States as my agent for this purpose" (Arthur Train and Robert Williams Wood, *The Man Who Rocked the Earth* [New York: Doubleday, Page & Co., 1915], 11).

26. For example, G. Powell, *All Things New* (1926); B. Newman, *Armoured Doves* (1931); and H. Edmonds, *The Professor's Last Experiment* (1935).

27. H. G. Wells, *The Open Conspiracy* (1928), *The Shape of Things to Come* (1933), and *World Brain* (1938), among others.

28. Warren Wagar, *Terminal Visions: The Literature of Last Things* (Bloomington: Indiana University Press, 1982), 27, 110.

29. *Flash Gordon* was created by Alex Raymond, who drew the strip from 1934 to 1944. It was later taken over by other artists, including Austin Briggs, Mac Rabay, Dan Barry, and Al Williamson.

30. Broad, *Star Warriors*.

31. "You [scientists] do not make the decision as to the rightness or wrongness [of my predictions], but only the inexorable judge, Nature herself. Therefore I ask all those who commit themselves with me, to put their names down on this sheet, so that their names will not be damned by humanity along with those of the others" (Rudolf Heinrich Daumann, *Protuberanzen. Ein utopischer Roman* [Berlin: Schützen-Verlag, 1940], 122).

32. Robert Millikan's *Science and the New Civilisation* [1930] was a national best-seller.

33. Don A. Stuart [John W. Campbell, Jr.], "Atomic Power," *Astounding Stories*, Dec. 1934.

34. W. Kaempffert, one of this group, declared, "Religion may preach the brotherhood of man; science practises it" (*Science, Today and Tomorrow* [New York: Viking, 1945], 266).

35. "There isn't even any word for . . . an [anti-social] act in the Lithian language" (James Blish, *A Case of Conscience* [1959; reprint, Harmondsworth: Penguin, 1963], 74).

36. C. S. Lewis, *Out of the Silent Planet* [1946]; see chapter 12 below.

37. In a letter published with the novel, Einstein called it "ein wesentlicher Beitrag zur Überwindung der gefährlichen Gleichgültigkeit des Publikums in bezug auf die großen internationalen Probleme unserer Zeit" (a significant contribution towards overcoming the dangerous public indifference concerning the huge international problem of our time) (Oskar Maria Graf, *Die Erben des Untergangs. Roman einer Zukunft* [Frankfurt am Main: Nest Verlag, 1959]).

38. James Hogan, *The Genesis Machine* (New York: Ballantine Books, 1978), 284–85. The last pages of the novel raise the interesting question whether Clifford's statements about the J-machine were only a hoax, to stop the world's rearming. This ambiguity is never finally resolved in the novel, presumably because Hogan intends to suggest that the *will* to disarm is all that is really necessary.

39. Arthur Conan Doyle, "The Final Problem" [1893], in *A Treasury of Sherlock Holmes Stories*, ed. Adrian Conan Doyle (Garden City, N.Y.: Hanover House, 1955), 259.

40. Stuart Edward White, *The Sign at Six* (Indianapolis: Bobbs-Merrill, 1912), 250.

41. William MacHarg and Edwin Balmer, *The Achievements of Luther Trant* (Boston: Small, Maynard & Co., 1910). This may be seen as a development of the early use of a reputed mesmerist as the agent for extracting a confession from a murderer in Leopold Lewis's play *The Bells* [1871].

42. Lasswitz was greatly influenced by his teacher Gustav Theodor Fechner, who incorporated the idealistic philosophy of Schiller and Goethe into his own scientific studies in medicine, physics, and psychology to arrive at a particular neo-Romantic

vitalist stance (see Edwin M. J. Kretzman, "German Technological Utopias of the Pre-War Period," *Annals of Science* 3 [1938], 418).

43. "Wie die Naturerkenntnis uns das Vertrauen zum Erfolg unserer Arbeit gibt, so liefert sie uns auch zugleich durch die Beherrschung der Natur das einzige Mittel, die Lebensbedingungen der Menschheit zu vervollkommen. . . . Es ist doch klar, sollen die Lebensbedingungen verbessert werden, daß neue Güter geschaffen werden müssen. Die einzige Quelle dafür ist die Natur und das einzige Mittel ist die Technik in ihrer Verbindung mit der Natur" (As the knowledge of Nature gives us the confidence to pursue our work, so it also allows us the only means of dominating Nature in order to perfect the living conditions of Mankind. It is therefore clear that, if the living conditions are to be improved, new benefits will have to be achieved. The only source for that is Nature and the only method is technology in conjunction with Nature) (Kurd Lasswitz, "Über Zukunftsträume II," *Die Nation*, 20 May 1899).

44. See chapter 6 above.

45. "Gegen das Weltgesetz" is one of the two tales published collectively as *Bilder aus der Zukunft* (Pictures of the future) [1878]. Cf. Lasswitz's later and more famous novel, *Auf zwei Planeten* (Two planets) [1897], in which the scientists' reactions to the Martians vary from suspicion to interest, but Lasswitz clearly suggests that any hope for the future of mankind depends on cooperating with this superior society. See note 11 above.

46. John Perry, another spokesperson for this view, observed with disgust that "almost all the most important, the most brilliant, the most expensively educated people in England" remained uninformed about the principles and methods of science. "This ignorance of our great men tends to create ignorance in our future leaders; is hurtful to the strength of the nation now, and retards our development in all ways" (quoted in Turner, "Public Science in Britain, 1880–1919").

47. Ibid., 593.

48. H. G. Wells, *A Modern Utopia* [1905] (London: Unwin, Atlantic edition, 1926), 92.

49. See chapter 10 above.

50. H. G. Wells, *The World Set Free* [1914] (London: Unwin, Atlantic edition, 1926), 141–42. Wells gave this belief more than fictional expression in a series of essays advocating a world state organized on scientific grounds. See, e.g., *The Open Conspiracy* [1928], *The World Brain* [1938], and *The Fate of Homo Sapiens* [1939].

51. "Nützlich und dem Menschen dienstbar wird [die neue Energie] fließen müssen. Dann aber wird ihr reicher Strom ein neues Zeitalter befruchten. Ein Zeitalter, in dem der alte Erdball der Menschheit zu klein wird und sie ihren Pfad zu anderen Gestirnen lenkt" (Hans Dominik, *Die Macht der drei, Roman* [Berlin: Verlag Scherl, 1922], xiv).

52. B. F. Skinner, *Walden Two* (New York: Macmillan, 1948), 211–12.

53. Zapparoni creates automatons like glass bees but retains complete control over them and hence over mankind, which happily relies on their labor. See chapter 14 below.

54. Eugene Rabinowitch, "History's Challenge to Scientists," *Bulletin of the Atomic Scientists* 12 (1956), 238. This attitude is discussed in more detail in chapter 14 below.

55. The first U.S. artificial satellite, Explorer I, was launched in January of the following year (1958). The vilification of scientists began only after the first Russian

atomic test on August 29, 1949, a few months before Klaus Fuchs admitted to passing atomic secrets to Russia both during and after the Manhattan Project. For almost a decade fears about the Russians' catching up had been mitigated by the assumption that U.S. scientists were well in advance of their rivals. With the Sputnik launching, however, a shocked United States was forced to realize that the Soviet Union was no longer dependent on information gleaned from the West, but was leading the way in space research. The marked increase in "alien invasion" films during the late fifties and sixties was symptomatic of the paranoia associated with a potential Soviet invasion (see Carl Jung, *Flying Saucers: A Modern Myth of Things Seen in the Sky* [London: Routledge & Kegan Paul, 1977], 1–24).

56. Ian Fleming's creation James Bond, Agent 007, became the most famous prototype of this new hero in films of the sixties and seventies. The Bond films were especially popular in Britain, where they fuelled the idea that Britain, which in reality was losing its position as a world powerbroker, was still able to shape international affairs (see Mick Broderick, *Nuclear Movies: A Filmography* [Northcote, Victoria: Post-Modem, 1988], 22–23).

Chapter 12. Mad, Bad, and Dangerous to Know: Reality Overtakes Fiction

1. Andrew Tudor, "Seeing the Worst Side of Science," *Nature* 340 (24 Aug. 1989), 589–92.

2. Sir William Crookes, quoted in Frederick Soddy to Sir Ernest Rutherford, 19 Feb. 1903, Rutherford Papers, Cambridge.

3. Pierre Biquard, *Frédéric Joliot-Curie et l'énergie atomique,* Savants du Monde Entier 3 (Paris: Seghers, 1961), 14–15, quoted in Spencer R. Weart, *Scientists in Power* (Cambridge, Mass.: Harvard University Press, 1979). Weart comments: "Curie may have been thinking of radium more as a poison than as a source of energy, but the parallel with explosives is suggestive" (294 n. 11).

4. Frederick Soddy, *The Interpretation of the Atom* (London: John Murray, 1932), 4. Within ten years this latter sentence was proved wrong.

5. This was not the case everywhere. Many French physicists and chemists, despite their international reputation, were forced into industry because of the failure of the French government to provide adequate funding for research. By 1920 the Paris Faculty of Sciences could not meet its debts (see Weart, *Scientists in Power*, 14).

6. See chapter 16 below.

7. Spencer R. Weart, "The Physicist as Mad Scientist," *Physics Today,* June 1988, 35.

8. Freeman Dyson, *The Day after Trinity: J. Robert Oppenheimer and the Atomic Bomb* (Kent, Ohio: Transcript Library, 1981), 14, 30.

9. W. H. Rhodes, "The Case of Summerfield," in *Caxton's Book,* ed. Daniel O'Connell (Westport, Conn.: Hyperion, 1974), 19. Kurt Vonnegut's *Cat's Cradle* [1963] uses a similar scenario but substitutes ice for fire; the effect on life is much the same.

10. Not only are these terms characteristic of the post-Darwinian perception of nature but they are almost exactly the terms in which Wells arraigned the sentimental, romantic view of nature; see, e.g., *A Modern Utopia* (1905).

11. Arthur Conan Doyle, "The Disintegration Machine," in *The Complete Professor Challenger Stories* (London: John Murray & Jonathan Cape, 1976), 531.

12. Doyle's most famous villain, Moriarty, lifelong opponent of Sherlock Holmes,

does not invent devices of destruction, but relies on his native mathematical brilliance to run a more profitable international crime ring. Holmes describes him as one "endowed by nature with a phenomenal mathematical faculty. At the age of twenty-one he wrote a treatise upon the binomial theorem, which has had a European vogue. On the strength of that he won the mathematical chair at one of our smaller universities and had, to all appearances, a most brilliant career before him. But the man had hereditary tendencies of the most diabolical kind" (Arthur Conan Doyle, "The Final Problem" [1893], in *A Treasury of Sherlock Holmes Stories* [Garden City, N.Y.: Hanover House, 1955], 258). Moriarty retains many of the stereotyped attributes of the academic, reinterpreted in a sinister light as he devises schemes for world domination. Thus: "His forehead domes out in a white curve, and his two eyes are deeply sunken in his head. . . . His shoulders are rounded from much study, and his face protrudes forward and is forever slowly oscillating from side to side in a curiously reptilian fashion" (260). One of the most striking things about Moriarty, though, is that this cold and satirical character is in many respects similar to Holmes. This estimate is mutual, and as suggested in chapter 10, the two characters are quite similar.

13. Murray Leinster, "The Storm That Had to be Stopped" and "The Man Who Put Out the Sun," *Argosy*, 1 Mar. 1930, 454–89, and 14 June 1930, 24–58.

14. We saw a parallel situation in Stewart White's novel *The Sign at Six* [1912], in which a mad, evil scientist holds a city at ransom (see chapter 11 above).

15. Georg Kaiser, *Gas I*, trans. Hermann Scheffauer, in *Twenty-five Modern Plays*, ed. S. M. Tucker (New York: Harper, 1953), 605.

16. Georg Kaiser, *Gas II*, trans. Winifred Katzin, in Tucker, *Twenty-five Modern Plays*, 639.

17. Priestley may have had in mind Einstein's resistance to the statistical implications of quantum theory and his much-quoted dictum, "God does not play dice."

18. J. B. Priestley, *The Doomsday Men* (London: Heinemann, 1938), 289.

19. F. H. Rose, *The Maniac's Dream* (London: Duckworth, 1946), 158–59.

20. L. Sprague De Camp, "Judgement Day," *Astounding Science Fiction* 55, no. 6 (1955), 72, 73.

21. See Paul Brians, *Nuclear Holocausts: Atomic War in Fiction, 1895–1984* (Kent, Ohio: Kent State University Press, 1987), 29.

22. Spencer Weart points out that this was a reference to the similar feature of a robot-making scientist in the classic 1926 film *Metropolis* (Spencer Weart, *Nuclear Fear: History of Images* [Cambridge, Mass.: Harvard University Press, 1988], 277).

23. Philip K. Dick, *Dr. Bloodmoney* (Boston: Gregg, 1977), 165.

24. See, e.g., Fredric Jameson, "After Armageddon: Character Systems in *Dr. Bloodmoney*," *Science Fiction Studies* 2, no. 1 (1975), 40.

25. J. Tyndall, "The Belfast Address," in *Fragments of Science*, vol. 2 (New York: Appleton, 1892), 201.

26. Another Faustian biologist, who in human breeding experiments intends to outdo God by producing stronger types, is Lambacher in Hans Henny Jahnn's novel *Trümmer des Gewissens (Der staubige Regenbogen)* (ed. Walter Muschg [1959; reprint, Frankfurt am Main: Europäische Verlagsanstalt, 1961]). Undeterred by the human suffering entailed in his experimental program, Lambacher considers individuals expendable in the cause of science and his own personal aggrandizement: "Es geht

um mehr als einen einzelnen Menschen" (It is a question of more than one single person) (66).

27. André Gide, *The Vatican Cellars* [1914], trans. Dorothy Bussy (London: Cassell & Co., 1952), 5.

28. A. Hyatt Verrill, "The Plague of the Living Dead," *Amazing Stories*, Apr. 1927, 20.

29. A. Hyatt Verrill, "The Ultra-Elixir of Youth," ibid., Aug. 1927, 483.

30. Alexander Snyder, "Blasphemers' Plateau," ibid., Oct. 1926, 663.

31. For a discussion of this point see Leland Sapiro, "The Faustus Tradition in the Early Science Fiction," *Riverside Quarterly* 1, no. 2 (1964), 53–55.

32. For a full list see chapter 7 above.

33. Alfred Döblin, *Berge, Meere und Giganten. Roman* (Berlin: S. Fischer Verlag, 1924).

34. Obsessed with demonstrating their power over nature by making man independent of nature, the scientists have also, by the twenty-sixth century, set up experimental centers where, as part of the *künstliche Lebensmittelsyntheseproject,* unsuspecting human guinea pigs are given *Schweinspeise* and *Schweingetränke.*

35. Sydna Stern Weiss points out that in Döblin's novel, "the centuries in which the scientists and technocrats control the earth are periods of impoverishment of human physical and mental capacities and of exploitation and rape of the earth and nature. The successive commitments of these men (1) to total independence of Nature . . . (2) to materialism and construction of useless weaponry for its own sake, and (3) to attempts at re-designing the earth and transforming themselves into such creatures as emerge from the world's past . . . lead to catastrophes. These functionaries of technology cannot govern because they provide man with neither spiritual nor physical nourishment" ("Scientists in Late Nineteenth-Century and Twentieth-Century German Literature: Their Projects, Personalities, and Problems" (Ph.D. diss., Princeton University, 1975).

36. Aldous Huxley, *Jesting Pilate* (London: Chatto & Windus, 1926), 273.

37. In the introduction to the 1947 edition of "The Machine Stops," Forster wrote that the story was "a counterblast to one of the heavens of H.G. Wells"; and Kuno, Forster's spokesman, affirms his belief that "man is the measure," a principle Forster believed the Wellsian utopia had discarded.

38. Huxley prefaced *Brave New World* with a quotation from Nicholas Berdiaeff: "Les utopies apparaissent bien plus réalisables qu'on ne le croyait autrefois. Et nous nous trouvons actuellement devant une question bien autrement angoissante: Comment éviter leur réalisation définitive?"

39. So influential were these anti-utopias in inculcating a general fear of psychological conditioning that few later writers have been brave enough to defend the idea as a means of introducing and maintaining a community-oriented attitude. One of the few exceptions was the psychologist B. F. Skinner, whose Wellsian-style novel *Walden Two* [1948] was written as fictional propaganda for his own behaviorist theories.

40. Examples of other derivative treatments include Bernard Wolfe, *Limbo* [1952]; Kurt Vonnegut, *Player Piano* [1952]; Isaac Asimov, *The Caves of Steel* [1954]; Ray Bradbury, *Fahrenheit 451* [1954]; and Frederick Pohl, *Drunkard's Walk* [1960].

41. See, e.g., J.B.S. Haldane, "Auld Hornie, F.R.S.," reprinted in *Shadows of Imag-*

ination, ed. M. R. Hillegas (Carbondale: Southern Illinois University Press, 1969), 15–25.

42. C. S. Lewis, *Perelandra* [1943] (New York: Macmillan, 1944), 81.

43. C. S. Lewis, *Out of the Silent Planet* [1938] (London: John Lane, the Bodley Head, 1946), 21–22.

44. Cf. Gulliver's account of Europe to the King of Brobdingnab and Cavor's to the Grand Lunar in H. G. Wells's *First Men in the Moon* (see chapter 10 above).

45. *Perelandra*, 81.

46. H. G. Wells, "Discovery of the Future," *Nature* 65 (6 Feb. 1902), 331. Cf. *The Food of the Gods* [1904] (London: Unwin, 1926), bk. 3, chap. 5, sec. 3, 305–6, and *Marriage* [1912] (London: Unwin, 1926), bk. 3, chap. 4, sec. 3, 316.

47. C. S. Lewis, *That Hideous Strength* (London: John Lane, the Bodley Head, 1946), 45.

48. In *The Abolition of Man* [1943] Lewis set out the philosophical position underlying his trilogy.

49. Christa Wolf, "Ein Brief," in *Mut zur Angst. Schriftsteller für den Frieden,* ed. Ingrid Krüger (Darmstadt: Luchterhand, 1982), 54–55, 157.

Chapter 13. The Impersonal Scientist

1. See chapter 6 note 1.

2. Thus the collapse of the Eastern bloc in Europe occurred primarily for economic reasons, which translate into efficiency of production, transportation, and communication.

3. Louis Macneice, "The Kingdom," in *Collected Poems*, ed. E. R. Dodds (London: Faber & Faber, 1979), 252–53.

4. D. H. Lawrence, *The Rainbow* (Harmondsworth: Penguin, 1966), 350.

5. During eighteen months in England (1916–17) supervising the construction of Russian icebreakers, Zamyatin became familiar with Wells's work. In his book *Herbert Wells* [1922] he praises Wells's scientific romances as "social-scientific fantasy," the new, urban fairy tales of the industrial age.

6. See, e.g., Alex M. Shane, *The Life and Works of Evgenji Zamjatin* (Berkeley: University of California Press, 1968).

7. Yevgeny Zamyatin, *We*, trans. Gregory Zilborg (New York: E. P. Dutton, 1924), 4. The work in Russian was probably completed in 1921, but it was first published in the United States in translation in 1924.

8. Patricia Warrick has pointed out that in addition to his obvious debt to Wells, Zamyatin has also drawn on Dostoevsky's *Notes from the Underground*. She suggests that Zamyatin was influenced by Dostoevsky's reservations about nineteenth-century rationalism, with its optimistic picture of a perfect society based on enlightened self-interest, and his recognition of the strong element of irrationality in the human psyche, even at the cost of unhappiness (see Patricia Warrick, "The Sources of Zamiatin's *We* in Dostoevsky's *Notes from the Underground*," *Extrapolation* 17 [1975], 63–77).

9. It is difficult to believe that Zamyatin was not inspired in this aspect by Coleridge's poem "Kubla Khan."

10. Algis Budrys, *Who?* (London: Quartet, 1975). The novel was expanded from a

short story of the same name, published in *Fantastic Universe*, 1955.

11. See chapter 14 below.

12. Thomas Pynchon, *Gravity's Rainbow* (New York: Viking, 1973), V434/B506.

13. Walter Gratzer suggests that Dr. Obispo bears a certain resemblance to Serge Voronoff, the leader of the monkey testis transplanters, but that Huxley may also have had in mind Casimir Funk, who caused a diplomatic incident when he tried to procure from Abyssinian tribesmen the castrated testicles of Italian prisoners in order to isolate testosterone (Walter Gratzer, *The Longman Literary Companion to Science* [Harlow, Essex: Longman, 1989], 245).

14. Aldous Huxley, *Point Counter Point* (London: Chatto & Windus, 1947), 26. Dr. Poole, the botanist of *Ape and Essence* [1948], is another of Huxley's emotionally retarded scientists, still dependent on his mother for advice and direction.

15. Aldous Huxley, *The Genius and the Goddess* (London: Chatto & Windus, 1955), 23.

16. The exception occurs in *Island* (see chapter 16 below).

17. C. P. Snow, *Death Under Sail* (London: Heinemann, 1959), 121–22. Snow published this detective story anonymously lest it adversely affect his career as a scientist.

18. C. P. Snow, *The Search*, rev. ed. (1938; reprint, Harmondsworth: Penguin, 1979), 69.

19. Snow takes some pains to emphasize Miles's coldness: when his mother dies he feels almost nothing except relief that he will no longer have to provide for her.

20. "If false statements are to be allowed, if they are not to be discouraged by every means we have, science will lose its one virtue, truth. The only ethical principle which has made science possible is that the truth shall be told all the time. If we do not penalise false statements in error, we open up the way, don't you see, for false statements by intention. And of course a false statement of fact, made deliberately, is the most serious crime a scientist can commit" (ibid., 257).

21. The fictional incident parallels a similar one in Snow's own career. Having worked with P. V. Bowden on the chemistry of vitamin A and its derivatives, Snow published a paper that was well received. However, it was later found that the experimental work was seriously flawed, and the conclusions drawn from it unfounded. Snow published no retraction and gave no satisfactory reason for the errors, but he did not publish another scientific paper.

22. Irving Fineman, *Doctor Addams* (London: Cresset, n.d.), 206–7.

23. Max Frisch, *Don Juan*, in *Max Frisch: Four Plays*, trans. Michael Bulloch (London: Methuen, 1969), 118–19.

24. Erich Nossack, "Die Schalttafel," in *Spirale* (Frankfurt: Suhrkamp Verlag, 1956), 47–113.

25. "Ich liebe die Ordnung. Ich bin ein wissenschaftlicher Mensch, nicht nur Wissenschaftler. Das heißt, mein Beruf hat von mir Besitz ergriffen, an Leib und Seele. Dennoch hat Ordnung für mich auch einen metaphysichen Sinn. Ein Grenzwert, jenseits dessen sich so etwas wie ein Weltprinzip ausdrückt. Ein Schöpfungsprinzip" (Heinz von Cramer, "Aufzeichnungen eines ordentlichen Menschen," *Leben wie im Paradies* [Hamburg: Hoffmann & Campe, 1964], 15).

26. "Du hast keine Macht—Ich bin frei—Du kannst mich nicht beherrschen—Du wirst mich nicht beherrschen— . . . Ich werde nicht tun, was du willst—gib es auf, gib es auf—versuche nicht noch einmal, mich zu überfallen, zu überlisten—" (You

have no power; I am free. You can't control me—you won't control me. I won't do it. Give up, give up. Don't try again to surprise me, to outwit me) (ibid., 27).

27. "Ich muß die Formel errechnen, die gewaltige Formel, welche nicht nur den Erdball zersprengt, sondern das All, die ganze Schöpfung, auseinanderreißt. . . . Ich bin eine zu komplizierte Maschine. Ich kann den mathematischenn Prozess nicht durch führen, wenn ich mich an das gewöhnliche Stromnetz anschließe . . . Ich muß hinaus." (I must calculate the formula, the powerful formula that will blow up not only the globe but the universe, the whole creation, rip everything apart, I am too complicated a machine. I can't carry through the mathematical process if I connect myself to the usual power supply system. . . . I must get out) (40–41).

28. "Alle Lebenserscheinungen zu umfassen, mathematisch und berechenbar zu umfassen, weil er durch diese Fülle mathematisierter Weltkenntnis zur Totalität des eigenen Lebens gelangen werde" (Hermann Broch, *Die unbekannte Große*, in *Gesammelte Werke*, ed. Ernest Schönwiese, vol. 10 [Zurich: Rhein-Verlag, 1961], 76).

29. "Innerhalb des Mathematischen bloß von jenem Menschen etwas geleistet werden könne, der sich von allem Sündigen oder wie man es sonst nennen wollte, *rein* hielt" (ibid., 91).

30. "Die rationale und wissenschaftliche Erkenntnis bloß einen Teil einer größeren und zugleich einfältigeren Erkenntnis darstellt, einer wahrhaft mystischen Erkenntnis, die beweislos und doch evident ist, wo sie Leben und Tod, Rationales und Irrationales umschließt" (Rational and scientific knowledge describes merely a part of a greater and, at the same time, simple knowledge, a truly mystical perception which is unprovable and yet evident, where life and death, the rational and the irrational come together) (Hermann Broch, "Grundzüge des Romans *Die unbekannte Große*," in *Gesammelte Werke*, vol. 10, 170).

31. Compare, for example, the following passage in Broch's essay on James Joyce: "The theory of relativity has discovered, however, that there is a source of error beyond the error in principle: namely the act of viewing itself, the observation in itself. It concluded, therefore, that in order to avoid this source of error, . . . a theoretical unity of physical object and physical subject must be created" (Herman Broch, "James Joyce und die Gegenwart," in *Hermann Broch, Dichten und Erkennen—Essays*, vol. 1, ed. Hannah Arendt [Zurich: Rhein-Verlag, 1955], 197).

32. Stephen Sewell, *Welcome the Bright World* (Sydney: Alternative Publishing Co-op, 1983), 24.

33. "Wir wähnen eine Schatzgrube wunderbarer Schätze entdeckt zu haben, und wenn wir wieder ans Tageslicht kommen, haben wir nur falsche Steine und Glasscherben mitebracht; und trotzdem schimmert der Schatz im Finstern unverändert" (M. Maeterlinck, "Moralité mystique," from *Le Trésor des humbles*, pt. 4, quoted in translation in Robert Musil, *Die Verwirrungen des Zöglings Törleß* [Reinbek bei Hamburg: Rowalt Taschenbuch Verlag, 1959]).

34. To some extent this was true also in Musil's first play, *Die Schwärmer* (The visionaries), in which the protagonist, Thomas, was a cold, rational scientist married to a passionate, sensuous woman, Maria.

35. Robert Musil, *Der Mann ohne Eigenschaften* [1930–78], translated as *The Man without Qualities* by Eithne Wilkins and Ernst Kaiser, 3 vols. (London: Secker & Warburg, 1953–60), 1:296.

36. It clearly concerned Musil greatly, since it demanded not only seven hundred

pages of *Der Mann ohne Eigenschaften* but the major part of his fiction for its elucidation.

37. Asimov was considerably influenced at the beginning of his career by John W. Campbell, the editor of *Astounding Science Fiction*, where most of his robot stories were published (see Isaac Asimov, *The Early Asimov, or Eleven Years of Trying* [1972] [London: Gollancz, 1973]; for a discussion of Campbell's views see chapter 11).

38. Isaac Asimov, "On Computers," *DP Solutions*, Apr. 1975, 2. A similar view had been put forward even earlier by Mary Boole, the wife of the mathematician George Boole. She regarded Babbage and Jevons as the two greatest benefactors of humanity the century had produced, for they demonstrated "that calculation and reasoning, like weaving and ploughing, are work, not for human souls, but for clever combinations of iron and wood" (quoted in John Cohen, *Human Robots in Myth and Science* [London: George Allen & Unwin, 1966], 113).

39. The Three Laws of Robotics prescribe that (1) a robot may not injure a human being, or, through inaction, allow a human being to come to harm; (2) a robot must obey the orders given it by human beings except where such orders would conflict with the First Law; and (3) a robot must protect its own existence as long as such protection does not conflict with the First or Second Law.

40. Isaac Asimov, "Catch that Rabbit" [1944], in *I, Robot* (St. Albans, Hertfordshire: Panther, 1968), 90.

41. Isaac Asimov, "Evidence," in ibid., 171.

42. At the end of *Vulcan's Hammer* one of the characters poses the problem in precisely these terms: "We humans . . . were pawns of those two things [machines]. They played us off against one another like inanimate pieces. The things became alive and the living organisms were reduced to things. Everything was turned inside out, like some terrible morbid view of reality" (Philip K. Dick, *Vulcan's Hammer* [New York: Ace Books, 1960], 153).

43. For a detailed discussion of this aspect of Dick's work see Patricia S. Warrick, *The Cybernetic Imagination in Science Fiction* (Cambridge, Mass.: MIT Press, 1980), 206–30.

44. Christa Wolf, "Self-Experiment: Appendix to a Report" [1973], trans. Jeanette Clausen, *New German Critique* 13 (Winter 1978), 115.

45. "Dies waren ja Menschen auf einer Isolierstation, ohne Frauen, ohne Kinder, ohne Freunde, ohne andere Vergnügen als ihre Arbeit, strengsten Sicherheits- und Geheimhaltungsvorschriften unterworfen; . . . Sie kennen, habe ich gelesen, nicht Vater noch Mutter. Nicht Bruder noch Schwester. Nicht Frau noch Kind (es gibt dort keine Frauen, . . . ! Ist diese beklemmende Tatsache Grund für die Computerliebe der jungen Leute? Oder ihre Folge?)" (Christa Wolf, *Störfall. Nachrichten eines Tag* [Frankfurt am Main: Luchterhand, 1988], 70).

46. The uncertainty principle, discovered by Werner Heisenberg in 1927, states that it is impossible to measure both the position and the momentum of a particle simultaneously with more than strictly limited precision. (The uncertainty in position multiplied by the uncertainty in momentum must always exceed Planck's constant.) This is a special case of the complementarity principle, enunciated by Nils Bohr in 1927, which states that an experiment on one aspect of a system of atomic dimensions destroys the possibility of learning about a "complementarity" aspect of the same system.

47. The parallels between Wolters's story and that of Fuchs as described by Alan

Moorehead in *The Traitors* (New York: Charles Scribner's Sons, 1952) have been detailed by J. H. Talbot in "The Theme of the Scientist's Responsibility in the Nuclear Age in Contemporary German Drama" (Ph.D. diss., Boston University, 1968). Talbot traces the similarities not only in the events but also in the character and background of the two men.

48. Carl Zuckmayer, *The Cold Light*, trans. Elizabeth Montagu (typescript held by Columbia University Library, n.d.), act 1, sc. 1, p. 7. This statement must be received with some reservation, since it is voiced by Northon, the British intelligence agent whose task it is to persuade Wolters to confess to his espionage activities, who therefore has a strong ulterior motive.

49. The other scientists in the play span a wide spectrum of types and attitudes. The militaristic Ketterick is troubled by no scruples about the use of the atomic bomb against any enemy, real or potential; the sensitive Löwenschild, a man of conscience, profoundly regrets the destructive potential of the research he does but believes that this is the lesser of two evils and that adhering to one's conscience is the only morality.

50. It is in the cold light of dawn that the announcement of the Hiroshima bombing is made.

Chapter 14. *Scientia Gratia Scientiae:* The Amoral Scientist

1. Shiv Visvanathan, "Atomic Physics: The Career of an Imagination," *Alternatives: A Journal of World Policy* 10, no. 2 (1984), 208.

2. In his letter (drafted by Sachs) Einstein first mentioned the work of Fermi and Szilard and went on to link the chain reaction that could be initiated from the breakdown of uranium with resultant bomb production. He concluded with the warning: "I understand that Germany has actually stopped the sale of uranium from the Czechoslovakian mines which she has taken over. That she should have taken such early action might perhaps be understood on the ground that the son of the German Under-Secretary of State, von Weizsäcker, is attached to the Kaiser-Wilhelm-Institut in Berlin where some of the American work on uranium is now being repeated" (quoted in David Irving, *The German Atomic Bomb: The History of Nuclear Research in Nazi Germany* [New York: Simon & Schuster, 1967], 67).

3. See Brian Easlea, *Science and Sexual Oppression: Patriarchy's Confrontation with Women and Nature* (London: Weidenfeld & Nicolson, 1981); and idem, *Fathering the Unthinkable: Masculinity, Scientists, and the Nuclear Arms Race* (London: Pluto, 1983).

4. Russell McCormmach, *Night Thoughts of a Classical Physicist* (Cambridge, Mass.: Harvard University Press, 1982), 148–49.

5. Stanislaw Lem, "Robots in Science Fiction," in *SF: The Other Side of Realism*, ed. Thomas D. Clareson (Bowling Green, Ky.: Bowling Green University Popular Press, 1971), 320.

6. A term such as *morally* is, of course, a misnomer in relation to robots, who lack the power of choice. However, Asimov contrives to suggest a moral parameter by repeated and favorable comparison of his robots' actions with the generally selfish behavior of their attendants. This is a deliberate strategy on the part of Asimov, whose intention is to overcome his readers' fears and resistance to the notion of widespread introduction of robots.

7. Karel Capek, *R.U.R.: A Fantastic Melodrama*, trans. Paul Selver (New York: Dou-

bleday, Page & Co., 1923). Strictly speaking, the robots in Capek's play are not robots (mechanical figures) but creatures constructed from a chemical compound resembling protoplasm. They are thus, in modern science fiction parlance, androids.

8. The name Rossum is derived from the Czech word *rozum* (reason).

9. H. G. Wells's *The Sleeper Awakes* and "A Story of the Days to Come" and Evgeny Zamyatin's *We* had drawn on this theme.

10. "Ein Wort, das zu einer unserer gängigen Phrasen geworden ist, nämlich 'Wissen ist Macht,' gewann hier einen neuen, unmittelbaren, gefährlichen Sinn" (A saying that has become one of our current phrases, namely, "Knowledge is power," acquired here a new, immediate, and dangerous meaning) (Ernst Jünger, *Gläserne Bienen* [Stuttgart: Ernst Klett Verlag, 1957], 79).

11. He is explicitly called at various times "magician," "Faust," "enchanter," "prophet," "wizard," "sorcerer," "creator," "benefactor of dogs," "priest," "shaman," "robber," "chieftain," "bank robber," and "vampire" (Mikhail Bulgakov, *The Heart of a Dog* [London: Collins & Harvill, 1968]).

12. Diana Burgin remarks: "These details suggest his participation in an operatic, pagan rite, which contradicts the scientific seriousness of the occasion. [He] is presented here not as a genuine godhead, but as an actor playing the role of high priest in an opera" (Diana L. Burgin, "Bulgakov's Early Tragedy of the Scientist-Creator: An Interpretation of *The Heart of a Dog*," *Slavic and Eastern European Journal* 22 [1978], 499).

13. He lectures his assistant, who thinks that the experiment failed only because the brain used was that of a criminal and that it would have been a triumph if he had used the brain of a Spinoza: "This, doctor, is what happens when a researcher, instead of keeping in step with nature, tries to force the pace and lift the veil. Result— Sharikov. We have made our bed and now we must lie on it. . . . I, Philip Preobrazhensky, could perform the most difficult feat of my whole career by transplanting Spinoza's, or anyone else's pituitary and turning a dog into a highly intelligent being. But what in heaven's name for? That's the point. Will you kindly tell me why one has to manufacture artificial Spinozas when some peasant woman may produce a real one any day of the week? . . . Mankind, doctor, takes care of that. . . . My discovery . . . is worth about as much as a bent penny" (*Heart of a Dog*, 108).

14. Anatoly Dneprov, "Formula for Immortality," in *New Soviet Science Fiction*, ed. Theodore Sturgeon, trans. Helen Saltz Jacobson (New York: Collier Macmillan, 1979), 133–34. Cf. Dneprov's "The S*T*A*P*L*E Farm," in *Other Worlds, Other Seas*, ed. Darko Suvin (New York: Random House, 1970), which is also on the theme of genetic engineering; it is discussed below in chapter 15.

15. Albert Einstein, quoted in John S. Weltman, "Trinity: The Weapons Scientists and the Nuclear Age," *SAIS Review* 5, no. 2 (1985), 39.

16. Freeman Dyson, "Weapons and Hope," pt. 2, *New Yorker*, 13 Feb. 1984, 77.

17. Joseph Rotblat, "Leaving the Bomb Project," *Bulletin of the Atomic Scientists* 41 (1985), 18.

18. Carl Sandburg, "Mr. Attila," in *Complete Poems* (New York: Harcourt Brace, 1950).

19. Bertolt Brecht, *The Life of Galileo*, trans. Desmond I. Vesey, 2d ed. (London: Methuen, 1965), 8.

20. "Zum Schluß ist [Galileo] ein Förderer der Wissenschaften und ein sozialer Verbrecher. . . . das wissenschaftliche Werk wiegt das soziale Versagen nicht auf"

(quoted in Käthe Rülicke, "Bemerkungen zur Schluß-Szene," in Bertolt Brecht, *Materialen zu Brechts "Leben des Galilei"* [Frankfurt: Suhrkamp Verlag, 1963], 106, 97).

21. Bertolt Brecht, "Notes" on *The Life of Galileo*, in *Plays*, 2d ed., vol. 1 (London: Methuen, 1965), 340.

22. Arthur Koestler, *The Sleepwalkers* [1959] (London: Penguin, 1988).

23. Brecht held that the audience, far from suspending its disbelief, should never be encouraged to surrender its separation from the events taking place on the stage. He believed that this sense of alienation (*Verfremdungseffekt*) was essential if the audience was to be provoked into thinking objectively about the issues presented in the play.

24. C. P. Snow, *The New Men* (London: Macmillan, 1954), 46.

25. U.S. Atomic Energy Commission [USAEC], *In the Matter of J. Robert Oppenheimer* (Washington, D.C.: USGPO, 1954), 251. It was Oppenheimer who named the atomic test site "Trinity," after John Donne's Holy Sonnet, "Batter my heart, three-person'd God," and who, on witnessing the first atomic bomb demonstration, was moved to recall two passages from the *Bhagavad Gita:* "The radiance of a thousand suns / which suddenly illuminate the heavens / all in one moment—thus / the splendour of the Lord"; and "And I am Death, who taketh all, / who shatters worlds."

26. USAEC, *In the Matter of J. Robert Oppenheimer: Transcript of Hearing before Personnel Security Board and Text of Principal Documents and Letters* (Cambridge, Mass., 1971), lx.

27. Heinrich Schirmbeck, *Ärgert dich dein rechtes Auge*, translated as *The Blinding Light* by Norman Denny (London: Collins, 1960), 341–42.

28. Heinz von Cramer, *Die einzige Staatsvernunft in Konzession des Himmels* (Hamburg: Hoffmann & Campe, 1961). The major atomic tests were held at Emu Field in 1953 and at Maralinga in 1956. The bombs ranged in size from 1 kiloton to 25 kilotons, the largest being greater than the bomb dropped on Hiroshima.

29. Daniel Lang, "Ex-Oracles," *Harper's Magazine* 245 (Dec. 1972), 34.

30. Freeman Dyson, quoted in Spencer R. Weart's review of the film *The Day after Trinity*, in *Bulletin of the Atomic Scientists* 38 (1982), 41–42.

31. Pearl S. Buck, *Command the Morning* (New York: John Day, 1959). Buck also wrote *A Desert Incident* (1959), a play having the same theme and set in the same place.

32. Dexter Masters, *The Accident* (London: Cassell, 1955), 18.

33. Heiner Kipphardt, *In the Matter of J. Robert Oppenheimer*, trans. Ruth Speirs (London: Methuen, 1967), pt. 2, sc. 2, 82.

34. Martin Cruz Smith, *Stallion Gate* (Glasgow: Harper Collins, 1992), 167, 240.

35. Because of a Nazi decree, Domagk was required to decline the prize, but in 1947 he received a gold medal in lieu of the prize money.

36. Richard Gordon, *The Invisible Victory* (Harmondsworth: Penguin, 1979), 264.

37. Gordon may have been unaware that such moral schizophrenia was not confined to Nazi Germany. In 1943 Robert Oppenheimer had asked Enrico Fermi to explore the feasibility of using radioactive food poisoning as a weapon. Oppenheimer has been quoted as saying, "I think we should not attempt a plan unless we can poison food sufficient to kill a half million men" (see M. Sherwin, "How Well They Meant," *Bulletin of the Atomic Scientists* 41 [1985], 8).

38. Wernher von Braun (1912–77) attended the Technische Hochschule in Berlin

and was part of the team at Peenemünde that developed the V-2 rocket, the forerunner of all modern rockets. After the war he emigrated to the United States, and in 1950 he was named director of the U.S. Army Ballistic Missile Agency at Huntsville, Alabama. There von Braun and his team developed the launch vehicles for a whole series of rockets, from *Jupiter* to the *Saturn V*.

39. Kurt Vonnegut, *Cat's Cradle* (London: Gollancz, 1965). The title refers to a children's game in which the strings can form elaborate and ever-changing patterns or can become hopelessly knotted.

40. This is reminiscent of Summerfield's discovery that would progressively boil all the water on the planet (see chapter 12 above).

41. However, lest we should begin to gain faith in scientists, Vonnegut tells us that the humanitarian von Koenigswald has been an official at Auschwitz and thus has a "terrible deficit . . . in his kindliness account" (159).

42. Edward Albee, *Who's Afraid of Virginia Woolf?* (Harmondsworth: Penguin, 1979), 113.

43. William Broad, *Star Warriors* (New York: Simon & Schuster, 1985), 13, 85. Broad was a science journalist for the *New York Times*.

44. Christa Wolf, *Störfall. Nachrichten eines Tag* (Frankfurt am Main: Luchterhand, 1988), 70. "Die sich—getrieben . . . von der Hyperaktivität bestimmter Zentren ihres Gehirns—nicht dem Teufel verschrieben haben . . . , sondern der Faszination durch ein technisches Problem."

45. "Was sie kennen, diese halben Kinder mit den hochtrainierten Gehirnen, mit ihrer ruhelosen, Tag und Nacht fieberhaft arbeitenden linken Gehirnhälfte—was sie kennen, ist ihre Maschine. Ihr lieber geliebter Computer. . . . Was sie kennen, ist das Ziel, den atomgetriebenen Röntgenlaser zu konstruieren, das Kernstück jener Phantasie von einem total sicheren Amerika durch die Verlegung künftiger Atomwaffenschlachten in den Weltraum." Hagelstein's girlfriend, Josephine Stein, on the other hand, campaigned against the project and left him when he refused to stop his research. Wolf explores Stein as a new moral Gretchen figure, not all-accepting like Faust's Gretchen but opposing her Hagelstein-Faust. Wolf was also heartened to learn some five months later that Hagelstein had finally left the Star Wars project.

46. Saul Bellow, *Mr. Sammler's Planet* (Harmondsworth: Penguin, 1978), 176.

47. Subtitled "Blueprints for a World Revolution," Wells's treatise *The Open Conspiracy* (1928) offered a program for setting up a utopian society based on efficiency and order by means of a science-based education system.

48. Norbert Wiener's best-known books were *I Am a Mathematician* (1956), *The Human Use of Human Beings* (1950), and *Cybernetics* (1948).

49. Norbert Wiener, *The Human Use of Human Beings: Cybernetics and Society* (New York: Avon Books, 1971), 244.

50. Mathilde von Zahnd, for example, uses this argument in Dürrenmatt's play *Die Physiker*: "Alles Denkbare wird einmal gedacht" (Everything conceivable is conceived once).

Chapter 15. The Scientist's Science Out of Control

1. *Fail-Safe* (1964), by Eugene Burdick and Harvey Wheeler, also introduced the figure of the scientist-strategist, who advises the U.S. government to launch an all-out

attack on Russia, and another scientist who disregards all human considerations in order to develop military hardware.

2. Harle Owen Cummins, "The Man Who Made a Man," *Welsh Rarebit Tales* (Boston: Mutual Books, 1902), 10.

3. Cf. ibid., 5: "I am going to do what has never been done in the history of the world except by God himself."

4. For Hoffmann's "Der Sandmann" see chapter 6 above; for Poe's "Maelzel's Chess-Player" and Bierce's "Moxon's Master" see chapter 10.

5. Hamilton was almost certainly influenced by A. Merritt's story "The Metal Monster" (1920), in which the monsters are not made by the scientists but are found by four American explorers in Asia. They are interesting because their alleged technology—electric impulses passing through metals and crystals to produce artificial intelligence—prefigures that of the computer.

6. Some other examples are Karel Capek's *Krakatit* (1925) (see also chapter 12 above) and E. C. Large's *Sugar in the Air* (1937) in which Charles Pry, an idealistic, unemployed chemical engineer becomes involved in working for a company that aspires to photosynthesize sugar in vitro by means of short-wave radiation. Pry manages to devise a method for producing "sunsap," an artificial sugar solution, but soon finds himself caught up in a dubious market enterprise. Nevertheless, he is finally able to retain his idealism and control the irrational and immoral practices of the commercial world.

7. Karel Capek, "The Meaning of *R.U.R.*," *Saturday Review* 136 (21 July 1923), 79.

8. Anatoly Dneprov, "The S*T*A*P*L*E Farm," in *Other Worlds, Other Seas*, ed. Darko Suvin (New York: Random House, 1970), 188.

9. There are three collections of Lem's cybernetic science fiction: *Tales of Pirx the Pilot* (1968), about the encounters of an astronaut Pirx with near-human robots; *The Robot Fables* (1964), a retelling of animal fables from the point of view of robots; and *The Cyberiad: Fables for the Cybernetic Age* (1964).

10. Stanislaw Lem, *The Futurological Conference* (New York: Avon Books, 1974), 20.

11. Stanislaw Lem, *His Master's Voice*, trans. Michael Kandel (San Diego: Harcourt Brace Jovanovich, 1983). The parallel with the discovery of pulsars at Cambridge in 1967 (see the discussion in relation to *A for Andromeda* below) is striking, though it is impossible that Lem could have heard of this closely guarded secret at the time of writing *His Master's Voice*.

12. Norman Spinrad's novel *Songs from the Stars* (1980) also explores this theme of "white" versus "black" science.

13. Richard E. Ziegfeld, *Stanislaw Lem* (New York: Frederick Ungar, 1985), 111.

14. Philip K. Dick, "The Preserving Machine," in *The Preserving Machine and Other Stories* (London: Gollancz, 1971), 14.

15. Fred Hoyle and John Elliot, *A for Andromeda* (London: Corgi Books, 1962), 92.

16. Hoyle and Elliot's suggestion of the insidious means whereby apparently "pure" research—and what science could be more "pure" than radio astronomy?—may attain a political or military significance was paralleled in the words of another Cambridge astronomer, the Nobel prize winner Sir Martin Ryle, a contemporary of Hoyle's. At the end of his life Ryle, who had been involved in the development of radar, wrote: "At the end of World War II I decided that never again would I use my scientific knowledge for military purposes; astronomy seemed about as far removed

as possible. But in succeeding years we developed new techniques for making very powerful radio telescopes; these techniques have been perverted for improving radar and sonar systems. A sadly large proportion of the PhD students we have trained have taken the skills they have learnt in these and other areas into the field of defence. . . . To so many it is simply an intriguing scientific problem; the morality and the responsibility are pushed aside, and the politicians make the decisions. I am left at the end of my scientific life with a feeling that it would have been better to have become a farmer in 1946" (Ryle to Professor Carlos Chagas, of the Pontifical Academy of Sciences, 24 Feb. 1983, quoted with the permission of Lady Ryle).

17. Christa Wolf, *Störfall. Nachrichten eines Tag* (Frankfurt am Main: Luchterhand, 1988), discussed in relation to the amoral scientist in chapter 14.

18. "Ob er Optimist oder Pessimist gewesen ist, hat er gesagt: Nein. Keinesfalls wird der Reaktorkern durchsmelzen. Oder: Aber doch. Doch, doch: Auch das ist gar nicht ausgeschlossen. Dann wäre jene Erscheinung zu erwarten gewesen, die der Humor der Wissenschlaftler so anschaulich 'Chinasyndrom' getauft hat" (ibid., 11).

19. Ibid., 112.

20. As early as 1939, the work of Enrico Fermi and Leo Szilard on nuclear chain reactions was being repeated by von Weizsäcker and others at the Kaiser-Wilhelm-Institut in Berlin.

21. Hans Rehberg, *Johannes Keppler: Schauspiel in drei Akten* (Berlin: S. Fischer Verlag, 1933). Kepler is persecuted by the Protestants for supporting the Gregorian—and hence Catholic—calendar, and by the Catholics for advocating the Copernican view of the universe, which was interpreted as undermining religious authority.

22. "Gott hat mich geleitet und was der ewige Gott mir schenkte, soll ich fortwerfen eines Herzogs wegen?" (ibid., 78).

23. "Eine Wahrheit für die man nicht bereit ist zu sterben, ist keine. . . . Ein Mensch, der nicht bereit ist, für die Wahrheit zu sterben, ist keiner.—Die Wahrheit verlangt Opfer, wie alles Grosse Opfer von uns verlangt. Ich wollte epikuräisch um das Opfer herumkommen. Das geht aber nicht, wenn man die Sache ernst nimmt. Humoristischerweise wäre ich dabei trotzdem beinahe Märtyrer wider Willen für die Wahrheit geworden. . . . Jetzt ist es etwas anderes. Jetzt will ich!" (Max Brod, *Galilei in Gefangenschaft* [Winterthur: Mondial Verlag, 1948], 774).

24. In another sense, Zwillinger's play proved curiously prophetic, since in 1992 a papal edict "cleared" Galileo of heresy.

25. Nigel Balchin, *The Small Back Room* (London: Collins, 1949), 190.

26. Nigel Balchin, *A Sort of Traitors* (London: Collins, 1949). "Mine eyes are full of tears, I cannot see; / And yet salt water blinds them not so much / But they can see a sort of traitors here. / Nay, if I turn mine eyes upon myself, / I find myself a traitor with the rest; / For I have given here my soul's consent" (*King Richard II* 4.1.244–49).

27. Freedom to publish scientific results is regarded, of course, as a fundamental prerequisite of research, and this became a crucial issue during the Second World War. Only more recently has it been threatened from other directions, as, for example, by commercial secrecy requirements.

28. Upton Sinclair, *A Giant's Strength* (London: T. Werner Laurie, 1948), 20.

29. Sinclair's scenario predates the actual enquiry into the activities of J. Robert Oppenheimer in May 1954.

30. "Sie tun das Falsche! Sie hoffen!" (Hans Henny Jahnn, *Trümmer des Gewissens*

(Der staubige Regenbogen) (Ruins of conscience: The dusty rainbow), ed. Walter Muschg [Frankfurt am Main: Europäische Verlagsanstaldt, 1961], 199).

31. Buzzati suggests a certain pride in Einstein's desire to finish his work before he dies. Despite the devil's assurance that he will know "as soon as you get to the other side," he pleads, "Here my work has considerable interest" (Dino Buzzati, "Appointment with Einstein," in *Restless Nights*, trans. Lawrence Venuti [Manchester: Carcanet, 1984], 12).

32. Compare Douglas Stewart's presentation of Rutherford, aware of the destructive potential of his discovery of nuclear fission but persuading himself that it would be used only for good (see chapter 16, n. 31).

33. "Förderungen an eine imaginäre Welt, Förderungen, nicht zu sündigen nach dem Sündenfall" (Friedrich Dürrenmatt, *Theaterschriften und Reden* [Zurich: Im Verlag der Arche, 1966], 275).

34. Möbius's predicament is further symbolized by the fact that when the single Möbius loop is cut longitudinally one-third of the distance from one edge, it forms two interlocking Möbius loops, one twice the length of the other.

35. "Für ihn ist das Irrenhaus keine poetische Metapher, sondern die höchst normale Endstation einer realen Entwicklung. Er untersucht die verbreitetste Form unseres noch nicht diagnostizierten Wahnsinns, die Jagd nach Atomphysikern" (Walter Muschg, "Dürrenmatt und die Physiker," *Moderna Sprak* 56 [1961], 280).

36. Friedrich Dürrenmatt, *The Physicists*, trans. James Kirrup (London: Samuel French, 1963), 42.

37. As in the setting of the asylum, Dürrenmatt clearly intends this "mad" statement to have relevance to the real world.

38. The secret hearing lasted from 12 April to 14 May 1954, and the play is distilled from the three thousand pages of typed records published by the U.S. Atomic Energy Commission in May 1954. Oppenheimer himself accused Kipphardt of misinterpretation but approved Jean Vilar's stage version of the hearings, *Le Dossier Oppenheimer*, which is based on Kipphardt's play (see, e.g., Franz Vossen, "Jean Vilar in der Sache J. Robert Oppenheimer," *Süddeutsche Zeitung*, 15 Dec. 1964, 12).

39. Kipphardt does not make clear in the play that at least two petitions were drawn up by scientists working on the bomb projects urging that the bomb not be deployed, at least until a public demonstration of its power could be held. Oppenheimer's committee effectively ignored these petitions. At the time of the petitions, Oppenheimer did not feel that the scientists were qualified to determine whether, or how, the bomb should be used (see Robert Jungk, *Brighter Than a Thousand Suns*, trans. James Cleugh [New York: Harcourt, Brace & Co., 1958], 186).

40. Heiner Kipphardt, *In the Matter of J. Robert Oppenheimer*, trans. Ruth Speirs (London: Methuen, 1967), 102-3.

41. Kipphardt also explores the relative importance of the public and the private roles of scientists. Oppenheimer sees no moral inconsistency in having active communists as personal friends while being an adviser to the government on atomic matters, but it is clear that this constitutes an anomaly for most of those present. It leads Ward Evans, a member of the Personnel Security Board and a professor of chemistry, to ask a question that recurs strategically and with increasing pertinence during the enquiry: "What kind of people are physicists?" In his reply to this question Oppenheimer explains how he became interested in politics, especially Marxism,

during the Depression: "Physicists are interested in new things. They like to experiment, and their thoughts are directed towards changes. In their work, and also in political matters. . . . You cannot produce an atomic bomb with irreproachable, that is, conformist ideas. Yes-men are convenient but ineffectual" (ibid., 34, 36).

42. In 1983 there was a reunion of the Los Alamos scientists to commemorate their achievement of forty years before. Isidor Rabi, then eighty-six, presented a paper entitled "We meant well," in which he lamented that the "nations are now lined up like people before the ovens of Auschwitz while we are trying to make the ovens more efficient. . . . We meant well. . . . We gave it away. We gave the power away to people who didn't understand it and now it's gotten out of our hands" (I. Rabi, quoted in Gregg Herken, *Counsels of War* [New York: Knopf, 1985]).

43. Ibid., 88–89. Cf. Bethe's statement in 1954: "I am afraid my inner troubles stayed with me and are still with me and I have not resolved this problem. I still have the feeling that I have done the wrong thing. But I have done it" (quoted in Jungk, *Brighter Than a Thousand Suns*, 291).

44. Kipphardt, *In the Matter of J. Robert Oppenheimer*, 106–7.

45. Kipphardt admits that he has selected from the report of the hearing but claims that this was only in a manner necessary for artistic presentation. He denies that he has distorted the truth of the proceedings (see his preface in ibid.).

46. See, e.g., Evans's soliloquy, "Perhaps I should have turned down this appointment . . . I cannot reconcile these interrogations with my idea of science" (ibid., 24).

47. "Ich wollte das Leben erforschen, seinen Aufbau ergründen, seinen Geheimnissen nachgehen" (Friedrich Dürrenmatt, *Der Mitmacher. Komödie* [Basel: Reiss, 1973], 14).

48. His walk is a "run-and-shuffle to the door, bent over at an angle of forty-five degrees, centre of gravity in advance of his feet, like an old style music hall comedian coming on to the stage" (C. P. Snow, *In Their Wisdom* [London: Macmillan, 1974], 29).

49. Although Sedgwick's operation towards the end of the novel is a triumphant success surgically, there is little hope that it has accomplished anything of lasting value, since Sedgwick is already so old.

50. Snow's Godwin Lectures at Harvard University, 1960, "On the Essentials of Free Government and the Duties of the Citizen," were later published as *Science and Government* (Cambridge, Mass: Harvard University Press, 1961).

51. He falls further from grace with the university authorities when, during a public lecture at which he was expected to talk about his experiences in 1870 in order to promote patriotism and student enlistment, he reaffirms the nonpolitical nature of science and recalls how Max Planck also had refused to bolster nationalism and instead appealed to all scientists "to persevere at their scientific posts in the midst of the temporary upheavals" (Russell McCormmach, *Night Thoughts of a Classical Physicist* [Cambridge, Mass.: Harvard University Press, 1982], 14).

52. Howard Brenton, *The Genius* (London: Methuen, 1983), 21.

53. The responses are predictable: the vice-chancellor regards it as a student prank in poor taste; the bursar tries to persuade Lehrer to be respectable; and a left-wing student turns out to be ultraconservative and goes off to work for MI6.

54. Gilly posts her equations through the letter box of the Soviet Embassy, explaining to a passing cat that she has "done something at least."

55. The Antrobuses' maid Sabina, a realist, says in wonderment, "Mr. and Mrs.

Antrobus! Their heads are full of plans and they're as confident as the day they began" (Thornton Wilder, *The Skin of Our Teeth*, in *Three Plays by Thornton Wilder* [London: Longmans Green Co., 1958], 250).

56. A similar humorous affirmation of the ordinary individual in the face of the inadequacy of scientists emerges in "To Explain Mrs. Thompson," a story by Philip Latham (in real life the astronomer R. S. Richardson). Latham presents a group of internationally famous astronomers who pride themselves on their ability to produce explanations for any phenomena. However, confronted by the appearance in the Andromeda nebula of a group of new stars that slowly assume the lines of a female face, they are unable to produce a suitable theory to fit the facts. The face, reminiscent of a comic strip character, turns out to be identical with that of a recently deceased American woman, whose grieving husband sees the appearance of his wife's face in the sky as some tasteless joke. The astronomers fail to produce even the most rudimentary explanation. "'I often wonder if theoretical physics can ever really explain anything,' said Slater soberly. 'In the last analysis, I wonder if theoretical physics can ever do more than merely describe. . . . I can spin endless stories for you that describe, but when you ask me for one that *explains*—that is another matter entirely'" (Philip Latham, "To Explain Mrs. Thompson," in *The Secret Dreamers*, ed. F. Pohl [London: Gollancz, 1963], 103–4). Latham, like Wilder, suggests that science can never explain the sheer survival of the ordinary individual but that this hardly matters, since we can, he implies, do without the theories of science, thereby rendering the limitations of the scientists of little account.

57. "Waren wir Monster, als wir um einer Utopie willen—Gerechtigkeit, Gleichheit, Menschlichkeit, für alle—, die wir nicht aufschieben wollten, diejenigen bekämpften, in deren Interesse diese Utopie nicht lag (nicht liegt), und, mit unseren eigenen Zweifeln, diejenigen, die zu bezweifeln wagten, daß der Zweck die Mittel heiligt?" (Wolf, *Störfall*, 37).

58. "That we should have the choice only of living with radioactivity or with the death of the forests" (Daß wir nur die Wahl haben sollen, mit der Radioactivität oder mit dem Waldsterben zu leben) (ibid., 79).

59. The narrator's pet rat extrapolates from the present to the ultimate demise of humanity: "Vast plains infested with garbage, beaches strewn with garbage, valleys clogged with garbage . . . You and your works wrapped in clear plastic, sealed into vacuum bags, moulded in synthetic resin, you in chips and clips: the human race that was" (Günter Grass, *The Rat*, trans. Ralph Manheim [London: Secker & Warburg, 1987], 6).

Chapter 16. The Scientist Rehabilitated

1. See, e.g., Robert Millikan, *Evolution in Science and Religion* (New Haven: Yale University Press, 1927). Millikan received the Nobel prize for physics in 1923 for his measurement of the charge on the electron and for his work on the photoelectric effect.

2. For example, Mervyn LeRoy's film *Madame Curie* [1943] and John Glenister's *Marie Curie* [1977].

3. Naturalism involved the inclusion of copious material details relating to the milieu of the action and stressed psychological determinism in the characterization.

4. Émile Zola, *Doctor Pascal* [1893], trans. Vladimir Kean (London: Elek Books, 1957), 34.

5. Émile Zola, *Paris* [1898], trans. E. A. Vizetelly, 2d ed. (London: Chatto & Windus, 1898), 486, 116.

6. Sacré Coeur was still being built when Zola was writing the novel. It took from 1875 to 1914 to complete and was not in fact consecrated until after the end of the First World War.

7. Lewis wrote: "I thought of some day having a doctor hero. . . . one who could get beneath the routine practice into the scientific foundations of medicine—one who should immensely affect all life" (Sinclair Lewis, quoted in Mark Schorer, Afterword to *Arrowsmith: Sinclair Lewis* [1925] [New York: New American Library, 1961], 432).

8. De Kruif later made a list of the names of the actual persons who had suggested characters "as well as I can now remember them": for Martin Arrowsmith, R. G. Hussey, later professor of pathology at Yale; for Max Gottlieb, F. G. Novy, of the University of Michigan, and Jacques Loeb, of the Rockefeller Institute; for Terry Wickett, T. J. LeBlanc and J. H. Northrop; for DeWitt Tubbs, Simon Flexner; and for Almus Pickerbaugh, William de Kleine, medical director of the Red Cross.

9. The much-debated claims for cold fusion in 1988 and for the alleged discovery of a planet circling a pulsar in 1991 are two recent cases of the same pressure to publish prematurely.

10. One of the moral dilemmas that confront Arrowsmith occurs in the sequence set in the West Indies, when Arrowsmith goes to test the efficacy of his bacteriophage against an outbreak of bubonic plague. The experiment involves refusing inoculation to half the victims as a control, but Arrowsmith is too humane to adhere to this, and as a result the experiment is inconclusive. De Kruif's research, published in *Microbe Hunters*, also formed the basis of Sidney Howard's play *Yellow Jack* (1934), which celebrates the heroism of doctors fighting yellow fever in Cuba (see *Three Plays about Doctors*, ed. Joseph Merrand [New York: Washington Square, 1961]).

11. He achieves his highest moral statement when, in response to Gottlieb's credo, quoted above, he enunciates his own self-dedicatory prayer: "God give me the unclouded eyes and freedom from haste. God give me a quiet and relentless anger against all pretense and all pretentious work and all work left slack and unfinished. God give me a restlessness whereby I may neither sleep nor accept praise till my observed results equal my calculated results or in pious glee I discover and assault my error. God give me strength not to trust to God!" (Lewis, *Arrowsmith*, 269). The pretentiousness of this self-indulgent rhetoric is apparently not part of Lewis's estimate.

12. We hear also of Gottlieb's callousness towards his wife when his research assumes priority and of his injunction to Arrowsmith to ignore the pleas of the plague sufferers and inoculate only the numbers required for the statistical experiment. For an extended discussion of Loeb's dedication to quantitative method see Charles E. Rosenberg, "Martin Arrowsmith: the Scientist as Hero," *American Quarterly* 15 (1963), 452.

13. "The sense one has of exquisite and wonderful rhythms—just beyond sight and sound! . . . a different scheme of harmonies . . . as if the whole world was fire and crystal and a-quiver" (H. G. Wells, *Marriage*, 2d ed. [London: Unwin, 1927], 130).

14. "He saw no ultimate truth in this seething welter of human efforts, no tragedy

as yet in its defeats, no value in its victories. It had to go on, he believed, until the spreading certitudes of the scientific method pierced its unsubstantial thickets, burst its delusive films, drained away its folly" (ibid., 237).

15. For Leo Szilard's acknowledgment of a debt to Wells's novel for both information and moral guidance see above, chapter 11, n. 24; and L. Szilard, "Reminiscences," in *Perspectives in American History*, vol. 2 (Cambridge, Mass.: Harvard University Press, 1968), 102.

16. Robert Nichols and Maurice Browne, *Wings Over Europe* (New York: Corvici-Friede, 1929), 73.

17. Peter Kapitza, a Soviet citizen, worked with Rutherford in the Cavendish Laboratory, Cambridge, becoming a fellow both of Trinity College and of the Royal Society before returning to Moscow as director of the Institute of Physical Problems.

18. Bertolt Brecht's "Der Mantel des Ketzers," translated into English as "The Heretic's Coat" by Yvonne Kapp, Hugh Morrison, and Anthony Tallow in *Bertolt Brecht: Short Stories, 1921–1946* (London: Methuen, 1983), concerns the last months in the life of Giordano Bruno, who was executed by the church on charges of heresy that included his cosmic theory. Bruno is presented in an entirely heroic light. Not only is the charge against him shown to be false and motivated by the disappointment of those who had hoped to gain illicit magical powers from him but Bruno's humanity is the characteristic most stressed. Even in his extreme suffering, he is concerned to ensure that his coat, his one possession of any value, is returned to the tailor, to whom he owes a debt.

19. Bertolt Brecht, "Notes" on *The Life of Galileo*, in *Plays*, 2d ed., vol. 1 (London: Methuen, 1965), 347.

20. The plague scene is unhistorical, being set between 1610 and 1616, whereas the plague did not reach Tuscany until 1632 (See Alfred D. White, "Brecht's *Leben des Galilei*: Armchair Theatre?" *German Life and Letters*, n.s. 27 [Jan. 1972], 126). Its inclusion illustrates Brecht's determination to absolve Galileo from the charge of cowardice.

21. It is perhaps significant that Selby has become a vibrations specialist by way of her study of music, a further indication of her holistic approach.

22. Charles Morgan, *The Burning Glass* (London: Macmillan, 1953), 11.

23. Ironically, one of the stories Campbell published in March 1944 was Cleve Cartmill's "Deadline," which contained a description of an atomic bomb so close to the real thing that FBI agents "visited" both Campbell and Cartmill on the suspicion that they were privy to classified material.

24. An early story by Stuart, "Atomic Power" (1934), epitomized this role. In this story the earth is undergoing entropic death, expanding into quiescence, with consequent panic on the part of its inhabitants. Enter a brilliant physicist who discovers atomic power, recharges the energy of the planet, and restores it to functional equilibrium. "He had found the secret of the vast power that would warm the frozen peoples and power the industry as Earth thawed out once more" (Don A. Stuart [John W. Campbell, Jr.], "Atomic Power," *Astounding Stories*, Dec. 1934, 250).

25. Anthony Burgess, Review of *Terminal Visions*, by Warren Wagar, *Times Literary Supplement*, 18 Mar. 1983, 256.

26. James Hilton, *Nothing So Strange* (Melbourne: Macmillan, 1948), 295.

27. No suspicion attached to Fuchs until late in 1949, and he made his confession,

the only real evidence against him, only in 1950. Aldridge's novel was published early in 1949. Recent evidence suggests that in any case Fuchs could not have been responsible for transferring any crucial information to the Soviet Union. The "secrets" that Fuchs passed on were already outdated, and every important assumption on which they were based was wrong. The Soviet scientists could have ascertained all the essential information about the Teller-Ulam version of the H-bomb by analyzing the fallout from the Mike test, just as the British scientists built their H-bomb using information from Soviet fallout given them by the United States to analyze (see Daniel Hirsch and William G. Mathews, "The H-Bomb: Who Really Gave Away the Secret?" *Bulletin of the Atomic Scientists* 46 [1990], 22–30). "Mike" was the code name for the first real thermonuclear test of the H-bomb at Eniwetok, in the Marshall Islands, in November 1952 (see Norman Moss, *Men Who Play God: The Story of the H-Bomb* [London: Gollancz, 1968], 55).

28. Mitchell Wilson was a physicist until 1945, when he abandoned that career to become a novelist. His earlier novel *Live with Lightning* (1949) also deals with the moral issues raised by physics.

29. See, e.g., Erika Gröger, "Der bürgerliche Atomwissenschaftler im englisch-amerikanischen Roman von 1945 bis zur Gegenwart," *Zeitschrift für Anglistik und Amerikanistik* 16, no. 1 (1968): 25–48.

30. Heinrich Schirmbeck, *Ärgert dich dein rechtes Auge*, translated as *The Blinding Light* by Norman Denny (London: Collins, 1960), 228.

31. Douglas Stewart, "Rutherford," in *Selected Poems* (Sydney: Angus & Robertson, 1973), 179.

32. At sixteen Tice had helped an escaped German prisoner, and in Hiroshima after the war he rejected the military platitudes and experienced "a revelation, nearly religious, that the colossal scale of evil could only be matched or countered by some solitary flicker of intense and private humanity" (Shirley Hazzard, *The Transit of Venus* [1980] [Harmondsworth: Penguin, 1982], 53).

33. Aldous Huxley, *Island* [1962] (Harmondsworth: Penguin, 1968), 219.

34. The net effect is little more than a reworking of the thesis Huxley had presented in *Literature and Science* (1963).

35. Ursula LeGuin, *The Dispossessed* [1974] (London: Panther, 1985), 9.

Implications

1. Mary Shelley's introduction to *Frankenstein* describes the discussion by Byron, Shelley, and Polidori of recent scientific theories that led directly to her nightmare and hence to the novel. For a more detailed analysis of contemporary scientific issues referred to in Frankenstein see Marilyn Butler, "The First *Frankenstein* and Radical Science," *Times Literary Supplement*, 9 Apr. 1993, 12–14.

2. Theodore Roszak, "The Monster and the Titan: Science, Knowledge, and Gnosis," *Daedalus* 103, no. 3 (1974), 31.

BIBLIOGRAPHY

Part 1: Works of Fiction with Scientist Characters

Akenside, Mark. "The Virtuoso" [1737]. In *A Book of Science Verse*, edited by W. Eastwood. London: Macmillan, 1961.

Albee, Edward. *Who's Afraid of Virginia Woolf?* [1962]. Harmondsworth: Penguin, 1979.

Aldiss, Brian. *The Dark Light Years: A Science-Fiction Novel*. London: Faber & Faber, 1964.

———. *Frankenstein Unbound*. New York: Random House, 1973.

———. *Moreau's Other Island*. London: Jonathan Cape, 1980.

Aldridge, James. *The Diplomat* [1949]. London: Bodley Head, 1950.

Andres, Stefan. *Die Sintflut: der graue Regenbogen*. Munich: R. Piper & Co. Verlag, 1959.

Aristophanes. *The Clouds*.

Asimov, Isaac. *I, Robot* [1967]. St. Albans, Hertfordshire: Panther, 1968.

Atlas, Martin. *Die Befreiung: ein Zukunftsroman*. Berlin: F. Dümmler, 1910.

Bacon, Francis. *New Atlantis* [1626]. In *The Works of Francis Bacon*, edited by James Spedding, Robert Leslie Ellis, and Douglas Denon Heath, 3:119–66. 14 vols. London: Longman & Co., 1857–74. Facs. ed. Stuttgart: Friedrich Fromann Verlag, 1962–64.

Balchin, Nigel. *The Small Back Room* [1943]. London: Collins, 1949.

———. *A Sort of Traitors*. London: Collins, 1949.

Balmer, Edwin, and Philip Wylie. *When Worlds Collide and After Worlds Collide* [1932]. New York: J. B. Lippincott Co., 1933.

Balzac, Honoré de. *The Elixir of Life* [1830]. In *La Comédie humaine*, vol. 42. London: Caxton, n.d.

———. *Grandeur et decadence de César Birotteau*. Paris: Calmann Levy, 1891.

———. *The Quest of the Absolute* [1834]. Translated by Ellen Marriage. London: Newnes, n.d.

Banville, John. *Doctor Copernicus*. New York: W. W. Norton & Co., 1976.

Behn, Aphra. *The Emperor of the Moon* [1687]. In *The Works of Aphra Behn*, edited by M. Summers. London: Heinemann, 1915.

Bellamy, Edward. *Dr. Heidenhoff's Process* [1880]. In *Edward Bellamy: Works*. London: Frederick Warne & Co. n.d.

————. *Looking Backward, 2000–1887* [1888]. Edited by John L. Thomas. Cambridge: Belknap, 1967.

Bellow, Saul. *The Dean's December*. London: Secker & Warburg, 1982.

————. *Mr. Sammler's Planet* [1969]. Harmondsworth: Penguin, 1978.

Benford, Gregory. *Timescape*. London: Gollancz, 1980.

Bierce, Ambrose. "Moxon's Master" [1893]. In *Science-Fiction Thinking Machines*, edited by Groff Conklin. New York: Vanguard, 1954.

Binder, Eando. "Adam Link's Vengeance." *Amazing Stories*, Feb. 1940.

Blake, William. "An Island in the Moon" [1784–85]. In *The Complete Writings of William Blake*, edited by Geoffrey Keynes. London: Oxford University Press, 1966.

Blish, James. *A Case of Conscience* [1959]. Reprint. Harmondsworth: Penguin, 1963.

————. *The Seedling Stars* [1957]. London: Arrow Books, 1972.

Bond, Edward. "Passion." *New York Times*, 15 Aug. 1971.

Borchert, Wolfgang. *Lesebuchgeschichten* [1946]. In *Draussen vor der Tür und ausgewählte Enzählungen*. Reinbek bei Hamburg: Rowohlt Verlag, 1956.

Braun, Günther, and Johanna Braun. *Ein objektiver Engel. Roman*. Berlin: Verlag Neuesheben, 1967.

Brecht, Bertolt. "The Experiment" [1939]. Translated by Yvonne Kapp. In *Short Stories, 1921–1946*, edited by J. Willett and R. Manheim. London: Methuen, 1983.

————. "The Heretic's Coat" [1939]. Translated by Yvonne Kapp. In *Short Stories, 1921–1946. See* Brecht, "The Experiment."

————. *The Life of Galileo* [1938–47]. In *Plays*, vol. 1. Translated by D. I. Vesey. 2d ed. London: Methuen, 1965.

Breggin, Peter. *After the Good War*. New York: Stein & Day, 1973.

Brenton, Howard. *The Genius* [1983]. Royal Court Writers. London: Methuen, 1985.

Brereton, Jane. "Merlin: A Poem by a Lady." In *The Rhetoric of Fiction*, edited by W. P. Jones. London: Routledge & Kegan Paul, 1966.

Broch, Hermann. *Die unbekannte Große* [1933]. In *Gesammelte Werke*, edited by Ernst Schönweise, vol. 10. Zurich: Rhein-Verlag, 1961.

Brod, Max. *Galilei in Gefangenschaft* . Winterthur: Mondial Verlag, 1948.

————. *Tycho Brahes Weg zu Gott*. Leipzig: Kurt Wolff Verlag, 1916.

Brown, Charles Brockden. *Wieland or the Transformation* [1798]. New York: Hafner, 1958.

Browning, Robert. "Paracelsus" [1835]. In *The Poetical Works of Robert Browning*. London: Oxford University Press, 1953.

Brunt, Samuel. *A Voyage to Cacklogallinia: With a Description of the Religion, Policy, Customs, and Manners of that Country* [1727]. New York: Columbia University Press, 1940.

Buck, Pearl S. *Command the Morning*. New York: John Day, 1959.

————. *A Desert Incident*. New York: New York Studio Duplicating Service, 1959.

Budrys, Algis. *Who?* [1958]. London: Quartet, 1975.

Bulgakov, Mikhail. *The Heart of a Dog*. Translated by Michael Glenny. London: Collins & Harvill, 1968.

Burdick, Eugene, and Harvey Wheeler. *Fail-Safe*. London: Hutchinson, 1963.

Butler, Samuel. *Erewhon* [1872]. New York: Airmont Classics, 1967.

————. *The Poetical Works of Samuel Butler*. Edited by J. Mitford. London: Bell & Daldy.

Buzzatti, Dino. "Appointment with Einstein" [1958]. In *Restless Nights,* translated by Lawrence Venuti. Manchester: Carcanet, 1984.

Byron, George Gordon, Lord. *Manfred* [1817]. In *Poetical Works,* edited by Frederick Page. London: Oxford University Press, 1973.

Campbell, Clyde C. "The Avatar." *Astounding Stories,* July 1935.

Capek, Karel. *Krakatit* [1923]. Translated by Laurence Hyde. London: Geoffrey Bles, 1925.

———. *R.U.R. (Rossum's Universal Robots: a Fantastic Melodrama)* [1921]. Translated by Paul Selver. Garden City, N.Y.: Doubleday, Page & Co., 1923.

Carlyle, Thomas. "Signs of the Times" [1829]. In *Thomas Carlyle: Selected Writings,* edited by Alan Shelston. Harmondsworth: Penguin, 1971.

Cavendish, Margaret, Duchess of Newcastle. *The Description of a New World Called the Blazing World.* London: A. Maxwell, 1666.

Centlivre, Susannah. *The Basset-Table* [1705]. In *Plays.* New York: Garland Publishers, 1982.

———. *A Bold Stroke for a Wife* [1718]. Edited by Thalia Stathas. London: Edward Arnold, 1969.

Chambers, Robert William. *Police.* New York: Appleton, 1915.

Chatterton, Ruth. *The Betrayers.* Boston: Houghton Mifflin, 1953.

Chaucer, Geoffrey. *The Canon's Yeoman's Prologue and Tale* [ca. 1391]. In *The Canterbury Tales,* translated by Nevil Coghill. Harmondsworth: Penguin, 1957.

Chekhov, Anton. "A Boring Story (From an Old Man's Notebook)" [1889]. In *Lady with Lapdog and Other Stories,* translated by David Magarshack. Harmondsworth: Penguin, 1980.

———. "Ward No. 6" [1892]. In *Lady with Lapdog and Other Stories. See* Chekhov, "A Boring Story."

Chesterton, G. K. *The Man Who Was Thursday* [1908]. Beaconsfield: Darwen Finlayson, 1963.

———. "The Unthinkable Theory of Professor Green." In *Tales of the Long Bow.* New York: Dodd, Mead & Co., 1925.

Clarke, Arthur C. *2001: A Space Odyssey.* London: Hutchinson, 1968.

———. *2010: Odyssey Two.* London: Granada, 1982.

Clough, Arthur Hugh. "The New Sinai." In *A Choice of Clough's Verse,* edited by Michael Thorpe. London: Faber & Faber, 1969.

Cobb, Weldon J. *A Trip to Mars.* New York: Street & Smith, 1901.

Colby, Merle. *The Big Secret.* New York: Viking, 1949.

Collins, Wilkie. *Heart and Science.* London: Chatto & Windus, 1893.

Conrad, Joseph. *Lord Jim* [1900]. New York: Bantam, 1958.

———. *The Secret Agent* [1907]. Cambridge: Cambridge University Press, 1990.

Conrad, Michael Georg. *In purpurner Finsternis: Roman-Improvisation aus dem dreissigsten Jahrhundert.* Berlin: Verein für freies Schriftthum, 1895.

Cooper, William. *Memoirs of a New Man: A Novel.* London: Macmillan, 1966.

Cowley, Abraham. "To the Royal Society." Frontispiece to Thomas Sprat, *History of the Royal Society* [1667]. London: Routledge, 1959.

Cowper, William. "The Task" [1784]. In *The Poetical Works of William Cowper,* edited by H. S. Milford. London: Oxford University Press, 1959.

Cramer, Heinz von. *Aufzeichnungen eines ordentlichen Menschen.* In *Leben wie in Para-dies.* Hamburg: Hoffmann & Campe, 1964.

———. *Die einzige Staatsvernunft in Konzession des Himmels.* Hamburg: Hoffmann & Campe, 1961.

Crichton, Michael. *The Andromeda Strain.* London: Corgi, 1976.

Cromie, Robert. *The Crack of Doom.* London: Digby, Long & Co. 1895.

———. *A Plunge into Space* [1891]. Westport, Conn.: Hyperion, 1976.

Cummings, Ray. *The Man Who Mastered Time.* Chicago: A. C. McClung & Co., 1929.

Cummins, Harle Owen. "The Man Who Made a Man." In *Welsh Rarebit Tales.* Boston: Mutual Book Co., 1902.

Dacré, Charlotte. *Zofloya: or The Moor.* London: Longman, Hurst, Rees & Orne, 1806.

Darwin, Erasmus. *Zoonomia or The Laws of Organic Life.* London: J. Johnson, 1794–96.

Daumann, Rudolf Heinrich. *Protuberanzen : ein utopischen Roman.* Berlin: Schützen-Verlag, 1940.

De Harsanyi, Zsolt. *The Star-gazer.* New York: Putnam's, 1939.

de Kruif, Paul. *The Sweeping Wind.* London: Rupert Hart-Davis, 1962.

de l'Isle-Adam, Villiers. *L'Eve Future* [1886]. Edited by N. Satiat. Paris: Flammarion, 1992.

Desaguliers, J. T. *The Newtonian System of the Universe, the Best Model of Government: an Allegorical Poem.* London: John Sexex, 1729.

Dick, Philip K. *Do Androids Dream of Electric Sheep?* New York: Signet, 1968.

———. *Dr. Bloodmoney—or How We Got Along after the Bomb* [1965]. Boston: Gregg, 1977.

———. *The Man in the High Castle.* London: Gollancz, 1975.

———. "The Preserving Machine" [1953]. In *The Preserving Machine and Other Stories.* London: Gollancz, 1971.

———. *Vulcan's Hammer.* New York: Ace Books, 1960.

Dickens, Charles. "Chemical Contradictions." *Household Words,* 14 Sept. 1850.

———. "Full Report of the First Meeting of the Mudfog Association for the Advance-ment of Everything." *Bentley's Miscellany* 2 (1837).

———. "The Haunted Man" [1848]. In *Christmas Books.* London: Chapman & Hall, n.d.

Disraeli, Benjamin. *Tancred: a Young and High Born Visionary.* London: Henry Colburn, 1847.

Dneprov, Anatoly. "Formula for Immortality." Translated by Helen Saltz Jacobson. In *New Soviet Science Fiction,* edited by Theodore Sturgeon. London: Collier Mac-millan, 1979.

———. "The Island of the Crabs." In *Other Worlds, Other Stories,* edited by Darko Suvin. New York: Random House, 1970.

———. "The S*T*A*P*L*E Farm." In *Other Worlds, Other Stories,* edited by Darko Suvin. New York: Random House, 1970.

Döblin, Alfred. *Berge, Meere und Giganten: Roman.* Berlin: S. Fischer Verlag, 1924.

Dominik, Hans. *Atomgewicht 500.* Berlin: Verlag Scherl, 1935.

———. *Die Macht der Drei: Roman.* Berlin: Verlag Scherl, 1922.

Doyle, Arthur Conan. "The Beetle-Hunter. " In *The Conan Doyle Stories.* London: John Murray, 1929.

——. "The Disintegration Machine." In *The Complete Professor Challenger Stories.* London: John Murray & Jonathan Cape, 1976.

——. "The Final Solution." In *A Treasury of Sherlock Holmes.* New York: Hanover House, 1955.

——. "The Great Keinplatz Experiment." In *The Conan Doyle Stories. See* Doyle, "Beetle-Hunter."

——. *The Land of Mist* [1926]. In *The Complete Professor Challenger Stories. See* "Disintegration Machine."

——. *The Lost World* [1912]. London: John Murray & Jonathan Cape, 1979.

——. *The Maracot Deep* [1928]. London: Pan, 1977.

——. "The Physiologist's Wife." In *The Conan Doyle Stories. See* Doyle, "Beetle-Hunter ."

——. *The Poison Belt.* London: Hodder & Stoughton, 1913.

——. "The Ring of Thoth." In *The Conan Doyle Stories. See* Doyle, "Beetle-Hunter."

——. "A Study in Scarlet." In *A Treasury of Sherlock Holmes. See* Doyle, "Final Solution."

——. "When the World Screamed." In *The Complete Professor Challenger Stories. See* "Disintegration Machine."

Dryden, John. "Annus Mirabilis" [1666]. In *The Poems and Fables of John Dryden,* edited by J. Kinsley. London: Oxford University Press, 1962.

Du Maurier, Daphne. "The Breakthrough" [1966]. In *Don't Look Now and Other Stories.* Harmondsworth: Penguin, 1974.

Du Maurier, George. *Trilby* [1894]. London: Harper, 1912.

Dürrenmatt, Friedrich. *Der Mitmacher: Kommodie.* Basel: Reiss, 1973.

——. *The Physicists* [1962]. London: French's Acting Edition, 1963.

Dwinger, Edwin Erich. *Es geschah im Jahre 1965.* Salzburg: Pilgram Verlag, 1957.

Eastwood, W., ed. *A Book of Science Verse: The Poetic Relations of Science and Technology.* London: Macmillan, 1961.

Edmonds, H. *The Professor's Last Experiment.* London: Rich & Cowan, 1935.

Ehrlich, Max, *The Big Eye.* Garden City, N.Y.: Doubleday, 1949.

Eliot, George. *Middlemarch: A Study of Provincial Life.* 2 vols. [1871–72]. London: Dent, 1959.

——. "The Spanish Gypsy." In *The Spanish Gypsy, The Legend of Jubal and Other Poems, Old and New.* Edinburgh: Blackwood, 1868.

Ellanby, Boyd. "Chain Reaction." In *The Expert Dreamers. See* Pohl.

Ellison, Harlan. *I Have No Mouth and I Must Scream: Stories.* New York: Pyramid Books, 1967.

England, George Allan. *Darkness and Dawn.* Boston: Small, Maynard & Co., 1914.

——. *The Golden Blight.* New York: H. K. Fly Co., 1916.

Fielding, Henry. *The History of Tom Jones* [1749]. London: Dent, 1957.

Fineman, Irving. *Doctor Addams.* London: Cresset, n.d.

Flaubert, Gustave. *Madame Bovary* [1856]. Harmondsworth: Penguin, 1965.

Forster, E. M. "The Machine Stops" [1909]. In *Collected Short Stories.* Harmondsworth: Penguin, 1977.

Franklin, H. B., ed. *Future Perfect: American Science Fiction of the Nineteenth Century.* New York: Oxford University Press, 1966.

Frayn, Michael. *The Tin Men.* London: Collins, 1965.

Freeman, Richard Austin. *The Famous Cases of Doctor Thorndyke*. London: Hodder & Stoughton, 1929.

Frisch, Max. *Don Juan, or the Love of Geometry* [1962]. Translated by Michael Bullock. In *Max Frisch, Four Plays*. London: Methuen, 1969.

———. *The Great Wall of China* [1947]. Translated by Michael Bullock. In *Max Frisch, Four Plays*. See Frisch, *Don Juan*.

Gaskell, Elizabeth. "Cousin Phillis" [1865]. In *Cousin Phillis and Other Tales*, edited by Angus Easson. Oxford: Oxford University Press, 1981.

———. *The Letters of Mrs. Gaskell*. Edited by J.A.V. Chapple and Arthur Pollard. Manchester: Manchester University Press, 1966.

———. *Wives and Daughters* [1866]. Harmondsworth: Penguin, 1969.

George, Peter. *Dr. Strangelove or: How I Learned to Stop Worrying and Love the Bomb* [1963]. Oxford: Oxford University Press, 1988.

Gibson, William, and Bruce Sterling. *The Difference Engine* [1991]. New York: Bantam Spectra, 1992.

Gide, André. *The Vatican Cellars* [1914]. Translated by Dorothy Bussy. London: Cassell, 1952.

Gissing, George. *Born in Exile* [1892]. London: Gollancz, 1970.

Godfrey, Hollis. *The Man Who Ended War*. Boston: Little, Brown & Co., 1908.

Godwin, Francis. *The Man in the Moone: or A Discourse of a Voyage thither by Domingo Gonsales "the Speedy Messenger"* [1638]. Reprint. Menston, Yorks.: Scolar, 1971.

Godwin, William. *Caleb Williams* [1794]. London: Oxford University Press, 1970.

———. *St. Leon: A Tale of the Sixteenth Century* [1799]. New York: Garland Publishing Co., 1974.

Goethe, Johann Wolfgang von. *Faust*. [Pt. 1, 1808; pt. 2, 1832]. Translated by Sir Theodore Martin. London: Dent, 1971.

Gordon, Richard. *The Invisible Victory* [1977]. Harmondsworth: Penguin, 1979.

Gosse, Edmund. *Father and Son* [1907]. Edited by Peter Abbs. Harmondsworth: Penguin, 1986.

Graf, Oskar Maria. *Die Erben des Untergangs: Roman einer Zukunft*. Frankfurt: Nest Verlag, 1959.

Greene, Robert. *The Honourable History of Friar Bacon and Friar Bungay* [1594]. Edited by J. A. Lavin. London: Benn, 1969.

Grover, Richard. "Poem on Sir Isaac Newton." In *View of Sir Isaac Newton's Philosophy*, by Henry Pemberton [1728]. New York: Johnson Reprint Corp., 1972.

Hacks, Peter. *Eröffnung des indischen Zeitalters: Schauspiel in Fünf Stücke* [1955]. Frankfurt am Main: Sührkamp Verlag, 1965.

Halley, Edmund. "Ode to the Illustrious Man, Isaac Newton." In *Mathematical Principles of Natural Philosophy and System of the World*, by Isaac Newton [1729]. Translated by Andrew Motte. Edited by R. T. Crawford. Berkeley: University of California Press, 1934.

Hamilton, Edmond. "The Metal Giants." *Weird Tales*, Dec. 1926.

Hanstein, Otfrid von. "Utopia Island." *Wonder Stories* 2 (May and June 1931).

Hardy, Thomas. *A Pair of Blue Eyes* [1873]. Edited by Roger Ebbatson. Harmondsworth: Penguin, 1986.

———. *Two on a Tower* [1882]. London: Macmillan, 1922.

———. *The Woodlanders* [1887]. London: Macmillan, 1923.

Hauptmann, Helmut. *Ivi*. Halle an der Saale: Mitteldeutscher Verlag, 1969.

Hawthorne, Nathaniel. "The Birthmark" [1845]. In *The Complete Short Stories of Nathaniel Hawthorne*. New York: Hanover House, 1959.

———. "Dr. Grimshawe's Secret" [1883]. In *The Complete Short Stories of Nathaniel Hawthorne*. See Hawthorne, "The Birthmark."

———. "Dr. Heidegger's Experiment" [1837]. In *The Complete Short Stories of Nathaniel Hawthorne*. See Hawthorne, "The Birthmark."

———. "Ethan Brand" [1851]. In *The Complete Short Stories of Nathaniel Hawthorne*. See Hawthorne, "The Birthmark."

———. "The Haunted Quack" [1831]. In *The Complete Short Stories of Nathaniel Hawthorne*. See Hawthorne, "The Birthmark."

———. "Rappacini's Daughter" [1844]. In *The Complete Short Stories of Nathaniel Hawthorne*. See Hawthorne, "The Birthmark."

———. *The Scarlet Letter* [1837]. Harmondsworth: Penguin, 1984.

———. *Septimius Felton, or the Elixir of Life* [1871]. In *The Dolliver Romance, Fanshawe and Septimius Felton*. Boston: Houghton Mifflin, 1886.

Hazzard, Shirley. *The Transit of Venus* [1980]. Harmondsworth: Penguin, 1982.

Heard, Gerald. *Doppelgängers* [1948]. London: Science Fiction Books, 1965.

Heath-Stubbs, J., and Salman Phillips, eds. *Poems of Science*. Harmondsworth: Penguin, 1984.

Hersey, John. *The Child Buyer* [1960]. Harmondsworth: Penguin, 1964.

Hilton, James. *Nothing So Strange* [1947]. Melbourne: Macmillan, 1948.

Hoffmann, E.T.A. "Klein Zaches genannt Zinnober: ein Märchen" [1819]. In *Sämtliche poetischen Werke*. Berlin: Der Tempel-Verlag, 1963. Vol. 1.

———. "The Sandman" [1816]. Translated by J. T. Bealby. In *Isaac Asimov Presents the Best Science Fiction of the Nineteenth Century*, edited by I. Asimov, C. G. Waugh, and M. Greenberg. London: Gollancz, 1983.

Hogan, James P. *The Genesis Machine*. New York: Ballantine, 1978.

Hospital, Janette Turner. *Charades*. Brisbane: Queensland University Press, 1988.

Howard, Sidney. *Yellow Jack* [1934]. In *Three Plays about Doctors*, edited by Joseph Mersand. New York: Washington Square, 1961.

Hoyle, Fred. *The Black Cloud*. London: Heinemann, 1957.

Hoyle, Fred, and John Elliott. *A for Andromeda*. London: Corgi, 1962.

———. *Andromeda Breakthrough*. London: Corgi, 1966.

Hudson, William Henry. *Far Away and Long Ago: A History of my Early Life*. London: Dent, 1918.

Hunt, Robert. *Panthea, The Spirit of Nature*. London: Reeve, Benham & Reeve, 1849.

Huxley, Aldous. *After Many a Summer* [1939]. London: Chatto & Windus, 1953.

———. *Antic Hay*. London: Chatto & Windus, 1923.

———. *Ape and Essence* [1948]. London: Chatto & Windus, 1951.

———. *The Genius and the Goddess*. London: Chatto & Windus, 1955.

———. *Island* [1962]. Harmondsworth: Penguin, 1968.

———. *Jesting Pilate*. London: Chatto & Windus, 1926.

———. *Point Counter Point* [1928]. London: Chatto & Windus, 1947.

Ibsen, Henrik. *An Enemy of the People* [1882]. Translated by Michael Meyer. London: Rupert Hart-Davis, 1963.

Jacob, Heinrich Eduard. *Die Physiker von Syrakus: Ein Dialog.* Berlin: Ernst Rowohlt Verlag, 1920.

Jahnn, Hans Henny. *Trümmer des Gewissens (Der staubige Regenbogen).* Edited by Walter Muschg. Frankfurt am Main: Europäische Verlagsanstalt, 1961.

Jakobs, Karl-Heinz. *Eine Pyramide für mich.* Berlin: Verlag Neues Leben, 1971.

James, Henry. *The Bostonians* [1886]. Harmondsworth: Penguin, 1971.

Jameson, Malcolm. *The Giant Atom* [1944]. New York: Bond-Charteris, 1945.

Jarry, Alfred. *Exploits and Opinions of Doctor Faustroll, Pataphysician* [1911]. In *Selected Works of Alfred Jarry,* edited by Roger Shattuck and Simmon Watson-Taylor. London: Methuen, 1965.

Johannsen, Olaf. *The Great Computer: A Vision* [1966]. Translated by N. Walford. London: Gollancz, 1968.

Johnson, Samuel. "The History of Rasselas, Prince of Abissinia" [1759]. In *Rasselas, Poems and Selected Prose,* edited by B. H. Bronson. New York: Holt, Reinhart & Windsor, 1958.

Jones, R. V. *The Most Secret War.* London: Hamish Hamilton, 1978.

Jonson, Ben. *The Alchemist* [1610]. In *The Complete Plays of Ben Jonson,* edited by G. A. Wilkes. Oxford: Clarendon, 1982.

Jünger, Ernst. *Gläserne Bienen.* Stuttgart: Ernst Klett Verlag, 1957.

Kafka, Franz. "In the Penal Colony." In *The Penal Colony, Stories and Short Pieces,* translated by Willa Muir and Edwin Muir. New York: Schocken, 1948.

Kaiser, Georg. *Gas I* [1918]. Translated by Hermann Scheffauer. In *Twenty-Five Modern Plays,* edited by S. M. Tucker and A. S. Downer. New York: Harper, 1953.

———. *Gas II* [1920]. Translated by Winifred Katzin. In Tucker and Downer, *Twenty-Five Modern Plays. See* Kaiser, *Gas I.*

Keller, Gottfried. *Das Sinngedicht* [1881]. In *Sämtliche Werke,* edited by Jonas Fränkel. Bern: Verlag Benteli-AG, 1934. Vol. 11.

Kellermann, Bernhard. *Der Tunnel.* Berlin: S. Fischer Verlag, 1913.

Ketterer, David. *New Worlds for Old.* New York: Anchor, Doubleday, 1974.

King, William. *The Transactioneer and Some of his Philosophical Fancies in Two Dialogues.* London: Booksellers of London & Westminster, 1700.

Kingsley, Charles. *Glaucus* [1855]. London: Dent, 1949.

———. *Two Years Ago* [1857]. London: Macmillan, 1889.

———. *The Water Babies* [1863]. London: Dent, 1949.

Kingsley, Sidney. *Men in White* [1933]. In *Three Plays About Doctors,* edited by J. Mersand. New York: Washington Square, 1961.

Kipling, Rudyard. "As Easy as A.B.C." [1912]. In *A Diversity of Creatures.* London: Macmillan, 1952.

———. "With the Night Mail: A Story of 2000 A.D." In *Actions and Reactions.* New York: Doubleday, Page & Co., 1909.

Kipphardt, Heinar. *In the Matter of J. Robert Oppenheimer* [1964]. Translated by Ruth Speirs. London: Methuen, 1967.

Kirst, Hans Hellmut. *Keiner Kommt davon: Bericht von den letzten Tagen Europas.* Vienna: Verlag Kurt Desch, 1957.

Kneale, Nigel. *Quartermass II.* Harmondsworth: Penguin, 1960.

Königsdorf, Helga. *Ein sehr exakter Schein.* Frankfurt am Main: Luchterhand Literaturverlag, 1990.

Kornbluth, C. M. "The Altar at Midnight" [1952]. In *Best Science Fiction Two*, edited by Edmund Crispin. London: Faber & Faber, 1968.

———. "Gomez" [1955]. In *Best Science Fiction Stories of C. M. Kornbluth*. London: Faber, 1968.

La Mettrie, Julien Offray de. *Man a Machine* [1747]. Translated by Gertrude Carmen Bussey. La Salle, Ill.: Open Court, 1912.

Landor, Walter Savage. "Gebir." In *The Works and Life of Walter Savage Landor*. London: Chapman & Hall, 1876. Vol. 7.

Large, E. C. *Sugar in the Air*. London: Jonathan Cape, 1937.

Lasswitz, Kurd. *Bilder aus der Zukunft*. Breslau: Verlag von S. Schottländer, 1878.

———. *Two Planets* [1897]. Translated by Hans H. Rudnick. Edited by Mark R. Hillegas. Carbondale: Southern Illinois University Press, 1971.

———. "Über Zukunftsträume." Parts 1 and 2. *Die Nation*, 13 and 20 May 1899.

Latham, Philip. "To Explain Mrs. Thompson." In *The Expert Dreamers. See* Pohl.

Lawrence, D. H. *The Rainbow* [1915]. Harmondsworth: Penguin, 1966.

LeGuin, Ursula. *The Dispossessed* [1974]. London: Panther, 1985.

———. *The Lathe of Heaven* [1971]. New York: Avon, 1973.

Leinster, Murray. "The Man Who Put Out the Sun." *Argosy*, 14 June 1930.

———. "The Storm That Had to Be Stopped." *Argosy*, 1 Mar. 1930.

Lem, Stanislaw. *The Cyberiad: Fables for the Cybernetic Age*. Translated by Michael Kandel. London: Secker & Warburg, 1975.

———. *The Futurological Congress*. New York: Avon Books, 1974.

———. *His Master's Voice* [1964]. Translated by Michael Kandel. San Diego: Harcourt Brace Jovanovich, 1983.

———. *The Invincible*. Translated by Wendayne Ackerman. New York: Seabury, 1973.

———. *Solaris*. Translated by Joanna Kilmartin and Steve Cox. New York: Berkeley University Press, 1971.

Lessing, Gotthold Ephraim. *Faust* (fragment). In *Gotthold Ephraim Lessings Sämtliche Schriften*, edited by Karl Lachmann, vol. 3. 23 vols. Stuttgart: G. J. Göschensche Verlagshandlung, 1886–1924.

Levi, Primo. *The Periodic Table* [1975]. Translated by Raymond Rosenthal. New York: Schocken, 1984.

Lewis, Clive Staples. *The Abolition of Man*. London: Oxford University Press, 1943.

———. *Out of the Silent Planet* [1938]. London: John Lane, the Bodley Head, 1946.

———. *Perelandra* [1943]. New York: Macmillan, 1944.

———. *That Hideous Strength* [1945]. London: John Lane, the Bodley Head, 1946.

Lewis, Leopold. *The Bells* [1871]. In *Nineteenth-Century Plays*, edited by George Rowell. London: Oxford University Press, 1960.

Lewis, Sinclair. *Arrowsmith* [1925]. New York: Signet, 1961.

Locke, W. J. *Septimus*. London: John Murray, 1909.

Lytton, Edward Bulwer, Lord. *A Strange Story*. London: Routledge, 1861.

———. *What Will He Do with It?* London: Routledge, 1858.

———. *Zanoni* [1842]. London: Routledge, 1853.

McCormach, Russell. *Night Thoughts of a Classical Physicist* [1969]. Cambridge, Mass.: Harvard University Press, 1982.

MacHarg, William, and Edwin Balmer. *The Achievements of Luther Trant*. Boston: Small, Maynard & Co. 1910.

McMahon, Thomas. *Principles of American Nuclear Chemistry: A Novel*. Boston: Little, Brown, 1970.

Macneice, Louis. "The Kingdom" [1943]. Pt. 6. "The Scientist." In *Collected Poems of Louis Macneice*, edited by E. R. Dodds. London: Faber & Faber, 1979.

Mailer, Norman. *Of a Fire on the Moon*. Boston: Little, Brown & Co., 1970.

Mallett, David. *The Excursion: a Poem in Two Books* [1728]. In *The Works of the English Poets*, edited by Alexander Chalmers. London, 1810. Vol. 14.

Mann, Thomas. *Dr. Faustus* [1947]. Translated by H. T. Lowe-Porter. In *Thomas Mann*. London: Secker & Warburg, 1979.

———. *The Magic Mountain* [1924]. Translated by H. T. Lowe-Porter. Harmondsworth: Penguin, 1973.

Marlowe, Christopher. *The Tragical History of Dr. Faustus* [1604]. In *Marlowe's Plays and Poems*, edited by M. R. Ridley. London: Dent, 1963.

Maron, Monika. *Flugasche: Roman*. Frankfurt am Main: S. Fischer Verlag, 1981.

Masters, Dexter. *The Accident*. London: Cassell, 1955.

Maugham, Somerset. *The Magician* [1908]. London: Heinemann, 1956.

Meredith, George. "Melampus" [1883]. In *The Poetical Works of George Meredith*, edited by George M. Trevelyn. London: Constable, 1912.

Miller, James. *The Humours of Oxford: a Comedy* [1726]. 2d ed. London: J. Watts, 1730.

Miller, Walter. *A Canticle for Leibowitz* [1959]. London: Weidenfeld & Nicolson, 1961.

Moorehead, Alan. *The Traitors*. New York: Charles Scribner's Sons, 1952.

Morgan, Charles. *The Burning Glass*. London: Macmillan, 1953.

———. *The Flashing Stream* [1938]. London: Macmillan, 1942.

Musil, Robert. "Eine Geschichte aus drei Jahrhunderten, 1927." In *Nachlass zu Lebzeiten*. Berlin: Ernst Rowohlt Verlag, 1981.

———. *The Man without Qualities* [1930–43]. 3 vols. Translated by Eithne Wilkins and Ernst Kaiser. London: Secker & Warburg, 1953–60.

———. "Tonka" [1924]. Translated by Eithne Wilkins and Ernst Kaiser. In *Tonka and Other Stories*. London: Secker & Warburg, 1965.

———. *Young Törless* [1906]. Translated by Ernst Kaiser and Eithne Wilkins. New York: Pantheon, 1978.

Newcomb, Simon. *His Wisdom the Defender*. New York: Harper, 1900.

Newman, B. *Armoured Doves*. London: Jarrolds, 1931.

Nichols, Robert, and Maurice Browne. *Wings over Europe* [1928]. New York: Covici-Friede, 1929.

Nossack, Erich. "Die Schalttafel." In *Spirale*. Frankfurt: Suhrkamp Verlag, 1956.

O'Brien, Fitz-James. "The Diamond Lens" [1858]. In *The Diamond Lens and Other Stories*. New York: AMS, 1969.

———. "The Golden Ingot" [1858]. in *The Diamond Lens and Other Stories. See* O'Brien, "The Diamond Lens."

Paltock, Robert. *The Life and Adventures of Peter Wilkins* [1751]. London: Dulau & Co., 1925.

Piercy, Marge. *Body of Glass* [1991]. London: Penguin, 1992.

Poe, Edgar Allan. *Arthur Gordon Pym*. In *The Complete Works of Edgar Allan Poe*. New York: Hearsts's International, 1914.

———. *The Balloon-Hoax* [1844]. In *The Complete Works of Edgar Allan Poe. See* Poe, *Arthur Gordon Pym*.

———. *Hans Pfall—A Tale* [1835]. In *The Complete Works of Edgar Allan Poe. See* Poe, *Arthur Gordon Pym.*

———. "Maelzel's Chess Player" [1836]. In *Complete Tales and Poems of Edgar Allen Poe.* New York: Random House, 1938.

———. "The Murders in the Rue Morgue." In *Tales of Mystery and Imagination.* London: Nelson, 1909.

———. "The Mystery of Marie Rôget." In *Tales of Mystery and Imagination. See* Poe, "Murders in the Rue Morgue."

———. "A Tale of the Ragged Mountains" [1844]. In *Tales of Mystery and Imagination. See* Poe, "Murders in the Rue Morgue."

———. "Von Kempelen and His Discovery" [1849]. In *The Science Fiction of Edgar Allan Poe,* edited by Harold Beaver. Hammondsworth: Penguin, 1976.

Pohl, F., ed. *The Expert Dreamers.* London: Gollancz, 1963.

Pope, Alexander. *The Dunciad* [1728, 1742]. In *Selected Poetry and Prose,* edited by W. K. Wimsatt, Jr. New York: Holt, Reinhart & Winston, 1962.

———. *Essay on Man* [1732–34]. In *Selected Poetry and Prose. See* Pope, *Dunciad.*

Powell, G. *All Things New.* London: Hodder & Stoughton, 1926.

Priestley, J. B. *The Doomsday Men.* London: Heinemann, 1938.

Prior, Matthew. "Solomon: On the Vanity of the World." In *The Poetical Works of Matthew Prior,* edited by J. Mitford. 2 vols. London: Bell & Daldy, 1866.

Pynchon, Thomas. *The Crying of Lot 49.* New York: Bantam, 1966.

———. *Gravity's Rainbow.* New York: Viking, 1973.

Raabe, Wilhelm. *Pfisters Mühle* [1884]. In *Sämtliche Werke.* 3d ser., vol. 2. Berlin-Grunewald: Verlagsanstalt für Literatur und Kunst-Hermann Klemm, n.d.

Ramsay, Allan. "Ode to the Memory of Sir Isaac Newton: Inscribed to the Royal Society" [1731]. In *Poems by Allan Ramsay and Robert Fergusson,* edited by Alexander Manson Kinghorn and Alexander Law. Edinburgh: Scottish Academic Press, 1974.

Rankine, W.J.M. "The Mathematician in Love" [1874]. In *Songs and Fables,* edited by W. J. Millar. London: Griffin, 1881.

Ray, John. *The Wisdom of God* [1691]. London: W. Innys, 1743.

Rehberg, Hans. *Johannes Keppler: Schauspiel in drei Akten.* Berlin: S. Fischer Verlag, 1933.

Renard, Maurice. *New Bodies for Old* [1908]. New York: Macaulay & Co. 1923.

Rhodes, William Henry. "The Case of Summerfield" [1871]. In *Caxton's Book,* edited by Daniel O'Connell. Westport, Conn.: Hyperion, 1974.

Rolland, Romain, and Frans Masereel. *The Revolt of the Machines.* Zürich: Buchergilde Gutenberg, 1949.

Rutherford, Mark [William Hale White]. *Miriam's Schooling.* In *Miriam's Schooling and Other Papers.* London: T. Fisher Unwin, 1890.

Sandburg, Carl. "Mr. Attila." In *Complete Poems.* New York: Harcourt Brace, 1950.

Savage, Richard. "The Wanderer." In *Works of the English Poets,* edited by Samuel Johnson [1779–81]. Edited by G. B. Hill. Oxford: Oxford University Press, 1905. Vol. 45.

Schary, Dore. *The Highest Tree.* New York: Random House, 1960.

Schirmbeck, Heinrich. *The Blinding Light* [1957]. Translated by Norman Denny. London: Collins, 1960.

―――. "Die Nacht vor dem Duell." In *Die Nacht vor dem Duell: Erzählungen*. Frankfurt am Main: Fischerbücherei, 1964.

Schlegel, Friedrich. "To Ritter." In *Kritische Friedrich Schlegel Ausgabe* II, edited by E. Behler and H. Eichner, vol. 2. 35 vols. Munich: F. Schöningh, 1958–91.

Schnitzler, A. *Professor Bernhardi* [1912]. Translated by Louis Borell and Ronald Adam. London: Gollancz, 1936.

Serviss, Garrett P. *Edison's Conquest of Mars* [1898]. Los Angeles: Carcosa House, 1947.

―――. "The Second Deluge" [1912]. *Fantastic Novels Magazine* 2 (July 1948).

Sewell, Stephen. *Welcome the Bright World* [1982]. Sydney: Alternative Publishing Co-op. & Nimrod Theatre Press, 1983.

Shadwell, Thomas. *The Virtuoso* [1676]. Edited by M. H. Nicolson and D. S. Rodes. London: Edward Arnold, 1966.

Shaffer, Peter. *Equus*. London: Andre Deutsch, 1973.

Shakespeare, William. *The Tempest* [1611]. In *William Shakespeare: The Complete Works*, edited by Alfred Harbage. London: Allen Lane, the Penguin Press, 1969.

Shelley, Mary W. *Frankenstein, or the Modern Prometheus* [1818]. Edited by R. E. Dowse and D. J. Palmer. London: Dent, 1963.

―――. "The Mortal Immortal" [1834]. In *Isaac Asimov Presents the Best Science Fiction of the Nineteenth Century*. See Hoffmann, "Sandman."

Shelley, Percy Bysshe. *Alastor* [1816]. In *The Poetical Works of Percy Bysshe Shelley*, edited by Edward Dowden. London: Macmillan, 1890.

Shenstone, William. "The Beau to the Virtuosos; Alluding to a Proposal for the Publication of a Set of Butterflies." In *Eighteenth-Century English Literature*, edited by G. Tillotson, P. Fussell, Jr., and M. Waingrow. New York: Harcourt, Brace & World, 1969.

Shute, Nevil. *No Highway*. London: Heinemann, 1948.

―――. *On the Beach*. London: Heinemann, 1957.

Simmons, Henry Hugh. "The Automatic Apartment." *Amazing Stories*, Aug. 1927.

―――. "The Automatic Self-Serving Dining Table." *Amazing Stories*, Apr. 1927.

―――. "The Electro-Hydraulic Bank Protector." *Amazing Stories*, Dec. 1927.

Sinclair, Upton. *A Giant's Strength*. London: T. Werner Laurie, 1948.

―――. *O Shepherd, Speak!* New York: Viking, 1949.

Skinner, B. F. *Walden Two*. New York: Macmillan, 1948.

Smith, Martin Cruz. *Stallion Gate*. London: Collins Harvill, 1986.

Snow, Charles Percy. *The Affair* [1960]. Harmondsworth: Penguin, 1968.

―――. *The Corridors of Power*. London: Macmillan, 1964.

―――. *Death Under Sail* [1932]. Rev. ed. London: Heinemann, 1959.

―――. *In Their Wisdom*. London: Macmillan, 1974.

―――. *Last Things*. London: Macmillan, 1970.

―――. *New Lives for Old*. London: Gollancz, 1933.

―――. *The New Men*. London: Macmillan, 1954.

―――. *The Search* [1934, rev. ed. 1938]. Reprint. Harmondsworth: Penguin, 1979.

―――. *The Sleep of Reason*. London: Macmillan, 1968.

Snyder, Alexander. "Blasphemers' Plateau." *Amazing Stories*, Oct. 1926.

Solzhenitsyn, Alexander. *The First Circle*. Translated by T. P. Whitney. New York: Harper & Row, 1968.

Spinrad, Norman. *Songs from the Stars*. New York: Simon & Schuster, 1980.

Stapledon, Olaf. *Odd John, a Story between Jest and Earnest*. London: Science Fiction Books, 1954.

Steinbeck, John. *The Log of "The Sea of Cortez."* New York: Viking, 1951.

Steinmann, Hans-Jürgen. *Träume und Tage*. Halle an der Saale: Mitteldeutscher Verlag, 1970.

Stevenson, Robert Louis. "The Strange Case of Dr. Jekyll and Mr. Hyde" [1886]. In *Dr. Jekyll and Mr. Hyde and Other Stories*. New York: Magnum, 1968.

Stewart, Douglas. "Professor Picard." In *Selected Poems*. Sydney: Angus & Robertson, 1962.

———. "Rutherford." In *Selected Poems. See* Stewart, "Professor Picard."

Stifter, Adalbert. *Der Nachsommer* [1857]. Reprint. Leipzig: Insel Verlag, 1910.

Stockton, Frank R. *The Great Stone of Sardis*. London: Harper & Bros., 1898.

Stoker, Bram, *Dracula* [1897]. New York: Airmont, 1965.

Stuart, Don A. [John W. Campbell, Jr.]. "Atomic Power." *Astounding Stories*, Dec. 1934.

Swift, Jonathan. *Gulliver's Travels* [1726]. Edited by Harold Williams. London: Dent, 1956.

———. "The Mechanical Operation of the Spirit" [1710]. In *Collected Works*. 16 vols. Oxford: Basil Blackwell, 1957. Vol. 1.

Tasker, William. "An Ode to Curiosity." In *Poems*. London, 1779.

Tennyson, Alfred. "Locksley Hall" [1842]. In *Alfred Tennyson: Selected Poetry*, edited by Douglas Bush.

———. "In Memoriam A.H.H." [1850]. In *Alfred Tennyson: Selected Poetry. See* Tennyson, "Locksley Hall."

Thompson, Morton. *The Cry and the Covenant*. Garden City, N.Y.: Doubleday, 1949.

Thompson, Vance. *The Mouse-Colored Road*. New York: Appleton, 1913.

Thomson, James. "A Poem Sacred to the Memory of Sir Isaac Newton" [1727]. In Thomson, *The Castle of Indolence and Other Poems*, edited by A. D. McKillop. Lawrence: University of Kansas Press, 1961.

———. *The Seasons* [1726–30]. In *The Poetical Works of James Thomson*. 2 vols. London: Bell & Daldy, 1860. Vol. 1.

Thoreau, Henry David. *Walden, or Life in the Woods* [1854]. New York: W. W. Norton, 1966.

Tickner-Edwards, E. "The Man Who Meddled with Eternity" [1901]. In *Beyond the Gaslight: Science in Popular Fiction, 1895–1905*, edited by H. Evans and D. Evans. London: Frederick Muller, 1976.

Train, Arthur, and Robert William Wood. *The Man Who Rocked the Earth*. New York: Doubleday, Page & Co., 1915.

Trollope, Anthony. *Doctor Thorne* [1858]. Boston: Houghton Mifflin, 1959.

Twain, Mark. "Some Learned Fables for Good Old Boys and Girls" [1875]. In *Science and Literature: A Reader*, edited by J. V. Cadden and P. R. Brostowin. Boston: D. C. Heath, 1964.

Verne, Jules. *Hector Servadac* [1877]. Translated by Ellen E. Frewer. London: Sampson, Low, Marston & Co., 1878.

———. *Journey to the Centre of the Earth* [1864]. Translated by R. Baldick. Harmondsworth: Penguin, 1970.

———. *The Mysterious Island* [1875]. London: Corgi, 1976.

———. *Robur the Conqueror.* Paris: Hetzel, 1886.

———. *Twenty Thousand Leagues under the Sea* [1870]. Translated by H. Frith. London: Dent, 1968.

Verrill, A. Hyatt. "The Plague of the Living Dead." *Amazing Stories*, Apr. 1927.

———. "The Ultra-Elixir of Youth." *Amazing Stories*, Aug. 1927.

Voltaire. *Micromégas* [1752]. In *The Works of Voltaire*, vol. 2, pt. 1. New York: St. Hubert Guild, 1901.

Vonnegut, Kurt, Jr. *Cat's Cradle* [1963]. New York: Dell, 1976.

———. "Fortitude." In *Wampeters, Foma and Granfaloons.* New York: Delacarte, 1975.

———. *Slaughterhouse Five or The Children's Crusade: A Duty Dance with Death* [1969]. New York: Dell, 1971.

Waterloo, Stanley. *Armageddon.* Chicago: Rand McNally & Co., 1898.

Wells, Herbert George. "Argonauts of the Air" [1895]. In *Selected Short Stories.* Harmondsworth: Penguin, 1970.

———. *The Croquet Player.* London: Chatto & Windus, 1936.

———. *Early Writings in Science and Science Fiction.* Edited by R. M. Philmus and D. Y. Hughes. Berkeley and Los Angeles: University of California Press, 1975.

———. *The First Men in the Moon* [1901]. London: Nelson, n.d.

———. *The Food of the Gods, and How it Came to Earth.* London: Macmillan, 1904.

———. *The Invisible Man* [1897]. London: Collins, n.d.

———. *The Island of Dr. Moreau* [1896]. Harmondsworth: Penguin, 1967.

———. *Love and Mr. Lewisham* [1900]. London: Benn, 1927.

———. *Men Like Gods.* New York: Macmillan, 1923.

———. "The Plattner Story" [1897]. In *Selected Short Stories. See* Wells, "Argonauts of the Air."

———. *Star Begotten* [1937]. London: Sphere, 1975.

———. "The Time Machine" [1895]. In *Selected Short Stories. See* Wells, "Argonauts of the Air."

———. *Tono-Bungay* [1908]. London: Pan, 1909.

———. *The War of the Worlds* [1898]. Harmondsworth: Penguin, 1971.

———. *The World Set Free.* London: Macmillan, 1914.

Wenfu, Lu. "The Well." Translated by Yu Fanquin. In *Chinese Literature, Fiction, Poetry, and Art*, Spring 1987, 93–144.

West, Morris L. *The Heretic.* London: Heinemann, 1970.

White, Stewart Edward. *The Sign at Six.* Indianapolis: Bobbs Merrill, 1912.

White, Stewart Edward, and Samuel Hopkins Adams. *The Mystery* [1907]. New York: Arno, 1975.

Whitman, Walt. "When I Heard the Learn'd Astronomer" [1865]. *Leaves of Grass 1850–1880.* In *The Portable Walt Whitman*, edited by Mark van Doren. New York: Viking, 1958.

Wicks, Mark. *To Mars via the Moon.* Philadelphia: J. B. Lippincott, 1911.

Wilder, Thornton. *The Skin of Our Teeth* [1942]. In *Three Plays by Thornton Wilder.* London: Longmans Green & Co., 1958.

Wilson, Mitchell. *Live with Lightning.* New York: Little, Brown, 1949.

———. *Meeting at a Far Meridian.* London: Secker & Warburg, 1961.

Wolf, Christa. "Self-Experiment: Appendix to a Report" [1973]. Translated by Jeanette Clausen. *New German Critique* 13 (Winter 1978), 109–31.

――――. *Störfall* [1987]. Frankfurt am Main: Luchterhand Literaturverlag, 1988.

Woolf, Virginia. *To the Lighthouse* [1927]. London: Hogarth, 1955.

Wright, Thomas. *The Female Virtuoso's: A Comedy*. London: J. Wilde, 1693.

Wylie, Philip. *The Gladiator* [1930]. Westport, Conn.: Hyperion, 1974.

――――. *Murderer Invisible* [1931]. Westport, Conn.: Hyperion, 1976.

――――. *Tomorrow!* New York: Reinhart, 1954.

Zamyatin, Yevgeny. *We*. Translated by Gregory Zilborg. New York: E.P. Dutton, 1924.

Zola, Émile. *Doctor Pascal* [1893]. Translated by Vladimir Jean. London: Elek Books, 1957.

――――. *Paris*. London: Chatto & Windus, 1898.

Zuckmayer, Carl. *Cold Light* [1955]. Translated by Elizabeth Montagu. Typescript, Columbia University Libraries, n.d.

Zwillinger, Frank. *Galileo: Schauspiel* [1953]. Bayreuth: Reta Baumann Verlag, 1962.

Part 2: Criticism

Arnold, Matthew

Donovan, R. A. "Mill, Arnold and Scientific Humanism." *Annals of the New York Academy of Sciences* 360 (20 Apr. 1981).

Artaud, Antonin

Demaitre, Ann. "The Theatre of Cruelty and Alchemy: Artaud and le Grand Oeuvre." *Journal of the History of Ideas* 33 (1972).

Asimov, Isaac

Asimov, Isaac. *The Early Asimov, or Eleven Years of Trying*. London: Gollancz, 1973.

Patrouch, Joseph F., Jr. *The Science Fiction of Isaac Asimov*. Garden City, N.Y.: Doubleday, 1974.

Bacon, Francis

Adams, Robert P. "The Social Responsibilities of Science in *Utopia, New Atlantis*, and After." *Journal of the History of Ideas* 10 (1949).

Anderson, F. H. *The Philosophy of Francis Bacon*. Chicago: University of Chicago Press, 1948.

Blish, James. *Doctor Mirabilis*. London: Faber & Faber, 1964.

Blodgett, E. D. "Bacon's *New Atlantis* and Campanella's *Civitas Solis*." *PMLA* 46 (1931).

Bullough, G. "Bacon and the Defense of Learning." In *Seventeenth-Century Studies Presented to Sir Herbert Grierson*, edited by J. Dover Wilson. New York: Octagon, 1967.

Haydn, Hiram C. "The Science of the Counter-Renaissance." *The Counter-Renaissance*. New York: Charles Scribner's Sons, 1950.

Metz, Rudolph. "Bacon's Part in the Intellectual Movements of His Time." In *Seventeenth-Century Studies Presented to Sir Herbert Grierson. See* Bullough, "Bacon and the Defense of Learning."

Prior, Moody E. "Bacon's Man of Science." *Journal of the History of Ideas* 15 (1954).

Balzac, Honoré de

Luce, Louise Fiber. "Honoré de Balzac and the Voyant: A Recovered Alchemical Discourse." *L'Esprit Créateur* 18 (1978).
Murard, Jean. "Balzac, la médecine et les médecins." *Histoire de la Médicine,* Aug.–Sept. 1971, Oct. 1971, Nov. 1971.
Pritchett, V. S. *Balzac.* London: Chatto & Windus, 1933.

Blake, William

Ault, Donald. *Visionary Physics: Blake's Response to Newton.* Chicago: University of Chicago Press, 1974.

Brecht, Bertolt

Cohen, M. A. "History and Moral in Brecht's *The Life of Galileo.*" *Contemporary Literature* 11 (1970).
Demetz, P., ed. *Brecht: A Collection of Critical Essays.* Englewood Cliffs, N.J.: Prentice Hall, 1962.
Jevons, F. R. "Brecht's *Life of Galileo* and the Social Relations of Science." *Technology and Society* 4, no. 3 (1968).
Spalter, M. *Brecht's Tradition.* Baltimore: Johns Hopkins University Press, 1967.
White, A. D. "Brecht's *Leben des Galilei:* Armchair Theatre?" *German Life and Letters,* n.s. 27 (Jan. 1974).

Broch, Hermann

Ziolkowski, Theodore. "Hermann Broch and Relativity in Fiction." *Wisconsin Studies in Contemporary Literature* 8, no. 3 (1967).

Browne, Sir Thomas

Chalmers, G. K. "Sir Thomas Browne, True Scientist." *Osiris* 2 (1936).

Bulgakov, Mikhail

Burgin, Diana. "Bulgakov's Early Tragedy of the Scientist-Creator: An Interpretation of the *Heart of a Dog.*" *Slavic and East European Journal* 22 (1978).

Butler, Samuel

Bruun, S. V. "Who's Who in Samuel Butler's 'The Elephant in the Moon.'" *English Studies* 50 (1969).

Capek, Karel

Capek, Karel. "The Meaning of *R.U.R.*" *Saturday Review* 136 (21 July 1923).
Harkins, W. E. *Karel Capek.* New York: Columbia University Press, 1962.

Carlyle, Thomas

Turner, Frank M. "Victorian Scientific Naturalism and Thomas Carlyle." *Victorian Studies* 18, no. 3 (1975).

Chaucer, Geoffrey

Baum, Paul F. "The Canon's Yeoman's Tale." *MLN* 40 (1925).

Craik, T. W. "The Canon's Yeoman's Tale." In *The Comic Tales of Chaucer*. London: Methuen, 1964.

Damon, S. Foster. "Chaucer and Alchemy." *PMLA* 39 (1924).

Duncan, Edgar H. "The Literature of Alchemy and Chaucer's *Canon's Yeoman's Tale*: Framework, Theme, and Characters." *Speculum* 43 (1968).

Gardner, John. "*The Canon's Yeoman's Prologue and Tale*: An Interpretation." *Philological Quarterly* 46 (1967).

Grennan, Joseph Edward. "Chaucer's Characterisation of the Canon and his Yeoman." *Journal of the History of Ideas* 25 (1964).

———. "Chaucer's 'Secree of Secrees': An Alchemical Topic." *Philological Quarterly* 42 (1963).

Hamilton, Marie P. "The Clerical Status of Chaucer's Alchemist." *Speculum* 16 (1941).

Hartung, Albert E. "Inappropriate Pointing in *The Canon's Yeoman's Tale* G1236–1239." *PMLA* 77 (1962).

Young, Karl. "The 'Secree of Secrees' of Chaucer's Canon's Yeoman." *MLN* 58 (1943).

Coleridge, Samuel Taylor

Barfield, Owen. *What Coleridge Thought*. Middleton, Conn.: Wesleyan University Press, 1971.

Coburn, Kathleen. "Coleridge, a Bridge between Science and Poetry: Reflections on the Bicentenary of His Death." *Proceedings of the Royal Institution of Great Britain* 46 (1973).

Potter, G. R. "Coleridge and the Idea of Evolution." *PMLA* 40 (1925).

Cowley, Abraham

Laprevotte, Guy. "To the Royal Society d'Abraham Cowley." *Études Anglaises* 26 (1973) 129–44.

Dick, Philip K.

Lem, Stanislaw. "A Visionary among the Charlatans." *Science Fiction Studies* 2, no. 1 (1975).

Suvin, Darko. "P. K. Dick's Opus: Artifice as Refuge World View." *Science Fiction Studies* 2, no. 1 (1975).

Dickens, Charles

Metz, N. A. "Science in *Household Words*." *Victorian Periodicals Newsletter*, 1979.

Wilkinson, Ann Y. "*Bleak House*: From Faraday to Judgement Day." *English Literary History* 34 (1967).

Donne, John

Coffin, C. M. *John Donne and the New Philosophy*. London: Routledge & Kegan Paul, 1937.

Doyle, Arthur Conan

Higham, Charles. *The Adventures of Conan Doyle: Life of the Creator of Sherlock Holmes*. London: Hamish Hamilton, 1976.

Rauber, D. F. "Sherlock Holmes and Nero Wolfe: The Role of the 'Great Detective' in Intellectual History." *Journal of Popular Culture* 6 (1972).
Rose, Phyllis. "Huxley, Holmes, and the Scientist as Aesthete." *Victorian Newsletter* 38 (1970), 22–24.

Dryden, John

Bredvold, Louis I. "Dryden, Hobbes, and the Royal Society." *Modern Philology* 25 (May 1928).
Lloyd, Claude. "John Dryden and the Royal Society." *PMLA* 45 (1930).

Dürrenmatt, Friedrich

Jenny, Urs. *Dürrenmatt: A Study of His Plays.* London: Methuen, 1978.
Muschg, W. "Dürrenmatt und *Die Physiker.*" *Moderna Sprak* 56 (1961).

Eliot, George

Adam, Ian. "A Huxley Echo in *Middlemarch.*" *Notes and Queries* 2 (1964).
Adam, Ian, ed. *This Particular Web: Essays on Middlemarch.* Toronto: University of Toronto Press, 1975.
Briggs, Asa. "*Middlemarch* and the Doctors." *Cambridge Journal,* Sept. 1948.
Cline, C. L. "Qualifications of Medical Practitioners of *Middlemarch.*" In *Nineteenth Century Perspectives: Essays in Honor of Lionel Stevenson,* edited by C. de L. Ryals. Durham, N.C.: Duke University Press, 1974.
Derrow, H. A. "*Middlemarch* and the Physician." *Annals of Medical History* 9 (1927).
Greenberg, R. A. "Plexuses and Ganglia: Scientific Allusion in *Middlemarch.*" *Nineteenth-Century Fiction* 30 (1976).
Harvey, W. J. "The Intellectual Background of the Novel." In *Middlemarch–Critical Approaches to the Novel,* edited by Barbara Hardy. London: Athlone, 1967.
Hulme, Hilda M. "*Middlemarch* as Science-Fiction: Notes on Language and Imagery." *Novel* 2 (Fall 1968).
Kitchel, Anna, ed. "Quarry for *Middlemarch.*" *Nineteenth-Century Fiction* 4 (1949–50), suppl.
Knoepflmacher, U. C. *Religious Humanism and the Victorian Novel–George Eliot, Walter Pater, and Samuel Butler.* Princeton: Princeton University Press, 1970.
Levine, George. "Determinism and Responsibility in Works of George Eliot." *PMLA* 77 (1962).
———. "George Eliot's Hypothesis of Reality." *Nineteenth-Century Fiction* 35 (1980).
McCarthy, P. J. "Lydgate, 'The New, Young Surgeon' of Middlemarch." *Studies in English Literature, 1500–1900* 10 (1970).
Mason, Michael York. "*Middlemarch* and Science: Problems of Life and Mind." *Review of English Studies,* n.s. 22, no. 116 (1971).
Newton, K. M. "George Eliot, George Henry Lewes, and Darwinism." *Durham University Journal* 65 (June 1974).
Paris, Bernard. *Experiments in Life: George Eliot's Quest for Values.* Detroit: Wayne State University Press, 1965.
Shuttleworth, Sally. *George Eliot and Nineteenth-Century Science.* Cambridge: Cambridge University Press, 1984.

———. "The Language of Science and Psychology in *Daniel Deronda*." *Annals of the New York Academy of Sciences* 360 (20 Apr. 1981).

Emerson, Ralph Waldo

Haugrud, Raychel A. "John Tyndall's Interest in Emerson." *American Literature* 41 (1970).

Fielding, Henry

Eales, Nellie B. "A Satire on the Royal Society Dated 1743, Attributed to Henry Fielding." *Notes and Records of the Royal Society of London* 23 (1968).

Frisch, Max

Gontrum, P. "Max Frisch's Don Juan: A New Look at a Traditional Hero." *Comparative Literary Studies* 2, no. 2 (1965).

Frost, Robert

Yogi, L. L. "The Scientist in Robert Frost's Poetry." *Rajasthan University Studies in English* 10 (1977).

Gaskell, Elizabeth Cleghorn

Lucas, John. "Mrs. Gaskell: The Nature of Social Change." In *The Literature of Change*. Brighton, Sussex: Harvester, 1980.

Gissing, George

Korg, Jacob. "The Spiritual Theme of *Born in Exile*." In *Collected Articles on George Gissing*, edited by Pierre Coustillas. London: Frank Cass, 1968.

Goethe, Johann Wolfgang von

Cottrell, Alan P. *Goethe's View of Evil and the Search for the New Image of Man in Our Time*. Edinburgh: Edinburgh University Press, 1982.
Fink, Karl J. *Goethe's History of Science*. Cambridge: Cambridge University Press, 1992.
Gray, Ronald D. *Goethe, the Alchemist: A Study of Alchemical Symbolism in Goethe's Literary and Scientific Work*. Cambridge: Cambridge University Press, 1952.
———. *Goethe—A Critical Introduction*. Cambridge: Cambridge University Press, 1967.
Heller, Otto. *Faust and Faustus: A Study of Goethe's Relation to Marlowe*. New York: Cooper Square, 1972.

Hardy, Thomas

Ingham, Patricia. "Hardy and *The Wonders of Geology*." *Review of English Studies*, n.s. 31 (Feb. 1980).
Millgate, Michael. *Thomas Hardy: His Career as a Novelist*. London: Bodley Head, 1971.
Wickens, G. G. "Literature and Science: Hardy's Response to Mill, Huxley, and Darwin." *Mosaic* 14 (1981).

Hawthorne, Nathaniel

Bales, Kent. "Hawthorne's Prefaces and Romantic Perspectivism." *ESQ: A Journal of American Renaissance* 23 (1977).

———. "Sexual Exploitation in Rappacini's Garden." *ESQ: A Journal of American Renaissance* 24 (1978).

Crews, Frederick C. *The Sins of the Fathers: Hawthorne's Psychological Themes.* London: Oxford University Press, 1966.

Fogle, Richard Harter. *Hawthorne's Fiction: The Light and the Dark.* Norman: University of Oklahoma Press, 1964.

Fryer, Judith. *The Faces of Eve: Women in the Nineteenth-Century American Novel.* New York: Oxford University Press, 1978.

Gollin, Rita. *Nathaniel Hawthorne and the Truth of Dreams.* Baton Rouge: Louisiana State University Press, 1979.

Heilman, R. B. "'The Birthmark': Science as Religion." *South Atlantic Quarterly* 48 (1949).

Kaul, A. N., ed. *Hawthorne: A Collection of Critical Essays.* Englewood Cliffs, N.J.: Prentice Hall, 1966.

Laser, Marvin. "Head, Heart, and Will in Hawthorne's Psychology." *Nineteenth-Century Fiction* 10 (1955).

Martin, Terence. *Nathaniel Hawthorne.* Boston: Twayne, 1983.

Noble, David W. "The Analysis of Alienation by Twentieth-Century Social Scientists and Nineteenth-Century Novelists: The Example of Hawthorne's *The Scarlet Letter.*" *Festschriften* 112 (1972).

Pollin, Burton R. "'Rappacini's Daughter'—Sources and Names." *Names* 14, no. 1 (1966).

Ringe, Donald A. "Hawthorne's Psychology and the Head and Heart." *PMLA* 65 (1950).

Schroeder, John W. "That Inward Sphere: Notes on Hawthorne's Heart Imagery and Symbolism." *PMLA* 65 (1950).

Stein, W. B. *Hawthorne's Faust: A Study in Mythmaking.* Gainesville: University of Florida Press, 1953.

Stoehr, Taylor. *Hawthorne's Mad Scientists: Pseudoscience and Social Science in Nineteenth-Century Life and Letters.* Hamden, Conn.: Archon, 1978.

Thompson, W. R. "Aminadab in Hawthorne's 'The Birthmark.'" *MLN* 52 (1955).

Turner, Arlin. *Nathaniel Hawthorne, a Biography.* New York: Oxford University Press, 1980.

Wentersdorf, Karl P. "The Elements of Witchcraft in *The Scarlet Letter.*" *Folklore* 83 (1972).

Hoffmann, E.T.A.

Brantly, Susan. "A Thermographic Reading of E.T.A. Hoffmann's *Der Sandmann.*" *German Quarterly* 55, no. 3 (1982).

Ellis, J. M. "Clara, Nathanael, and the Narrator: Interpreting Hoffmann's *Der Sandmann.*" *German Quarterly* 54, no. 1 (1981).

Lawson, Ursula D. "Pathological Time in E.T.A. Hoffmann's *Der Sandmann.*" *Monatshefte* 60 (1968).

Prawer, S. S. "Hoffmann's Uncanny Guest: A Reading of *Der Sandmann.*" *German Life and Letters* 18 (1965).

von Matt, Peter. *Die Augen der Automaten: E.T.A. Hoffmann's Imaginationslehre als Prinzip seiner Einzählkunst.* Tübingen: Niemeyer, 1971.

Huxley, Aldous

Birnbaum, M. *Aldous Huxley's Quest for Values.* Knoxville: University of Tennessee Press, 1971.

Brander, Lawrence. *Aldous Huxley—A Critical Study.* London: Rupert Hart-Davis, 1969.

Clareson, Thomas D. "The Classic: Aldous Huxley's *Brave New World.*" *Extrapolation* 2 (1961).

James, Henry

Long, R. E. "Source for Dr. Mary Prance in *The Bostonians.*" *Nineteenth-Century Fiction* 19 (1964).

Purdy, S. B. *The Hole in the Fabric: Science, Contemporary Literature, and Henry James.* Pittsburgh: Pittsburgh University Press, 1977.

Johnson, Samuel

Boswell, James. *Life of Johnson* [1791]. Edited by R. W. Chapman. 3d ed. London: Oxford University Press, 1970.

Eberwein, Robert. "The Astronomer in [Samuel] Johnson's Rasselas." *Michigan Academician* 5, no. 1 (1973).

Philip, J. R. "Samuel Johnson as Anti-scientist." *Notes and Records of the Royal Society of London* 29 (1975).

Schwartz, Richard B. *Samuel Johnson and the New Science.* Madison: University of Wisconsin Press, 1971.

Kingsley, Charles

Chitty, Susan. *The Beast and the Monk: A Life of Charles Kingsley.* London: Hodder & Stoughton, 1974.

Gillespie, H. R., Jr. "George Eliot's Tertius Lydgate and Charles Kingsley's Tom Thurnall." *Notes and Queries* 2 (1964).

Thorp, Margaret F. *Charles Kingsley, 1819–1875.* New York: Octagon, 1969.

Kipphardt, Heiner

Zipes, Jack D. "Documentary Drama in Germany: Mending the Circuit." *Germanic Review* 62 (Jan. 1967).

La Mettrie, Julien Offray de

Vartanian, Aram. *La Mettrie's L'Homme Machine: A Study in the Origins of an Idea.* Princeton: Princeton University Press, 1960.

Lasswitz, Kurd

Hillegas, Mark R. "The First Invasion from Mars." *Michigan Alumnus Quarterly Review* 66 (1959).

LeGuin, Ursula

Bierman, Judah. "Ambiguity in Utopia: The Dispossessed." *Science Fiction Studies* 2, no. 3 (1975).
Bittner, James W. *Approaches to the Fiction of Ursula K. LeGuin.* Ann Arbor: UMI Research Press, 1984.
Bucknall, Barbara J. *Ursula K. LeGuin.* New York: Frederick Ungar & Co., 1981.
Cummins, Elizabeth. *Understanding Ursula K. LeGuin.* Columbia: South Carolina University Press, 1990.
Moylan, Tom. "Ursula K. LeGuin, *The Dispossessed.*" In *Demand the Impossible.* London: Methuen, 1986.
Selinger, Bernard. *LeGuin and Identity in Contemporary Fiction.* Ann Arbor: UMI Research Press, 1958.
Spivak, Charlotte. *Ursula K. LeGuin.* Boston: Twayne, 1984.
Tavormina, M. Teresa. "Physics as Metaphor: The General Temporal Theory in *The Dispossessed.*" *Mosaic* 13 (1980).

Lem, Stanislaw

Jarzebski, Jerzy. "Stanislaw Lem: Rationalist and Sensualist." *Science Fiction Studies* 4, no. 2 (July 1977).
Kandel, Michael. "Lem in Review (June 2238)." *Science Fiction Studies* 4, no. 1 (March 1977).
———. "Stanislaw Lem on Man and Robots." *Extrapolation* 14 (1972).
Philmus, Robert. "*Futurological Congress* as Metageneric Text." *Science Fiction Studies* 13, no. 3 (1986).
Rodnianskaia, Irina. "Two Faces of Stanislaw Lem: On *His Master's Voice.*" *Science Fiction Studies* 13, no. 3 (1986).
Ziegfield, Richard E. *Stanislaw Lem.* New York: Frederick Ungar, 1985.

Lewis, C. S.

Sammons, Martha C. *C. S. Lewis' Space Trilogy.* Winchester, Ill.: Cornerstone Books, 1980.

Lewis, Sinclair

Geismar, M. "Sinclair Lewis." In *The Last of the Provincials: The American Novel, 1915–1925.* London: Secker & Warburg, 1947.
Kazin, Alfred. "The New Realism: Sherwood Anderson and Sinclair Lewis." In *On Native Grounds: An Interpretation of Modern American Prose Literature.* New York: Harcourt, Brace & World, 1942.
Rosenberg, Charles E. "Martin Arrowsmith: The Scientist as Hero." *American Quarterly* 15 (1963).
Schorer, Mark. *Sinclair Lewis: An American Life.* New York: McGraw-Hill, 1961.
Schorer, Mark, ed. *Sinclair Lewis: A Collection of Critical Essays.* Englewood Cliffs, N.J.: Prentice Hall, 1962.

Lytton, Edward Bulwer

Christensen, Allan Conrad. *Edward Bulwer-Lytton: The Fiction of New Regions.* Athens: University of Georgia Press, 1976.

Mann, Thomas

Prusok, Rudi. "Science in Mann's *Zauberberg:* The Concept of Space." *PMLA* 88 (1973).

Marlowe, Christopher

Brown, Beatrice D. "Marlowe, Faustus, and Simon Magus." *PMLA* 54 (1939).
McAlindon, T. "The Ironic Vision: Diction and Theme in Marlowe's *Doctor Faustus.*" *Review of English Studies* 32, no. 126 (1981).
Szönyi, Gye. "The Quest for Omniscience: The Intellectual Background of Marlowe's Doctor Faustus." *Papers in English and American Studies* 1 (1980).

Meredith, George

Grabar, Terry H. "Scientific Education and Richard Feverel." *Victorian Studies* 14, no. 2 (1970).

Mill, John Stuart

Donovan, R. A. "Mill, Arnold, and Scientific Humanism." in *Annals of the New York Academy of Sciences* 360 (20 Apr. 1981).
Strong, Edward W. "William Whewell and John Stuart Mill: Their Controversy about Scientific Knowledge." *Journal of the History of Ideas* 16 (1955).

Musil, Robert

Goldgar, Harry. "Freud and Robert Musil's Törless." *Comparative Literature* 17 (1965).
Kirchberger, Lida. "Musil's Trilogy: An Approach to *Drei Frauen.*" *Monatshefte* 55 (1963).
Luft, David S. *Robert Musil and the Crisis of European Culture, 1880–1942.* Berkeley and Los Angeles: University of California Press, 1980.
Peters, Frederick G. *Robert Musil, Master of the Hovering Life: A Study of the Major Fiction.* New York: Columbia University Press, 1978.
Sokel, W. H. "Kleist's Marquise of O, Kierkegaard's Abraham, and Musil's Tonka: Three Stages of the Absurd as the Touchstones of Faith." *Wisconsin Studies in Contemporary Literature* 8, no. 4 (1967).

O'Brien, Fitz-James

Franklin, H. Bruce. "O'Brien and Science Fiction." In *Future Perfect: American Science Fiction of the Nineteenth Century.* New York: Oxford University Press, 1966.

Poe, Edgar Allan

Howarth, W., ed. *Twentieth-Century Interpretations of Poe's Tales.* Englewood Cliffs, N.J.: Prentice Hall, 1971.
Ketterer, D. "The Science Fiction Element in the Work of Poe." *Science Fiction Studies* 1 (1974).

Pope, Alexander

Nicolson, Marjorie Hope, and George S. Rousseau. *This Long Disease, My Life: Alexander Pope and the Sciences.* Princeton: Princeton University Press, 1968.

Prior, Matthew

Spears, Monroe K. "Matthew Prior's Attitude toward Natural Science." *PMLA* 63 (1948).

Proust, Marcel

Bottiger, L. E. "Remembrance of Disease Lifelong: Marcel Proust and Medicine." *British Medical Journal* 287 (3 Dec. 1983).

Pynchon, Thomas

Bloom, Harold, ed. *Thomas Pynchon.* New York: Chelsea House, 1986.
Friedman, Alan J. "Science and Technology." In *Approaches to Gravity's Rainbow,* edited by Charles Clerc. Columbus: Ohio State University Press, 1983.
Friedman, Alan J., and Manfred Puetz. "Science as Metaphor: Thomas Pynchon and *Gravity's Rainbow*." *Contemporary Literature* 15 (1974).
Morrison, Philip. Review of *Gravity's Rainbow*, by Thomas Pynchon. *Scientific American* 229 (Oct. 1973).
Tanner, Tony. *Thomas Pynchon.* London: Methuen, 1982.

Rabelais, François

Francis, K. H. "Some Popular Scientific Myths in Rabelais: A Possible Source." In *Studies in French Literature Presented to H. W. Lawton,* edited by J. C. Ireson, I. D. McFarlaine, and Carnet Rees. New York: Barnes & Noble, 1968.

Rimbaud, Arthur

Bouvet, Alphonse. "Rimbaud et l'alchemie." *Revue des sciences humaines,* April–June 1967.

Ritter, Johann Wilhelm

Rehm, Else. "Johann Wilhelm Ritter und die Universität Jena." *Jahrbuch des Freien Deutschen Hochstifts.* Tübingen: Niemeyer, 1973.
———. "Über den Tod und den letzten Verfügungen des Physikers Johann Wilhelm Ritter." *Jahrbuch des Freien Deutschen Hochstifts.* Tübingen: Niemeyer, 1974.

Shadwell, Thomas

Gilde, Joseph M. "Thomas Shadwell and the Royal Society: Satire in *The Virtuoso*." *Studies in English Literature* 10 (1970).
Lloyd, Claude. "Shadwell and the Virtuosi." *PMLA* 44 (1929).

Shelley, Mary

Awad, Louis. "The Alchemist in English Literature, Part I: Frankenstein." *Bulletin of the Faculty of Arts* (Fuad I University, Cairo) 13 (May 1951).

Bennett, Betty T., and Charles E. Robinson, eds. *The Mary Shelley Reader*. New York: Oxford University Press, 1990.

Bloom, Harold. "Frankenstein, or the New Prometheus." *Partisan Review* 32, no. 4 (1965).

Brooks, P. "Godlike Science / Unhallowed Arts: Language and Monstrosity in *Frankenstein*." *New Literary History* 9 (1978).

Buchen, Irving H. "*Frankenstein* and the Alchemy of Creation and Evolution." *Wordsworth Circle* 8 (1977).

Butler, Marilyn. "The First *Frankenstein* and Radical Science." *Times Literary Supplement*, 9 Apr. 1993.

Callahan, P. J. "*Frankenstein*, Bacon, and the Two Truths." *Extrapolation* 14 (1972).

Cude, Wilfred. "Mary Shelley's Modern Prometheus: A Study in the Ethics of Scientific Creativity." *Dalhousie Review* 52 (1972).

Fleck, P. D. "Mary Shelley's Notes to Shelley's Poems and *Frankenstein*." *Studies in Romanticism* 6, no. 4 (1967).

Florescu, Radu. *In Search of Frankenstein*. Boston: New York Graphic Society, 1975.

Goldberg, M. P. "Moral and Myth in Mrs. Shelley's *Frankenstein*." *Keats-Shelley Journal* 8 (Winter 1959).

Hume, Robert D. "Gothic versus Romantic: A Revaluation of the Gothic Novel." *PMLA* 84 (1969).

Ketterer, David. *Frankenstein's Creation: The Book, the Monster, and Human Reality*. English Literary Studies, 16. Victoria, British Columbia: University of Victoria Press, 1979.

Kiely, Robert. "Frankenstein." In *The Romantic Novel in England*. Cambridge, Mass.: Harvard University Press, 1973.

Kreutz, C. "Mary Wollstonecraft Shelleys Prometheusbild." in *Das Prometheussymbol in der Dichtung der Englischen Romantik, Palaestra* 236 (1963).

Levine, George. "*Frankenstein* and the Tradition of Realism." *Novel* 7 (Fall 1973).

Levine, George, and U. C. Knoepflmacher, eds. *The Endurance of Frankenstein: Essays on Mary Shelley's Novel*. Berkeley and Los Angeles: University of California Press, 1979.

Lovell, Ernest J., Jr. "Byron and Mary Shelley." *Keats-Shelley Journal* 2 (Winter 1953).

Lund, Mary Graham. "Mary Godwin Shelley and the Monster." *University of Kansas City Review* 28 (June 1962).

―――. "Shelley as Frankenstein." *Forum* 4, no. 2 (1963).

McInerney, P. "Frankenstein and the Godlike Science of Letters." *Genre* 13, no. 4 (1980).

McLeod, Patrick G. "Frankenstein: Unbound and Otherwise." *Extrapolation* 21 (1980).

Mays, Milton A. "*Frankenstein*, Mary Shelley's Black Theodicy." *Southern Humanities Review* 3 (1969).

Millhauser, M. "The Noble Savage in *Frankenstein*." *Notes and Queries* 190 (1946).

Moers, Ellen. "Female Gothic." In *Literary Women*. London: Allen & Co., 1977.

―――. "Female Gothic: The Monster's Mother." *New York Review of Books*, 21 Mar. 1974.

Nelson, Lowry, Jr. "Night Thoughts on the Gothic Novel." *Yale Review* 52 (1962).

Nitchie, Elizabeth. *Mary Shelley, "Author of Frankenstein."* New Brunswick, N.J.: Rutgers University Press, 1953.

Palmer, D. J., and R. E. Dowse. *"Frankenstein:* A Moral Fable." *Listener,* 23 Aug. 1962.

Pollin, B. R. "Philosophical and Literary Sources of *Frankenstein." Comparative Literature* 17 (1965).

Railo, Eino. *The Haunted Castle.* New York: Humanities Press, 1964.

Reed, John R. "Will and Fate in *Frankenstein." Bulletin of Research in Humanities* 83 (1980).

Rieger, James. "Dr. Polidori and the Genesis of *Frankenstein." Studies in English Literature* 3 (1963).

Roszak, Theodore. "The Monster and the Titan: Science, Knowledge, and Gnosis." *Daedalus* 103, no. 3 (1974).

Scholes, Robert, and E. S. Rabkin. *Science Fiction.* New York: Oxford University Press, 1977.

Scott, Sir Walter. "Remarks on *Frankenstein." Blackwood's Edinburgh Magazine* 2, no. 12 (1818).

Seed, David. *"Frankenstein:* Parable or Spectacle?" *Criticism* 24, no. 4 (1982).

Shelley, Percy Bysshe. "On *Frankenstein."* In *The Complete Works of Percy Bysshe Shelley,* edited by Roger Ingpen and Walter E. Peck, vol. 6. 10 vols. New York: Benn, 1926–29.

———. "Preface to *Frankenstein; or the Modern Prometheus,* 1818." In *The Complete Works of Percy Bysshe Shelley,* vol. 6. *See* Shelley, "On *Frankenstein."*

Small, Christopher. *Ariel Like a Harpy: Shelley, Mary, and Frankenstein.* London: Gollancz, 1972.

———. *Mary Shelley's Frankenstein: Tracing the Myth.* Pittsburgh: Pittsburgh University Press, 1973.

Spark, Muriel. *Child of Light: A Reassessment of Mary Shelley.* Hadleigh, Essex: Tower Bridge, 1951.

———. "Mary Shelley: a Prophetic Novelist." *Listener,* 22 Feb. 1951.

Swingle, L. J. "Frankenstein's Monster and Its Romantic Relatives: Problems of Knowledge in English Romanticism." *Texas Studies in Literature and Language* 15, no. 1 (1973).

Vasbinder, S. H. *Attitudes in Mary Shelley's "Frankenstein": Newtonian Monism as a Basis for the Novel.* Ann Arbor: University Microfilms, 1984.

Walling, W. A. *Mary Shelley.* New York: Twayne, 1972.

Ziolkowski, T. "Science, Frankenstein, and Myth." *Sewanee Review* 89, no. 1 (1981).

Shelley, Percy Bysshe

Grabo, Carl. *A Newton among Poets: Shelley's Use of Science in Prometheus Unbound.* New York: Cooper Square, 1968.

———. *Prometheus Unbound: An Interpretation.* Chapel Hill: University of North Carolina Press, 1935.

Smart, Christopher

Greene, D. J. "Smart, Berkeley, the Scientists, and the Poets." *Journal of the History of Ideas* 14 (1953).

Snow, C. P.

Bergonzi, B. "The World of Lewis Eliot." *Twentieth Century* 167 (Mar. 1960).

Bezel, Nail. "Autobiography and the 'Two Cultures' in the Novels of C. P. Snow." *Annals of Science* 32 (1975).

Greacen, Robert. *The World of C. P. Snow.* London: Scorpion, 1962.

Karl, F. R. *C. P. Snow: The Politics of Conscience.* Carbondale: Southern Illinois University Press, 1963.

Ramanathan, Suguna. *The Novels of C. P. Snow: A Critical Introduction.* London: Macmillan, 1978.

Shusterman, David. *C. P. Snow.* Boston: Twayne, 1975.

Snow, C. P. "Interview." *Review of English Literature* 3 (July 1962).

Thale, Jerome. *C. P. Snow.* Edinburgh: Oliver & Boyd, 1964.

Waring, A. G. "Science, Love, and the Establishment in the novels of D. A. Granin and C. P. Snow." *Forum for Modern Language Studies* 14, no. 1 (1978).

Stendhal [Marie Henri Beyle]

Milhand, Gérard. "Le Visage scientifique de Stendhal." *Europe*, July–September 1972.

Muller, Maurice. "En marge de l'idéologie à propos de l'ouvrage de Jean Théodorides: Stendhal du côté de la Science." *Studia Philosophica* 33 (1973).

Théodorides, Jean. "Une Source de Lucien Leuven: le Dr Prunelle, modèle probable du Dr du Poirier." *Stendhal Club* 88 (15 July 1980).

———. *Stendhal du côté de la Science.* Aran: Editions du Grand Chene, 1972.

Stevenson, Robert Louis

Strathdee, R. B. "Robert Louis Stevenson as a Scientist." *Aberdeen University Review* 36 (1956).

Stoker, Bram

Hennelly, Mark M., Jr. "Dracula: The Gnostic Quest and Victorian Wasteland." *English Literature in Transition* 20 (1977).

Swift, Jonathan

Case, Arthur, E. *Four Essays on Gulliver's Travels.* Princeton: Princeton University Press, 1945.

Crane, Ronald S. "The Houyhnhnms, the Yahoos, and the History of Ideas." In *Reason and the Imagination,* edited by J. A. Mazzeo. New York: Columbia University Press, 1962.

Eddy, W. A. *Gulliver's Travels—A Critical Study.* Gloucester, Mass.: Peter Smith, 1963.

Foster, M. P., ed. *A Casebook on Gulliver among the Houyhnhnms.* New York: Crowell, 1961.

Jeffares, A. Norman, ed. *Fair Liberty Was All His Cry.* London: Macmillan, 1967.

Kiernan, Colin. "Swift and Science." *Historical Journal* 14, no. 4 (1971).

Korshin, Paul J. "The Intellectual Context of Swift's Flying Island." *Philological Quarterly* 50 (1971).

Merton, Robert C. "The 'Motionless' Motion of Swift's Flying Island." *Journal of the History of Ideas* 27 (1966).
Monro, John M. "Book III of *Gulliver's Travels* Once More." *English Studies* 49 (1968).
Nicolson, Marjorie Hope. "The Scientific Background of Swift's *Voyage to Laputa*." In *Science and Imagination*. Ithaca: Cornell University Press, 1956.
Nicolson, Marjorie Hope, and N. Mohler. "Swift's 'Flying Island' in the *Voyage to Laputa*." *Annals of Science* 2 (1937).
Ross, Angus. *Swift: Gulliver's Travels*. London: Edward Arnold, 1972.
Sutherland, John H. "A Reconsideration of Gulliver's Third Voyage." *Studies in Philology* 54 (1957).
Tuveson, Ernest, ed. *Swift: A Collection of Critical Essays*. Englewood Cliffs, N.J.: Prentice Hall, 1964.
Williams, Kathleen. "Gulliver in Laputa." In *Twentieth-Century Interpretations of Gulliver's Travels*, edited by F. Brady. Englewood Cliffs, N.J.: Prentice Hall, 1968.

Tennyson, Alfred, Lord

Gibson, Walker. "Behind the Veil: A Distinction between Poetic and Scientific Language in Tennyson, Lyell, and Darwin." *Victorian Studies* 2, no. 1 (1958).
Gliserman, Susan M. "Early Victorian Science Writers and Tennyson's 'In Memoriam': A Study in Cultural Exchange." *Victorian Studies* 18, nos. 3 and 4 (1975).
Millhauser, Michael. *Fire and Ice: The Influence of Science on Tennyson's Poetry*. Lincoln: Tennyson Society, 1971.
Wickens, G. G. "The Two Sides of Early Victorian Science and the Unity of 'The Princess.'" *Victorian Studies* 23, no. 3 (1980).

Thomson, James

Drennon, Herbert. "James Thomson's Contact with Newtonianism and His Interest in Natural Philosophy." *PMLA* 49 (1934).
———. "Newtonianism in James Thomson's Poetry." *Englische Studien* 70 (1936).
———. "Scientific Rationalism and James Thomson's Poetic Art." *Studies in Philology* 31 (1934).
Ketcham, Michael G. "Scientific and Poetic Imagination in James Thomson's "Poem Sacred to the Memory of Sir Isaac Newton." *Philological Quarterly* 61 (1982).
McKillop, Alan D. *The Background of Thomson's The Seasons*. Hamden, Conn.: Archon, 1961.
———. *James Thomson: The Castle of Indolence*. Lawrence: University of Kansas Press, 1961.

Thoreau, Henry David

Baym, Nina. "Thoreau's View of Science." *Journal of the History of Ideas* 26 (1965).
Griffin, David. "The Science of Henry David Thoreau." *Synthesis* 1, no. 4 (1973).
Harding, Walter. "Walden's Man of Science." *Virginia Quarterly* 57 (1981).

Verne, Jules

Barthes, Roland. "The *Nautilus* and the Drunken Boat." In *Mythologies*, translated by Annette Lavers. London: Jonathan Cape, 1972.

Butor, Michel. "Le Point suprême et l'Age d'Or: à travers quelques oeuvres de Jules Verne." *Répertoire* 1 (1960).
Costello, Peter. *Jules Verne, "Inventor of Science Fiction."* London: Hodder & Stoughton, 1978.
de la Fuye, Marguerite Allotte. *Jules Verne.* Translated by Erik de Maury. London: Staples, 1954.
Evans, Arthur B. *Jules Verne Rediscovered: Didacticism and the Scientific Novel.* New York: Greenwood, 1988.
Kylstura, Peter H. "Some Backgrounds of Jules Verne." *Janus* 57 (1970).
Suvin, Darko. "Communication in Quantified Space: Utopian Liberalism of Jules Verne's Science Fiction." *Clio* 4, no. 1 (1974).
Winandry, André, "The Twilight Zone: Imagination and Reality in Jules Verne's *Strange Journeys.*" *Yale French Studies* 43 (1969).

Voltaire

Wade, Ira O. *Voltaire's "Micromégas."* Princeton: Princeton University Press, 1950.

von Arnim, Achim

Riley, Helene M. "Scientist, Sorcerer, or Servant of Humanity: The Many Faces of Faust in the Work of Achim von Arnim." *Journal of Germanic Studies* (University of British Columbia) 13 (1977).

Vonnegut, Kurt

Klinkowitz, Jerome. *Kurt Vonnegut.* London: Methuen, 1982.
McNelly, Willis E. "Kurt Vonnegut as Science Fiction Writer." in *Vonnegut in America: An Introduction to the Life and Work of Kurt Vonnegut,* edited by Jerome Klinkowitz and Donald L. Lawler. New York: 1977.
Zins, Daniel L. "Rescuing Science from Technocracy: *Cat's Cradle* and the Play of the Apocalypse." *Science Fiction Studies* 13, no. 2 (1986).

Wells, H. G.

Bergonzi, Bernard. "Another Early Wells Item." *Nineteenth-Century Fiction* 13 (1958).
———. *The Early H. G. Wells: A Study of the Scientific Romances.* Manchester: Manchester University Press, 1961.
———. *H. G. Wells: A Collection of Critical Essays.* Englewood Cliffs, N.J.: Prentice Hall, 1976.
Bower, R. "Science, Myth, and Fiction in H. G. Wells's *Island of Dr. Moreau.*" *Studies in the Novel* 8, no. 3 (1976).
Eisenstein, A. "Very Early Wells: Origins of Some Major Physical Motifs in *The Time Machine* and *The War of the Worlds.*" *Extrapolation* 8 (1972).
Harris, Wilson, ed. *Arnold Bennett and H. G. Wells: A Record of a Personal and a Literary Friendship.* London: Rupert Hart-Davis, 1960.
Haynes, Roslynn D. *H. G. Wells: Discoverer of the Future: The Influence of Science on His Thought.* London: Macmillan, 1980.
———. "The Unholy Alliance of Science in *The Island of Doctor Moreau.*" *Wellsian* 11 (1988).

————. "Wells's Debt to Huxley and the Myth of Doctor Moreau." *Cahiers Victoriens et Édouardiens* 13 (Apr. 1981).

Hennelly, Mark M., Jr. "*The Time Machine:* A Romance of 'The Human Heart'." *Extrapolation* 20 (1979).

Hillegas, Mark R. "Cosmic Pessimism in H. G. Wells's Scientific Romances." *Papers of the Michigan Academy of Science, Arts and Letters* 46 (1961).

Kagarlitski, J. *The Life and Thought of H. G. Wells.* Translated by Moura Budberg. London: Sidgwick & Jackson, 1966.

Le Mire, E. D. "H. G. Wells and the World of Science Fiction." *University of Windsor Review* 2, no. 2 (1967).

Lindsay, C. B. "A Formalist Approach to Wells' Science Fantasies." *Rocky Mountain Review of Language and Literature* 34, no. 3 (1980).

Locke, G. "Wells in Three Volumes?" *Science Fiction Studies* 3, no. 3 (1976).

McConnell, Frank. *The Science Fiction of H. G. Wells.* Oxford: Oxford University Press, 1981.

Newell, K. B. *Structure in Four Novels by H. G. Wells.* The Hague: Mouton, 1968.

Parrinder, Patrick. "*The First Men in the Moon:* H. G. Wells and the Fictional Strategy of His 'Scientific Romances.'" *Science Fiction Studies* 7, no. 2 (1980).

————. "*News from Nowhere, The Time Machine,* and the Break-up of Classical Realism." *Science Fiction Studies* 3, no. 3 (1976).

————. *H. G. Wells.* Edinburgh: Oliver & Boyd, 1970.

Parrinder, Patrick, ed. *H. G. Wells: The Critical Heritage.* London: Routledge, 1972.

Philmus, R. "Time Machine or the Fourth Dimension as Prophecy." *PMLA* 84 (1969).

————. "Very Early Wells: Origins of Some Physical Motifs in *The Time Machine* and the *War of the Worlds.*" *Extrapolation* 13 (1972).

Platzner, Robert L. "H. G. Wells's 'Jungle Book': The Influence of Kipling on *The Island of Dr. Moreau.*" *Victorian Newsletter* 36 (1969).

Raknem, Ingvald. *H. G. Wells and His Critics.* Trondheim, Norway: Universitetsforlaget / Allen & Unwin, 1962.

Roemer, Kenneth M. "H. G. Wells and the 'Momentary Voices' of a Modern Utopia." *Extrapolation* 23 (1982).

Suvin, Darko. "*The Time Machine* versus *Utopia* as a Structural Model for Science Fiction." *Comparative Literature Studies* 10, no. 4 (1973).

Suvin, Darko, and Robert M. Philmus, eds. *H. G. Wells and Modern Science Fiction.* London: Associated University Presses, 1977.

Wagar, Warren. *H. G. Wells and the World State.* New Haven: Yale University Press, 1963.

Williamson, Jack. *H. G. Wells: Critic of Progress.* Baltimore: Mirage, 1973.

Whitman, Walt

Lindfors, Berndt. "Whitman's 'When I heard the learn'd astronomer.'" *Walt Whitman Revue* 10 (1964).

Wolf, Christa

Buehler, George. *The Death of Socialist Realism in the Novels of Christa Wolf.* Frankfurt am Main: Peter Lang, 1984.

Kuhn, Anna K. *Christa Wolf's Utopian Vision: From Marxism to Feminism*. Cambridge: Cambridge University Press, 1988.

Wordsworth, William

Gaull, M. "From Wordsworth to Darwin: 'On to the Fields of Praise.'" *Wordsworth Circle* 1 (1979).

Yeats, William Butler

Spivey, W. B. "Yeats and the 'Children of Fire': Science, Poetry, and Visions of the New Age." *Studies in the Literary Imagination* 14, no. 1 (1981).

Zamyatin, Yevgeny

Shane, Alex M. *The Life and Works of Evgenyi Zamyatin*. Berkeley: University of California Press, 1968.
Warrick, Patricia. "The Sources of Zamiatin's *We* in Dostoevsky's *Notes from the Underground*." *Extrapolation* 17 (1975).

Zola, Émile

Schmidt, Günther. *Die literarische Rezeption des Darwinismus: das Problem der Vergebung bei Emile Zola und in Drama des deutschen Naturalismus*. Berlin: Akademie-Verlag, 1974.

Zuckmayer, Carl

Glade, Henry. "Carl Zuckmayer's Theory of Aesthetics." *Monatshefte* 52 (1960).
Peppard, Murray B. "Carl Zuckmayer: Cold Light in a Divided World." *Monatshefte* 49 (1957).

INDEX